Fungal Biology

Series Editors

Vijai Kumar Gupta
School of Natural Sciences, National University of Ireland Galway
Galway, Ireland

Maria G. Tuohy
School of Natural Sciences, National University of Ireland Galway
Galway, Ireland

More information about this series at http://www.springer.com/series/11224

Ram Prasad
Editor

Advances and Applications Through Fungal Nanobiotechnology

Springer

Editor
Ram Prasad
Amity Institute of Microbial Technology
Amity University, Noida
Uttar Pradesh, India

ISSN 2198-7777 ISSN 2198-7785 (electronic)
Fungal Biology
ISBN 978-3-319-42989-2 ISBN 978-3-319-42990-8 (eBook)
DOI 10.1007/978-3-319-42990-8

Library of Congress Control Number: 2016953738

Printed on acid-free paper

This Springer imprint is published by Springer Nature
The registered company is Springer International Publishing AG
The registered company address is: Gewerbestrasse 11, 6330 Cham, Switzerland

Preface

The study of fungi has become a valuable science in the last 100 years as it has provided to control a number of infectious diseases. In this direction, nanotechnology has emerged as a potential candidate. Nanotechnology is the study and application of extremely small things (1–100 nm) and can be used across all the other science fields, such as chemistry, biology, physics, materials science, and engineering. Biologically prepared tailored nanoparticles from fungi are gaining attention due to their cost-effective, sustainable, resource efficient, simplicity and eco-friendly nature.

In this book entitled *Advances and Applications Through Fungal Nanobiotechnology*, the editor has accumulated various advanced approaches for studying the fungal system for the benefit of humankind. The book covers synthesis of nanoparticles by fungi, the mechanism involved in the biosynthesis and unique template for synthesis of tailored nanoparticles targeted at therapeutic and diagnostic platform technologies.

This book should be immensely useful for microbiologists, nanotechnologists, researchers and teachers of fungal biology and those who are interested in fungal nanobiotechnology. I am honored that the leading scientists who have extensive, in-depth experience and expertise in fungal system and nanobiotechnology took the time and effort to develop these outstanding chapters. Each chapter is written by internationally recognized scientists so the reader is given an up-to-date and detailed account of our knowledge of the nanobiotechnology and various applications of fungi.

We are indebted to the many people who helped to bring this book to light. I wish to thank series editors Dr. Vijai Kumar Gupta and Dr. Maria G. Tuohy; Eric Stannard, Senior Editor, Botany, Springer; and Hemalatha Gunasekaran and Jaspher Jasmine, Springer, for generous assistance, constant support and patience in initializing the volume. I particularly thank Dr. Ishan Barman, Biophotonics Laboratory, Whiting School of Engineering, Department of Mechanical Engineering, Johns Hopkins University, USA, for providing necessary facilities for editing the book during my visit to his institute. I am also thankful to UICC and American Cancer Society for financial assistance to visit the Barman Laboratory. Special thanks go to my lovely wife Dr. Avita Maurya for her constant support and motivations in putting everything together. Dr. Prasad in particular is very thankful to Professors Ajit Varma and Narendra Tuteja, Amity University, for the kind support and constant encouragement. Special thanks are due to my esteemed friend and well wisher Dr. Rishikesh Pandey, Laser Biomedical Research Center, Massachusetts Institute of Technology, USA; Drs. Vivek Kumar, Manoj Kumar, Devendra Kumar Choudhary, Neeraj Shrivastava, Amity Institute of Microbial Technology, all faculty colleagues of Amity University and my Ph.D. student Nafe Aziz.

Noida, UP, India Ram Prasad

Contents

1 **Understanding Mechanism of Fungus Mediated Nanosynthesis: A Molecular Approach** 1
Anal K. Jha and Kamal Prasad

2 **Innovation of Strategies and Challenges for Fungal Nanobiotechnology** 25
Necdet Saglam, Ozfer Yesilada, Ahmet Cabuk, Mesut Sam, Semran Saglam, Sedef Ilk, Ezgi Emul, Pınar Aytar Celik, and Ekrem Gurel

3 **Marine-Derived Fungi: Potential Candidates for Fungal Nanobiotechnology** 47
Anjana K. Vala, Hiral B. Trivedi, and Bharti P. Dave

4 **Green Synthesis of Metal Nanoparticles by Fungi: Current Trends and Challenges** 71
Luciano Paulino Silva, Cínthia Caetano Bonatto, and Vera Lúcia Perussi Polez

5 **Microbial Enzymes: Current Features and Potential Applications in Nanobiotechnology** 91
Mohammadhassan Gholami-Shabani, Masoomeh Shams-Ghahfarokhi, Zeynab Gholami-Shabani, and Mehdi Razzaghi-Abyaneh

6 **The Effect of Mycobiota on the Biointerface of Polyaniline Surface** 129
Elena Binkauskienė, Vitalija Jasulaitienė, Ausra Selskienė, and Albinas Lugauskas

7 **Synthesis Techniques and Evaluation Methods of Nanoparticles as Fungicides** 141
Youn Su Lee and Ahmed I. S. Ahmed

8 Plant Fungal Disease Management Using Nanobiotechnology as a Tool 169
Sandra Pérez Álvarez, Norma Elena Leyva López, Jesús Méndez Lozano, Edgar Antonio Rodríguez Negrete, and María Elena Santos Cervantes

9 Antifungal Products by Fungi in Food Nano-Packaging 193
Arash Babaei, Mohamad Javad Akbarian Mymand, and Samaneh Faraji Kafshgari

10 Fungal Nanoparticles: An Emerging Tool in Medical Biology 213
Anurag Yadav, Amit Verma, and Kusum Yadav

11 Intervention of Fungi in Nano-Particle Technology and Applications 241
Ramesh Raliya, Vinod Saharan, Ramprakash Saran, Kailash Choudhary, Jagadish C. Tarafdar, and Pratim Biswas

12 Microbial Laccases and Nanobiotechnology: Environmental Perspective 253
Sunita J. Varjani

13 Polymer Inorganic Nanocomposites: A Sustainable Antimicrobial Agents 265
Ayat F. Hashim, Mousa A. Alghuthaymi, Alexander Yu. Vasil'kov, and Kamel A. Abd-Elsalam

14 Advances in Bio-Botanicals Formulations with Incorporation of Nanotechnology in Intensive Crop Management 291
Pinki Bhandari, Megha Pant, P.K. Patanjali, and S.K. Raza

15 Nano-Biofungicides: Emerging Trend in Insect Pest Control 307
Atanu Bhattacharyya, Prabha Duraisamy, Marimuthu Govindarajan, Abdul A. Buhroo, and Ram Prasad

16 Nanocellulose Production Using Cellulose Degrading Fungi 321
Nadanathangam Vigneshwaran and Prasad Satyamurthy

Index 333

Contributors

Kamel A. Abd-Elsalam Agricultural Research Center, Plant Pathology Research Institute, Giza, Egypt

Unit of Excellence in Nano-Molecular Plant Pathology Research (ARC), Giza, Egypt

Ahmed I.S. Ahmed Plant Pathology Unit, Plant Protection Department, Desert Research Center, Cairo, Egypt

Mousa A. Alghuthaymi Biology Department, Science and Humanities College, Shaqra University, Al-Quway'iyah, Saudi Arabia

Sandra Pérez Álvarez Instituto Politécnico Nacional, CIIDIR-IPN, Unidad Sinaloa, Departamento de Biotecnología Agrícola, Guasave, Sinaloa, Mexico

Arash Babaei Biology Department, Faculty of Sciences, Malayer University, Malayer, Iran

Pinki Bhandari Institute of Pesticide Formulation Technology, Gurgaon, Haryana, India

Atanu Bhattacharyya Department of Biomedical Engineering, Nanotechnolgy Section, Rajiv Gandhi Institute of Technology and Research Centre, Bangalore, India

Elena Binkauskienė State Research Institute Center for Physical Sciences and Technology, Vilnius, Lithuania

Pratim Biswas Washington University, St. Louis, MO, USA

Cínthia Caetano Bonatto Embrapa Genetic Resources and Biotechnology, Laboratory of Nanobiotechnology (LNANO), Nanotechnology and Synthetic Biology Group, Brasilia, DF, Brazil

Institute of Biological Sciences, University of Brasilia, Brasilia, DF, Brazil

Abdul A. Buhroo Section of Entomology, P.G. Department of Zoology, University of Kashmir, Srinagar, India

Ahmet Cabuk Department of Biotechnology, Graduate School of Natural and Applied Science, Osmangazi University, Eskisehir, Turkey

María Elena Santos Cervantes Instituto Politécnico Nacional, CIIDIR-IPN, Unidad Sinaloa, Departamento de Biotecnología Agrícola, Guasave, Sinaloa, Mexico

Pınar Aytar Celik Department of Biology, Faculty of Science and Letters, Osmangazi University, Eskisehir, Turkey

Kailash Choudhary Lachoo Memorial College of Science and Technology, Jodhpur, Rajasthan, India

Bharti P. Dave Department of Life Sciences, Maharaja Krishnakumarsinhji Bhavnagar University, Bhavnagar, India

Prabha Duraisamy Department of Environmental Sciences, Bharathiar University, Coimbatore, Tamilnadu, India

Ezgi Emul Division of Nanotechnology and Nanomedicine, Graduate School of Science and Engineering, Hacettepe University, Ankara, Turkey

Mohammadhassan Gholami-Shabani Department of Mycology, Pasteur Institute of Iran, Tehran, Iran

Zeynab Gholami-Shabani Faculty of Aerospace, Science and Research Campus, Islamic Azad University, Tehran, Iran

Marimuthu Govindarajan Unit of Vector Control, Phytochemistry and Nanotechnology, Department of Zoology, Annamalai University, Chidambaram, Tamil Nadu, India

Ekrem Gurel Department of Biology, Faculty of Science and Letters, Abant Izzet Baysal University, Bolu, Turkey

Ayat F. Hashim Agricultural Research Center, Plant Pathology Research Institute, Giza, Egypt

Unit of Excellence in Nano-Molecular Plant Pathology Research (ARC), Giza, Egypt

Sedef Ilk Faculty of Agricultural Sciences and Technologies, Nigde University, Nigde, Turkey

Vitalija Jasulaitienė State Research Institute Center for Physical Sciences and Technology, Vilnius, Lithuania

Anal K. Jha Aryabhatta Centre for Nanoscience and Nanotechnology, Aryabhatta Knowledge University, Patna, India

Samaneh Faraji Kafshgari Department of Food Science and Technology, Agricultural Sciences and Resource University of Gorgan, Gorgan, Iran

Norma Elena Leyva López Instituto Politécnico Nacional, CIIDIR-IPN, Unidad Sinaloa, Departamento de Biotecnología Agrícola, Guasave, Sinaloa, Mexico

Youn Su Lee Division of Bioresource Sciences, Kangwon National University, Chuncheon, South Korea

Jesús Méndez Lozano Instituto Politécnico Nacional, CIIDIR-IPN, Unidad Sinaloa, Departamento de Biotecnología Agrícola, Guasave, Sinaloa, Mexico

Albinas Lugauskas State Research Institute Center for Physical Sciences and Technology, Vilnius, Lithuania

Mohamad Javad Akbarian Mymand Biology Department, Faculty of Sciences, Malayer University, Malayer, Iran

Edgar Antonio Rodríguez Negrete Instituto Politécnico Nacional, CIIDIR-IPN, Unidad Sinaloa, Departamento de Biotecnología Agrícola, Guasave, Sinaloa, Mexico

Megha Pant Institute of Pesticide Formulation Technology, Gurgaon, Haryana, India

P.K. Patanjali Institute of Pesticide Formulation Technology, Gurgaon, Haryana, India

Vera Lúcia Perussi Polez Embrapa Genetic Resources and Biotechnology, Laboratory of Prospection of Bioactive Compounds (LPCB), Bioactive Compounds Group, Brasilia, DF, Brazil

Ram Prasad Amity Institute of Microbial Technology, Amity University, Noida, Uttar Pradesh, India

Kamal Prasad Aryabhatta Centre for Nanoscience and Nanotechnology, Aryabhatta Knowledge University, Patna, India

Ramesh Raliya Washington University, St. Louis, MO, USA

S.K. Raza Institute of Pesticide Formulation Technology, Gurgaon, Haryana, India

Mehdi Razzaghi-Abyaneh Department of Mycology, Pasteur Institute of Iran, Tehran, Iran

Necdet Saglam Division of Nanotechnology and Nanomedicine, Graduate School of Science and Engineering, Hacettepe University, Ankara, Turkey

Semran Saglam Department of Atomic and Molecular Physics, Faculty of Science, Gazi University, Ankara, Turkey

Vinod Saharan Maharana Pratap University of Agriculture and Technology, Udaipur, Rajasthan, India

Mesut Sam Department of Biology, Faculty of Science and Letters, Aksaray University, Aksaray, Turkey

Ramprakash Saran J. N. Vyas University, Jodhpur, Rajasthan, India

Prasad Satyamurthy ICAR-Central Institute for Research on Cotton Technology, Mumbai, India

Ausra Selskienė State Research Institute Center for Physical Sciences and Technology, Vilnius, Lithuania

Masoomeh Shams-Ghahfarokhi Department of Mycology, Faculty of Medical Sciences, Tarbiat Modares University, Tehran, Iran

Luciano Paulino Silva Embrapa Genetic Resources and Biotechnology, Laboratory of Nanobiotechnology (LNANO), Nanotechnology and Synthetic Biology Group, Brasilia, DF, Brazil

Institute of Biological Sciences, University of Brasilia, Brasilia, DF, Brazil

Jagadish C. Tarafdar Central Arid Zone Research Institute, Jodhpur, Rajasthan, India

Hiral B. Trivedi Department of Life Sciences, Maharaja Krishnakumarsinhji Bhavnagar University, Bhavnagar, India

Anjana K. Vala Department of Life Sciences, Maharaja Krishnakumarsinhji Bhavnagar University, Bhavnagar, India

Sunita J. Varjani School of Biological Sciences and Biotechnology, University and Institute of Advanced Research, Gandhinagar, Gujarat, India

Alexander Yu. Vasil'kov A.N. Nesmeyanov Institute of Organoelement Compounds (INEOS) of Russian Academy of Sciences, Moscow, Russia

Amit Verma Department of Biochemistry, College of Basic Science and Humanities, S.D. Agricultural University, Banaskantha, Gujarat, India

Nadanathangam Vigneshwaran ICAR-Central Institute for Research on Cotton Technology, Mumbai, India

Anurag Yadav Department of Microbiology, College of Basic Science and Humanities, S.D. Agricultural University, Banaskantha, Gujarat, India

Kusum Yadav Department of Biochemistry, University of Lucknow, Lucknow, Uttar Pradesh, India

Ozfer Yesilada Department of Biology, Faculty of Science and Letters, Inonu University, Malatya, Turkey

About the Editor

Ram Prasad is Assistant Professor at the Amity Institute of Microbial Technology, Amity University, Uttar Pradesh. Dr. Prasad has completed his Ph.D. degree from the Department of Microbiology, Chaudhary Charan Singh University, Meerut, UP, India, in collaboration with School of Life Sciences, Jawaharlal Nehru University (JNU), New Delhi, India. Dr Prasad received his M.Sc. in Life Sciences at JNU and also qualified CSIR-NET, ASRB-NET, and GATE. His research interest includes plant-microbe-interactions, sustainable agriculture and microbial nanobiotechnology. Dr. Prasad has ninety-five publications to his credit, including research papers & book chapters and five patents issued or pending, and edited or authored several books. Dr. Prasad has eleven years of teaching experience and he has been awarded the Young Scientist Award (2007) and Prof. J.S. Datta Munshi Gold Medal (2009) by the International Society for Ecological Communications; FSAB fellowship (2010) by the Society for Applied Biotechnology; Outstanding Scientist Award (2015) in the field of Microbiology by Venus International Foundation; and the American Cancer Society UICC International Fellowship for Beginning Investigators (USA, 2014). In 2014-2015, Dr. Prasad served as Visiting Assistant Professor in the Department of Mechanical Engineering at Johns Hopkins University, USA.

Chapter 1
Understanding Mechanism of Fungus Mediated Nanosynthesis: A Molecular Approach

Anal K. Jha and Kamal Prasad

Abstract The chapter details different processes of biosynthesis of inorganic (metallic and oxide) nanoparticles mediated by the different members of fungi. The biosynthetic mechanism (at molecular level) has been discussed in detail. The nanosynthesis is broadly dependent upon the modulation of key parameters like temperature, pH and other medium conditions. It is conclusively found that although, the cellular level organization matters along with their metabolic fluxes/signal transduction pathways, it is the different stress shearing cues at different levels (ranging from cell wall to nucleus) that bestows a unique echelon to an individual genera in the phylogeny.

1 Introduction

Mother Nature inhabited endearing cohorts with the exponential leap of evolution. Most of them are amazingly diverse yet unique in terms of their cellular level organization and mode of nutrition. Fungi, can literally be perceived as an eukaryotic, heterotrophic, absorptive organisms having chitineous cell walls, reproducing mostly asexually or sexually by producing spores, and grow by budding or even by cell elongation of the hyphal tip. They are characterized by a distinctive, multinucleate vegetative (somatic) thallus called mycelium and survive by obtaining nourishment systemically or through rhizoids. They have got naturally bestowed property to depolymerise the complex molecules like proteins, carbohydrates and lipids as source of carbon and energy and due this, they are ecologically considered among primary decomposers. Due to microbial dimension and eukaryotic cell structure, they have developed enrapturing modes of adaptability in order to survive right from the stale decaying food to living human/animal body.

A.K. Jha • K. Prasad (✉)
Aryabhatta Centre for Nanoscience and Nanotechnology, Aryabhatta Knowledge University, Patna 800001, India
e-mail: k.prasad65@gmail.com

R. Prasad (ed.), *Advances and Applications Through Fungal Nanobiotechnology*, Fungal Biology, DOI 10.1007/978-3-319-42990-8_1

Materials with nano-sized scale have fascinated extensively to the researchers due to their exponential promises in about every part of our life. Nanotechnology deals with small structures/materials having the dimensions less than 100 nm. To demonstrate a few, the hydrogen atom is ~0.1 nm, a virus may be ~100 nm, a red blood corpuscle ~7000 nm in diameter while an average human hair is ~10,000 nm wide. Nano-science and technology is the area that which deals on (1) the development of synthetic methods and surface analytical tools for building structures and materials, (2) to understand the change in chemical and physical properties due to miniaturization, and (3) the use of such properties in the development of novel and functional materials and devices. It is well known that due to the quantum effect there is alteration in the properties of conventional materials at nano level and the performance of surfaces start to lead the performance of bulk materials. By amenably manipulating the size, composition and shape of the nanoscale materials; it is quite possible to tune their optical, electrical, mechanical, magnetic, and chemical properties (Daniel and Astruc 2004; Krolikowska et al. 2003; Schmid 1992; Suman et al. 2010; Valtchev and Tosheva 2013). Many synthetic procedures have emerged with time for the preparation of metals and/or oxide nanoparticles such as extensive ball milling, chemical, hydrothermal, solvothermal, flame combustion, emulsion precipitation, forming materials around/within templates, growth of secondary materials on a crystalline lattice in which the lattice parameter do not match (strained-layer growth), etc. (Flores et al. 2013; Kalishwaralal et al. 2008; Pingali et al. 2005; Shin et al. 2004; Wang et al. 2007). Each of these procedures are either capital and/or labour intensive or time consuming and having its own advantages and disadvantages. During the last two decades, the biosynthesis of noble metal nanoparticles (silver, gold, platinum and palladium) has been noticed significantly due to the growing need to develop environmentally sociable technologies in material synthesis (Chandran et al. 2006; Jia et al. 2009; Prasad 2014; Song et al. 2010; Xie et al. 2007). In recent years, for the synthesis of various types of nanomaterials, nanobiotechnology has emerged as an upcoming field (Ghodake et al. 2013; Liu et al. 2013, 2014; Tanvir et al. 2012). The growth of ecologically-benign, 'green' synthesis protocols is in consonance with the recent RoHS and WEEE legislation stipulated by the EU.

Since the beginning of life on earth biological entities and inorganic materials have been in constant interaction with each other. Because of this interaction cue, life could sustain on this planet with a well-organized deposit of minerals. Living cells are the best known examples of machines that operate at the nano-level and perform a number of jobs ranging from generation of energy to extraction of targeted materials with very high efficiency. Microorganisms are natural producers of nanoscale materials. Being the cell factories themselves, they have proved to be promising tools for the emerging technologies and biomedical applications. During recent years there has been much interest in research into the interaction between inorganic substances and biological systems and highlighted their prospective application for the fabrication of nanomaterials with interesting technological properties (Aziz et al. 2015; Beveridge and Murray 1980; Cunningham and Lundie 1993; Deplanche and Macaskie 2008; Deplanche et al. 2008; Du et al. 2007; Fortin and Beveridge 2000; Jha and Prasad 2010a; Jha et al. 2008a, 2009a, 2010; Labrenz et al. 2000; Mokhtari et al. 2009; Nair and Pradeep 2002; Prasad and Jha 2009;

Prasad et al. 2007, 2010, 2014; Saifuddin et al. 2009; Southam and Beveridge 1994). In order to authenticate the above discussed facts, an attempt was made to use microbes (*Lactobacilli* and fungal members like yeast, *Fusarium, Aspergillus*, etc.) for the purpose of synthesizing nanoparticles.

Because of being eukaryotes, fungi have remarkable advantage in terms of metabolic flux as well as cellular level organization. Yeast being a member of the kingdom fungi has been taken into regular use as media supplement in different culture procedures and this organism itself has-been a very good source of different enzymes and vitamins. They are handy, non-toxic and amenable in terms of culturing/handling. Baker's yeast (*Saccharomyces cerevisiae*) has also been considered among microbes in order to assay their potential as putative candidate fungal genera for the synthesis of metal and/or oxide nanoparticles principally due to their well-organized cellular system and metabolic fluxes. Though, microbial systems like *Fusarium* and *Aspergillus* were used earlier, recently *Pencillium*, *Volveria* and many others have been employed for this purpose (Narayanan and Sakthivel 2010; Ray et al. 2011; Haq et al. 2015; Philip 2009a).

Yeast (*Sachharomyces cerevisae, Schizosaccharomyces pombe, Candida glabrata, MKY3,* etc.) being a member of the kingdom fungi has been taken into regular use as media supplement in different culture procedures and this organism itself has been a very good source of different enzymes and vitamins. They are handy, non-toxic and quite amenable in terms of culturing. In the recent past, yeast and other fungal members have been considered for the synthesis of metal, oxide and chalcogenide nanoparticles (Agnihotri et al. 2009; Ahmad et al. 2002, 2003; Balaji et al. 2009; Bansal et al. 2005; Basavaraja et al. 2008; Bhainsa and D'Souza 2006; Bhambure et al. 2009; Dameron et al. 1989; Durán et al. 2005; Fayaz et al. 2009; Gharieb et al. 1999; Jha and Prasad 2010b; Jha et al. 2008b, 2009c; Kalishwaralal et al. 2008; Kathiresan et al. 2009; Kowshik et al. 2002, 2003; Kumar et al. 2008; Mukherjee et al. 2001a, b, 2002; Philip 2009b; Prasad et al. 2015; Reese and Winge 1988; Sanghi and Verma 2009; Sastry et al. 2003; Senapati et al. 2005; Shaligram et al. 2009; Shankar et al. 2003). Accordingly, efforts have been made to understand the mechanism of nano-transformation of accomplishing biosynthesis at the molecular level.

2 Methodology

Eukaryotes have comparatively lower requirements in culture media as they have a better level of cell organization and metabolism. The fungal mycelia of *Fusarium oxysporum* is inoculated in a 500 ml Erlenmeyer flask containing known volume of MGYP medium and incubated for 72 h under shaking conditions (200 rpm) at 27 °C. After incubation, the fungal mycelia are harvested and washed thoroughly under sterile conditions. The fungus *Aspergillus niger* NCIM 616 is obtained and maintained on potato-dextrose agar slants at 25 °C. Stock cultures are maintained by subculturing at monthly intervals. The fungus is grown at pH 5.5 and 25 °C for 7 days. *Aspergillus fumigatus* (NCIM 902) is maintained on potato dextrose agar slants. To prepare biomass for biosynthesis studies the fungus is grown aerobically in a liquid media containing (g/l) KH_2PO_4, 7.0: K_2HPO_4, 2.0; $MgSO_4 \cdot 7H_2O$, 0.1; $(NH_4)_2SO_4$, 1.0; yeast extract, 0.6; and glucose, 10.0. The flasks are inoculated, incubated on orbital shaker at 25 °C and agitated at 150 rpm. The biomass was harvested after 72 h of growth by sieving through a plastic sieve, followed by extensive washing with distilled water to remove any medium component from the biomass. Typically 20 g of biomass (fresh weight) is brought in contact with 200 ml of Milli-Q deionized water for 72 h at 25 °C in an Erlenmeyer flask and agitated in the same condition as described earlier. After the incubation, the cell filtrate is obtained by passing it through Whatman filter paper no. 1. The fungus culture, *Pencillium brevicompactum* WA 2315, is grown in seed medium containing (w/v) glucose, glycerol, peptone, $NaNO_3$, $MgSO_4$ and soybean meal, the pH was adjusted to 6.5–0.2 with 2 N H_3PO_4. The flasks are incubated in the incubator shaker at 180 rpm at 25 °C. After 3 days of incubation, the mycelium is separated by centrifugation (6000 rpm) and washed thrice with de-ionized water. The washed mycelia are suspended in deionized water and kept on rotary shaker for 96 h at 180 rpm. Edible mushroom, *Volvariella volvacea*, obtained from home surroundings is washed several times with de-ionised water. Sixty eight grams of finely cut mushroom is boiled for 2 min in 300 ml water and filtered. The filtrate is cooled to room temperature and used as reducing agent and stabilizer. In yet another recent study, about 5 g of dried mushrooms (*Agaricus bisporus*) were taken and washed thoroughly with double distilled water to remove dust and mud adhering to the surface of the mushrooms. The washed mushroom samples were kept in shadow conditions for drying. The mushroom samples are then cut into small pieces with sterile knife and then powdered into fine particles. The mushroom fragments were suspended into 350 ml of sterile distilled water and boiled for 10 min at 55 °C in Erlenmeyer's conical flask. The mushroom extract was filtered twice through Whatman's filter paper No. 1 and stored at 4 °C for further experiments. The filtrate was used as reducing agent for 1 mM of $AgNO_3$ (99.9 %). For the synthesis of silver nanoparticles the 350 ml of mushroom extract was added to 150 ml of 1 mm $AgNO_3$ solution incubated at room temperature for the reduction.

The mycelium of *Tricholoma crassum* was cultured in vitro and was used for production of silver nanoparticles. Tissue from basidiocarp was first cultured on potato dextrose agar (PDA). For liquid culture, the mycelium from solid substrate was inoculated in 50 ml potato dextrose broth in 250 ml flasks. The biomass was harvested after 72 h of growth in 28 °C by straining through a sieve. The biomass was washed with sterilized distilled water to remove medium component. 1 g of biomass (fresh weight) was added to 10 ml of deionized water in a 250 ml Erlenmeyer flask and agitated in the same condition for 72 h at 28 °C. The resulting cell filtrate was collected by passing it through Whatman filter paper no. 1. This filtrate was used for nanoparticles synthesis.

Yeast cells are allowed to grow as suspension culture in presence of suitable carbon (Glucose) and nitrogen (Glycine) source for 36 hours. This is treated as source culture. A small portion of it (25 ml) is filtered and diluted four times by adding 30 % Et-OH containing nutrients. This diluted culture is again allowed to grow for another 24 h until it attains a light straw colour/golden yellow colour. Now,

20 ml of molar metal salt solution is added to the culture solution and it is heated on steam bath up to 60 °C for 10–20 min. The pH of culture solution is suitably adjusted at this stage depending upon targeted task-synthesis of a metal or an oxide or chalcogenide.

3 Discussions

The procedure of nanotransformation involving synthesis of metal, oxide or chalcognide nanoparticles seemingly is the cumulative response of the biological system being taken in to use and its immediate chemical ambience along with basic metabolic fluxes/metabolite content/signal transduction of the organism being employed for the purpose. The biosynthetic mechanism for preparation of metal/oxide nanoparticles is illustrated in Fig. 1.1. The environments of most organisms are rarely constant, so some resilience to environmental stress is essential in order for organisms to persist. Chemical stressors such as organic and inorganic pollutants can have different modes of action, but one effect is common to many of these, as well as certain natural stressors such as radiation, and that is an association with oxidative damage in cells (Avery 2001; Limon-Pacheco and Gonsebatt 2009).

Today's knowledge on the molecular background of metal/metalloid toxicity in eukaryotes arises in great part from extensive genome, transcriptome, deletome, proteome, interactome and metabolome analyses having been done in baker's yeast cultures. *Saccharomyces cerevisiae* is an excellent model organism to address important biological questions because the available molecular biological, genetic

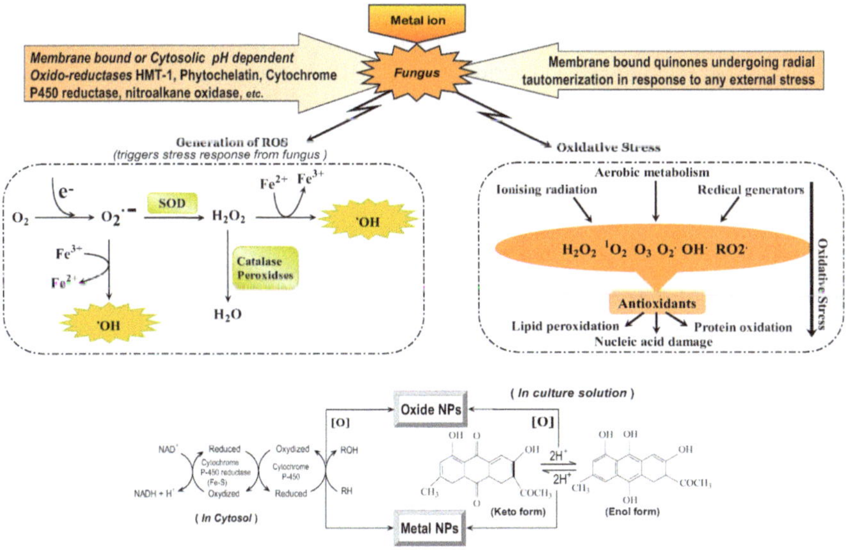

Fig. 1.1 Mechanism for the fungus mediated biosynthesis of metal/oxide nanoparticles

and bioinformatic tools are unprecedently sophisticated and versatile with this hemiascomycete. As shown by numerous examples, knowledge obtained from yeast-based models, e.g. on the mechanism of action of and the tolerance against toxic metals, can be transferred with high efficiency to higher eukaryotes including humans (González-Guerrero et al. 2009; Tamás et al. 2005). Challenged by metal/metalloid toxicity, fungal members trigger response at multiple levels originating from cell wall to nuclei that can be understood as follows:

3.1 Extracellular Chelation and Binding to Cell Wall Constituents (First Line response of Cellular Defense)

Production of nanoparticles using filamentous fungi has some advantages over other organisms. Filamentous fungi are easy to handle, require simple raw materials and has high wall-binding capacity (Haq et al. 2015; Prasad et al. 2015). Along with this, a number of simple hydroxy/methoxy derivatives of benzoquinones and toluquinones are elaborated by lower fungi (especially Penicillium and Aspergillus species) in response a metallic stress. Presence of these metabolites may trigger a redox reaction due to tautomerization leading to synthesis of a nanomaterial (Jha and Prasad 2010c). Cell membranes are often bestowed with small molecular mass metabolites like peptides and proteins are pivotally important element of almost all metal/metalloid detoxification processes and hence, the significance of extracellular and cytosolic chelation reactions cannot be underestimated (González-Guerrero et al. 2009; Tamás et al. 2005; Wysocki and Tamás 2010; Verbavatz et al. 2009). The membrane-bound (as well as cytosolic) oxidoreductases and quinones might have played an important role in the process. The oxidoreductases are pH sensitive and work in alternative manner. At a lower value of pH, oxidase gets activated while a higher pH value activates the reductase (Jha and Prasad 2010c). Amino acids and amino acid derived molecules have high significance in plant to adapt in heavy metal stress conditions. N-containing metabolites majorly proline is frequently synthesized under heavy metal stress such as Cd, Cu, Ni, and Zn. Proline has three major functions in metal detoxification namely metal binding, antioxidant defence, and signaling (Shanti and Karl 2006).

Glutathione (GSH) secretion is a very important element of the GSH-homeostasis in yeast under different environmental conditions (Perrone et al. 2005) and it is sensible that yeast cells intensify GSH-secretion under As(III) exposures to relieve the intracellular detoxification pathways (Wysocki and Tamás 2010). Elaboration of inorganic metal chelators is very documented among different members of fungi, and especially among brown-rot and white-rot fungi and this process seems to be stimulated under Cu(II) and Cd(II) stress (Clausen and Green 2003; Jarosz-Wilkołazka et al. 2006). The bulk formation of water-insoluble metal-oxalate crystals is undoubtedly an efficient way to circumvent toxic metal ions which tend to enter fungal cells (Jarosz-Wilkołazka and Gadd 2003). In addition, oxalate is primarily important to maintain the lignolytic system of white rot basidimycetes (Schlosser and Höfer 2002).

A wide range of fungi has been reported to produce extracellular mucilaginous materials (ECMM or "emulsifier") with excellent toxic metal binding capabilities. As demonstrated by Paraszkiewicz et al. (Paraszkiewicz et al. 2007, 2010), Ni(II) [unlike Cu(II), Pb(II) and Zn(II)] (Paraszkiewicz and Długoński 2009) did not trigger ECMM production by *Curvularia lunata.* These authors reported the saturation of cellular fatty acids, which is clearly attributable to Ni(II)-initiated lipid peroxidation processes. Importantly, the pullulan production by *Aureobasidium pullulans* was positively affected by Ni (II) and Cd(II) exposures (Breierová et al. 2004), and pullulan increased the Cd(II) tolerance of this industrially important species (Čertik et al. 2005). Vesentini et al. (Vesentini et al. 2006) also reported elevated ratio of ECMM in the biomass in Cu(II) exposed *Trametes versicolor* and *Gloeophyllum trabeum* cultures. Isolates of the arbuscular mycorrhizal fungi Glomus and Gigaspora species produce a soil glycoprotein called glomalin, which possesses a remarkable capability to sequester Cu(II) (Cornejo et al. 2008; González-Chávez et al. 2004; Wright et al. 1996). The production of various metabolites like citric acid, homogeneous proteins, heterogeneous proteins, peroxidases by fungi made them effective for detoxification of heavym metals from industrial effluents. White rot fungi are ubiquitous in nature and their enzymes producing activity makes them effective decolourizers and remove toxic metals by biosorbtion ultimately rendering the effluents more eco-friendly (Tripathi et al. 2007). Exposure to elevated heavy metals concentration in mycorrhizal species of *Pinus sylvestris* made it to produce organic acids. Among different acid production, the level of oxalic acid is significantly high compared to other acids like malonic acid, citric acid, shikimic acid, lactic acid, acetic acid, propionic acid, fumaric acid, formic acid, iso-butyric acid and butyric acid are found in variable concentrations (Ulla et al. 2000). Along with this, *Fusarium Oxysporum, Aspergillus niger and Aspergillus fumigatus and Pencillium brevicompactum* being a eukaryotes have a better organized cellular system compared to the bacteria, therefore a drastic reduction in particle size of nanomaterials which has been reported is natural. *Fusarium* and other members belong to the kingdom fungi. Its adaptability is prodigal which is manifested in the form of its metabolic treasures beginning from membrane bound cellulases, nitrate reductases galactosides, quinones to cytosolic oxidoreductases like Cytochrome P450 (Durán et al. 2005) and an active FAD-containing form of nitroalkane oxidase is also present (Gadda and Fitzpatrick 1998). All of these cumulatively ensure a redial extra-cellular transformation of both metallic as well as oxide nanoaprticles (Gharieb et al. 1999; Karbasian et al. 2008). In yet another study for the purpose of optimization of the procedure, the biomass obtained by the growth of *Fusarium oxysporium* PTCC 5115 in MGYP medium was able to convert silver nitrate to nano silver through the enzyme nitrate reductase under the conditions of temperature, pH and time of the reaction, agitation rate and fungal biomass (Jha et al. 2009b). *Volvariella volvacea* and *Agaricus bisporus* (edible mushroom) have also been found to synthesize metals like Ag and Au and candidate metabolites are ascertained to be amino acids and proteins (Kalishwaralal et al. 2008).

3.2 *Transport, Intracellular Chelation and Compartmentalization (Second Line Response of Cellular Defense)*

Organisms are constantly challenged by ever-changing variables in their environment, including fluctuating nutrient levels, osmotic imbalance, exposure to toxic molecules, and non-optimal temperatures. While multicellular or motile organisms can usually alter these conditions by a change in location or physiology, single-celled organisms such as yeast are at the mercy of their situation, and must adapt or perish (Morano et al. 2012). Heavy metals enter cells through channels and transporters, which normally facilitate the uptake of essential transition metal micronutrients like Fe, Mn and Zn, anions including phosphate and sulphate as well as sugars (glucose) and sugar derivatives (glycerol) (Tamás et al. 2005; Wysocki and Tamás 2010). In theory, one of the most simple and most effective way to keep off toxic metals/metalloids outside the cell is to eliminate the channel or transporter responsible for the uptake of a given toxic metal/metalloid ion. These ions may be channeled through multiple transporters into the cytoplasm and, aggravating the situation, the absence of even one of these transport routes may disturb the normal metabolism of the cells which in due course may come across the different stress shearing cues of the cells and give a desired nanomaterial. However, the elimination of thea few plasma membrane channels and transporters has been demonstrated to confer metal tolerance to metal/metalloid exposed S. cerevisiae cells (Tamás et al. 2005; Wysocki and Tamás 2010) like Pho87p low-affinity phosphate transporters for As(V); Fps1p aquaglyceroporin channel for As (III) and Sb(III); Zrt1p Zn(II) and Smf1p and Smf2p for Cd (II) and Mn(II); Sul1p and Sul2p sulphate transporters for Cr(VI) and Se(VI), etc.

Importantly, yeast cells lacking either Pho86p, which is required in trafficking Pho84p from the ER to the cytoplasma membrane, or Gtr1p, a cytoplasmic GTP binding protein, a regulator of phosphate transport through Pho84p, displayed As(V)-tolerant phenotypes (Bun-ya et al. 1992, 1996; Yompakdee et al. 1996). Furthermore, the elimination of *BSD2* encoding an ER protein trafficking Smf1p and Smf2p transporters to the vacuoles for degradation resulted in Cd(II) and Cu(II) hypersensitivities (Liu and Culotta 1999; Liu et al. 1997). However, other important cellular transporters leading to subsequent chelation may be as under: GSH complexes of Cd(II), As(III), Hg(II), Pb(II): Ycf1p ATP-binding cassette family vacuolar GSH S-conjugate transporter, a strong induction of which can be achieved by overexpressing Yap1p, the master regulator transcription factor of oxidative stress response (Sharma et al. 2002; Song et al. 2003; Wemmie et al. 1994); Yor1p plasma membrane ABC transporter (Nagy et al. 2006), phytochelatin complexes of Cd(II): HMT1 vacuolar membrane transporter of ATP-binding cassette-type (*S. Pombe*) (Ortiz et al. 1995), although HMT1 is likely to be a GSH S-conjugate transporter in S. pombe according to most recent data published by Prévéral et al. (Prévéral et al. 2009). Importantly, vacuolar Zn(II) (GintZnT1; (González-Guerrero et al. 2005) and Cd(II)/Cu(II) (GintABC1; (González-Guerrero et al. 2009; González-Guerrero et al. 2010) transporters have been characterized in the endomycorrhizal fungus *Glomus intraradices* with high homology and functional orthology to well-characterized baker's yeast proteins, which clearly demonstrates the applicability of yeast-based heavy metal sequestration models in this case. Not surprisingly, the expression of GintABC1 was up regulated when the fungus was exposed to Cd(II) and Cu(II) (González-Guerrero et al. 2010). GSH is a key player in both heavy metal tolerance and oxidative stress defense (Hegedűs et al. 2007; Mendoza-Cózatl et al. 2005; Pócsi et al. 2004; Stephen and Jamieson 1997; Tamás et al. 2005; Westwater et al. 2002; Wu and Moye-Rowley 1994; Wysocki and Tamás 2010) and it's over expression increases the toxic metal/metalloid tolerance. When yeast γ-glutamylcysteine synthetase GSH1 and garlic phytochelatin synthase AsPCS1 were expressed either alone or simultaneously in *Arabidopsis thaliana* the transgenic plants accumulated and tolerated Cd and As remarkably well (Guo et al. 2008). Intracellular GSH levels can also be elevated considerably in baker's yeast by overexpressing the Hgt1p GSH-transporter but resulted in some GSH-induced cell toxicity (Srikanth et al. 2005), indicating the limitation of this approach. Considering GSH turnover, vacuolar γ-glutamyltranspeptidase and aminopeptidase are needed to recycle GSH-derived amino acids into the cytoplasm (Pócsi et al. 2004). Interestingly, a yeast strain with LAP4 mutation accumulated three times more Cd(II) than the control strain with an unexpected decrease in the Cd(II)elicited oxidative stress (Adamis et al. 2009). *S. pombe* deposits Cd(II) in the form of vacuolar high molecular mass phytochelatin-Cd(II) aggregates with CdS crystallites in their cores (Kowshik et al. 2002; Mendoza-Cózatl et al. 2005).

Numerous studies have revealed that all yeast genera can accumulate different heavy metals, but particularly interesting is the capability of accumulating significant amounts of highly toxic metals (Breierová et al. 2002). Cells, which are able to grow in media with high metal ion concentrations is called resistant. The yeast cell capacity to transform absorbed metal ions into complex polymer compounds that are nontoxic for the cell is perceived as resistance ad this might be acting as crux for synthesis of nanomaterials. The major molecules that contribute to the detoxification mechanisms in yeast cells are glutathione (GSH) and two groups of metal-binding ligands: metallothioneins and phytochelatins, both well characterized. GSH with the structure *c*-Glu-Cys-Gly is an important tripeptide involved in various metabolic processes in bacteria, yeasts, plants and animals. The unique redox and nucleophilic properties classify this compound as a detoxificator, actively taking part in the bioreduction and defense against free radicals and xenobiotics. GSH is also a structural unit in phytochelatin molecules, which is one of its major functions. Metallothioneins are low-molecular-weight, cysteine-rich metal-binding proteins, derived by mRNA translation (Kagi 1993). The great number of cysteine residues binds different metals by S–S bonds. Metallothioneins are classified according to the arrangement of these residues (Cherian and Chan 1993).

The low molecular mass, the ability to bind metal ions and the specific amino acid composition with high levels of cysteine and low levels of aromatic amino acid residues are common among all metallothionein classes (Kagi and Schaffer 1988). The synthesis of these proteins is determined by a family of genes. Although the model has been created on the basis of detoxification of Cu^{2+} ions, it is also accurate for other metals in yeast (Butt and Ecker 1987). In animals the function of metallothioneins is associated with the defense against cadmium toxicity and resistance towards Cu^{2+} and Zn^{2+}. Nevertheless, there is evidence that metallothioneins play a similar role in some plants and yeast species, e.g. in *S. cerevisiae* and *C. glabrata* (Mehra and Winge 1991; Mehra et al. 1988). Phytochelatins were discovered in the yeast *Schizosaccharomyces pombe* (Murasugi et al. 1983), but are common for most plants (Grill et al. 1985) and also some animal species (Cobbett 2000). Although, initially, they were described as cadmium binding peptides (Grill et al. 1985; Kondo et al. 1983), phytochelatin formation is induced by a large number of elements such as Cd^{2+}, Pb^{2+}, Zn^{2+}, Sb^{3+}, Ag^{+}, Ni^{2+}, Hg^{2+}, Cu^{2+}, Sn^{2+}, Au^{+}, Bi^{3+}, Te^{4+}, and W^{6+}, when supplemented to the medium (Rauser 1995). They have the general structure (*c*-Glu-Cys)n-Gly, where $n=2$–11, and a multitude of structural variants has been described in the scientific literature (Huang et al. 2007; Rauser 1995, 1999; Zenk 1996). Structurally, phytochelatins are related to GSH, serving as a substrate for their enzymatic biosynthesis, which has been confirmed by a number of physiological, biochemical and genetic studies (Huang et al. 2007; Rauser 1995, 1999; Zenk 1996). The enzyme phytochelatin synthase or *c*-Glu-Cys dipeptidyl transpeptidase (EC 2.3.2.15) catalyzes the reaction of transpeptidation of *c*-Glu-Cys dipeptide from a GSH molecule to a second molecule of GSH, resulting in phytochelatin PC2, or to a phytochelatin molecule, resulting in an n+1 oligomer (Grill et al. 1989). Phytochelatin synthesis begins within minutes after exposing yeast cells to cadmium ions and is regulated by enzyme activation in the presence of metal ions. The best activators are cadmium ions, followed by ions of Ag, Bi, Pb, Zn, Cu, Hg, and Au (Clemens et al. 1999; Ha et al. 1999; Vatamaniuk et al. 1999). In plants and yeasts, Cd-phytochelatin complexes are formed in the cytosol but accumulate in vacuoles (Salt and Wagner 1993). Detailed studies of the fission yeast *S. pombe* revealed that nearly the whole cadmium and phytochelatin amounts are located in vacuoles (Clemens et al. 1999). Compared to metallothioneins, phytochelatins feature many advantages, derived from their unique structure and especially the repeated *c*-Glu-Cys units and they have better metal-binding capacity (Mehra and Mulchandani 1995). In addition, phytochelatins can incorporate large amounts of inorganic sulfur, resulting in increased capacity of these peptides to bind cadmium (Mehra et al. 1994). The biomineralization of cadmium by the yeasts *S. pombe* and *C. glabrata* is a metal-triggered biotransformation, in which metal ions are consequently chelated with small selective peptides and co-precipitated with inorganic sulfur, resulting in nontoxic CdS clusters (Perego and Howell 1997). Dameron and Winge (Dameron and Winge 1990) discovered that the CdS crystals isolated from *S. pombe* and *C. glabrata* are monodisperse, with an average diameter of 2 nm. The crystal lattice consists of 85 CdS pairs covered by approximately 30 (*c*-Glu-Cys)n-Gly peptides, with $n=3$–5. Vacuoles represent the primary sites of intracellular metal/metalloid sequestration and storage in fungi (Tamás et al. 2005; Wysocki and Tamás 2010). All these observations in nut shell suggest that intracellular protein trafficking systems represent relevant and promising targets for engineering new fungal strains with an altered toxic metal/metalloid tolerance and may contribute towards large scale production of nanomaterials subject to modulation of experimental cues.

3.3 The Antioxidative System (Third Line of Cellular Defense)

Fungi exposed to toxic metal/metalloid stress commonly face oxidative cell injuries caused by reactive oxygen species (Avery 2001). Fungal cells possess a wide array of antioxidants to cope with different kinds of oxidative stress. For example, GSH independent and GSH-dependent enzyme activities are able to neutralize reactive oxygen species with remarkable efficiency (Pócsi et al. 2004). While elevated temperature represents the primary insult during heat shock (as described above), one of the major secondary consequences involves production of reactive oxygen species (ROS). All organisms are exposed to ROS during the course of normal aerobic metabolism or following exposure to radical-generating compounds (Halliwell 2006).

Molecular oxygen is relatively unreactive and harmless in its ground state, but can undergo partial reduction to form a number of ROS, including the superoxide anion and hydrogen peroxide (H_2O_2), which can further react to produce the highly reactive hydroxyl radical. ROS are toxic agents that can damage a wide variety of cellular components resulting in lipid peroxidation, protein oxidation, and genetic damage through the modification of DNA. *S. cerevisiae* responds to an oxidative stress using a number of cellular responses that ensure the survival of the cell following exposure to oxidants. These include defense systems that detoxify ROS, reduce their rate of production, and repair the damage caused by them. Many responses are ROS specific, but there are also general stress responses that are typically invoked in response to diverse stress conditions (Morano et al. 2012). ROS are continuously produced in actively metabolizing cells. However, *S. cerevisiae*, like all organisms, contains effective antioxidant defense mechanisms, which detoxify ROS as they are generated and maintain the intracellular redox environment in a reduced state. An oxidative stress is said to occur when ROS overwhelm these defenses, resulting in genetic degeneration and physiological dysfunction, leading eventually to cell death.

Antioxidant defenses include a number of protective enzymes that are present in different subcellular compartments and can be upregulated in response to ROS exposure. Non-enzymic defenses typically consist of small molecules that can act as free radical scavengers; to date, only ascorbic acid and GSH have been extensively characterized in yeast (Morano et al. 2012). Antioxidants, such as ascorbic acid can reduce transition metals giving these compounds a pro-oxidant character (Poljšak et al. 2005). Therefore, any overload of fungal cells with antioxidants (e.g. Tempol, Trolox, Melatonin) might exacerbate oxidative cell damages in the presence of redox active metals/metalloids (Lewinska and Bartosz 2007). The different enzymes involved in protection against ROS exposures may be as under:

3.3.1 Catalases

Catalases are ubiquitous heme-containing enzymes that catalyze the dismutation of H_2O_2 into H_2O and O_2. Yeast has two such enzymes: the peroxisosomal catalase A encoded by CTA1, and the cytosolic catalase T encoded by CTT1. The CTA1 expression is coordinated with peroxisomal fatty acid metabolism, suggesting that

Cta1 may function in the detoxification of H_2O_2 generated from fatty acid b-oxidation (Hiltunen et al. 2003). Catalases are important for the acquisition of peroxide resistance following pretreatment with low doses of H_2O_2 and upon entry into stationary phase, indicating a role during adaptive responses (Culotta et al. 2006).

3.3.2 Superoxide Dismutases

Superoxide dismutases (SODs) convert the superoxide anion to hydrogen peroxide, which can then be reduced to water by catalases or peroxidises. SODs are ubiquitous antioxidants, which differ in their intracellular location and metal cofactor requirements between different organisms. Enzyme activity is dependent on redox cycling of the bound metal cofactor. Yeast contains a cytoplasmic Cu,Zn-SOD (Sod1) and a mitochondrial matrix Mn-SOD (Sod2), which appear to play distinct roles during oxidative stress conditions (Culotta et al. 2006).

3.3.3 Methionine Sulfoxide Reductase

Amino acids are susceptible to oxidation by ROS (Stadtman and Levine 2003). Methionine residues are particularly susceptible, forming a racemic mixture of methionine-S-sulfoxide (Met-S-SO) and methionine-R-sulfoxide (Met-R-SO) in cells (Dean et al. 1997). Most organisms contain methionine sulfoxide reductases (MSRs), which protect against methionine oxidation by catalyzing thiol-dependent reduction of oxidized Met residues. This is particularly important because it means that methionine oxidation is readily reversible and can play an antioxidant role in scavenging ROS (Stadtman et al. 2003). Yeast contains three MSR enzymes that are required for resistance against oxidative stress (fRMsr/MsrA/MsrB) (Le et al. 2009). fRMsr is thought to be the main enzyme responsible for the reduction of free Met-R-SO, whereas MsrA and MsrB are active with Met-S-SO and Met-R-SO in proteins. A triple fRMsr/MsrA/MsrB mutant is viable on media containing methionine, but cannot grow if methionine is substituted with Met-SO (Le et al. 2009).

3.3.4 Thioredoxins

S. cerevisiae, like most eukaryotes, contains a cytoplasmic thioredoxin system, which functions in protection against oxidative stress. This comprises two thioredoxins (TRX1 and TRX2) and a thioredoxin reductase (TRR1) (Gan 1991). As in most organisms, yeast thioredoxins are active as antioxidants and play key roles in protection against oxidative stress induced by various ROS (Izawa et al. 1999; Kuge and Jones 1994). A major part ofthe antioxidant function of thioredoxins is mediated by peroxiredoxins. Oxidized thioredoxins (Trx1/Trx2) are rapidly observed (15 s) following exposure to hydrogen peroxide and are detected for 0.1 h before returning to the reduced form (Okazaki et al. 2007). Yeast also contains a complete mitochondrial thioredoxin system, comprising a thioredoxin (Trx3) and a thioredoxin reductase (Trr2) (Pedrajas et al. 1999). The redox states of the cytoplasmic and mitochondrial thioredoxin systems are independently maintained and cells can survive in the absence of both systems (Trotter and Grant 2005).

3.3.5 Peroxiredoxins

Peroxiredoxins (Prx) have multiple roles instress protection, acting as antioxidants, molecular chaperones, and in the regulation of signal transduction (Wood et al. 2003). They use redox-active Cys residues to reduce peroxides and have been divided into two classes, the 1-Cys and 2-Cys Prx's, on the basis of the number of Cys residues directly involved in catalysis. Typical 2-Cys Prx's are active as a dimer and contain two redox-active Cys residues that are required for enzyme activity (Chae et al. 1994; Park et al. 2000). During catalysis, the peroxidatic cysteine residue of one subunit is oxidized to a sulfenic acid, which condenses with the resolving cysteine from the other subunit to form a disulfide that is reduced by thioredoxin been described in yeast (Morgan and Veal 2007). All three display thioredoxin peroxidase activity, but appear to play distinct physiological roles. Tsa1 has best been characterized as an antioxidant in the detoxification of hydroperoxides (Garrido and Grant 2002; Wong et al. 2004), but has also been shown to act as a chaperone that promotes resistance to heat and reductive stresses (Jang et al. 2004; Rand and Grant 2006). Tsa2 is highly homologous to Tsa1 and possesses similar peroxidase and chaperone activities, but is expressed at significantly lower levels than Tsa1 (Jang et al. 2004).

3.3.6 The Glutathione System

The oxidation of sulfhydryl groups is one of the earliest observable events during ROS mediated damage. This underlies the importance of GSH (g-glutamylcysteinylglycine) which is typically found as the most abundant low molecular-weight sulfhydryl compound (mM concentrations) in most organisms. Many roles have been proposed for GSH in a variety of cellular processes including amino acid transport; synthesis of nucleic acids and proteins; modulation of enzyme activity; and metabolism of carcinogens, xenobiotics, and ROS (Schafer and Buettner 2001). Not surprisingly therefore, GSH is an essential metabolite in eukaryotes, and for example, mice that are deficient in GSH biosynthesis die rapidly (Shi et al. 2000). Similarly, GSH is an essential metabolite in yeast where it appears to be required as a reductant during normal growth conditions (Grant et al. 1996a). Oxidative stress converts glutathione to its oxidized disulfide form (GSSG). However, glutathione is predominantly present in its reduced GSH form in yeast and other eukaryotes due to the constitutive action of glutathione reductase (Glr1). Glr1 is an NADPH-dependent oxidoreductase, which converts GSSG to GSH using reducing power generated by the pentose phosphate pathway (López-Barea et al. 1990). Yeast GLR1 is not essential for normal aerobic growth, but is required for viability during exposure to oxidative stress and following starvation conditions (Grant et al. 1996b, c).

3.3.7 Glutaredoxins

Glutaredoxins (Grx) are small heat-stable oxidoreductases, which were first discovered in E. coli as GSH-dependent hydrogen donors for ribonucleotide reductase (Holmgren 1989). Classical cellular glutaredoxins contain a conserved dithiol active site (Cys-Pro-Tyr-Cys) and form part of the glutaredoxin system, in which glutathione reductase transfers electrons from NADPH to glutaredoxins via GSH. They have proposed roles in many cellular processes including protein folding and regulation, reduction of dehydroascorbate, and protection against ROS and sulphur metabolism (Holmgren 1989). Two yeast genes encode classical dithiol glutaredoxins (GRX1 and GRX2) (Luikenhuis et al. 1997). Grx1 and Grx2 are active as GSH-dependent oxidoreductases, but appear to have distinct cellular functions.

3.3.8 Glutathione Peroxidases

Eukaryotic glutathione peroxidises (Gpx's) are thought to provide the major enzymatic defense against oxidative stress caused by hydroperoxides. They reduce hydrogen peroxide and other organic hydroperoxides, such as fatty acid hydroperoxides, to the corresponding alcohol, using reducing power provided by GSH (Michiels et al. 1994). Mammalian cells also contain phospholipid hydroperoxide Gpx's (PHGpx's), which are able to reduce membrane phospholipid hydroperoxides (Roveri et al. 1994). Interestingly, yeast does not contain any classical Gpx's, but expresses three PHGpx's encoded by GPX1-3 (Avery and Avery 2001; Inoue et al. 1999). These PHGpx enzymes have activity with phospholipid hydroperoxides as well as nonphospholipid hydroperoxides and are able to protect membrane lipids against peroxidation.

3.3.9 Glutathione Transferases

Glutathione transferases (GSTs) are a major family of proteins, which are involved in the detoxification of many xenobiotic compounds (Sheehan et al. 2001). They catalyze the conjugation of electrophillic substrates to GSH prior to their removal from cells via glutathione conjugate pumps. Two genes encoding functional GSTs, designated GTT1 and GTT2, have been identified in yeast (Choi et al. 1998).

3.3.10 Ascorbic Acid

Ascorbic acid is a water soluble antioxidant, which commonly acts in a redox couple with glutathione in many eukaryotes (Winkler et al. 1994). However, the relevance of ascorbate to the yeast oxidative stress response is unclear since yeast contains a 5-carbon analog, erythroascorbate, which may have limited importance as an antioxidant. Conclusively, it can be said that the response of different fungi with respect to the metal ion is an interesting cascade of events subject to modulation of experimental cues in vitro. It may look to get accomplished in a jiffy but encompasses a very well-orchestrated conglomerate of metabolites and pathways. Future may find a series of genetically engineered fungal species for the purpose of nanomaterial synthesis with astonishing amenability of scaling up.

References

Adamis PD, Mannarino SC, Riger CJ, Duarte G, Cruz A, Pereira MD, Eleutherio EC (2009) Lap4, a vacuolar aminopeptidase I, is involved in cadmium-glutathione metabolism. Biometals 22:243–249

Agnihotri M, Joshi S, Kumar AR, Zinjarde S, Kulkarni S (2009) Biosynthesis of gold nanoparticles by the tropical marine yeast *Yarrowia lipolytica* NCIM 3589. Mater Lett 63:1231–1234

Ahmad A, Mukherjee P, Mandal D, Senapati S, Khan MI, Kumar R, Sastry M (2002) Enzyme mediated extracellular synthesis of CdS nanoparticles by the fungus, *Fusarium oxysporum*. J Am Chem Soc 124:12108–12109

Ahmad A, Mukherjee P, Senapati S, Mandal D, Khan MI, Kumar R, Sastry M (2003) Extracellular biosynthesis of silver nanoparticles using the fungus *Fusarium oxysporum*. Col Surf B Biointerfaces 28:313–318

Avery SV (2001) Metal toxicity in yeasts and the role of oxidative stress. Adv Appl Microbiol 49:111–142

Avery AM, Avery SV (2001) *Saccharomyces cerevisiae* expresses three phospholipid hydroperoxide glutathione peroxidases. J Biol Chem 276:33730–33735

Aziz N, Faraz M, Pandey R, Sakir M, Fatma T, Varma A, Barman I, Prasad R (2015) Facile algae-derived route to biogenic silver nanoparticles: synthesis, antibacterial and photocatalytic properties. Langmuir 31:11605–11612

Balaji DS, Basavaraja S, Deshpande R, Mahesh DB, Prabhakar BK, Venkataraman A (2009) Extracellular biosynthesis of functionalized silver nanoparticles by strains of *Cladosporium cladosporioides* fungus. Col Surf B Biointerfaces 68:88–92

Bansal V, Rautaray D, Bharde A, Ahire K, Sanyal A, Ahmad A, Sastry M (2005) Fungus mediated biosynthesis of silica and titania particles. J Mater Chem 15:2583–2589

Basavaraja S, Balaji SD, Lagashetty A, Rajasab AH, Venkataraman A (2008) Extracellular biosynthesis of silver nanoparticles using the fungus *Fusarium semitectum*. Mater Res Bull 43:1164–1170

Beveridge TJ, Murray RGE (1980) Sites of metal deposition in the cell wall of *Bacillus subtilis*. J Bacteriol 141:876–887

Bhainsa KC, D'Souza SF (2006) Extracellular biosynthesis of silver nanoparticles using the fungus *Aspergillus fumigatus*. Col Surf B Biointerfaces 47:160–164

Bhambure R, Bule M, Shaligram NS, Kamat M, Singhal R (2009) Extracellular biosynthesis of gold nanoparticles using *Aspergillus niger*–its characterization and stability. Chem Eng Technol 32:1036–1041

Breierová E, Vajczikova I, Sasinkova V, Stratilova E, Fisera M, Gregor T, Sajbidor J (2002) Biosorption of cadmium ions by different yeast species. Z Naturforsch C 57:634–639

Breierová E, Gregor T, Juršíková P, Stratilová E, Fišera M (2004) The role of pullulan and pectin in the uptake of Cd^{2+} and Ni^{2+} ions by *Aureobasidium pullulans*. Ann Microbiol 54:247–255

Bun-ya M, Harashima S, Oshima Y (1992) Putative GTP-binding protein, Gtr1, associated with the function of the Pho84 inorganic phosphate transporter in *Saccharomyces cerevisiae*. Mol Cell Biol 12:2958–2966

Bun-ya M, Shikata K, Nakade S, Yompakdee C, Harashima S, Oshima Y (1996) Two new genes, *PHO86* and *PHO87*, involved in inorganic phosphate uptake in *Saccharomyces cerevisiae*. Curr Genet 29:344–351

Butt TR, Ecker DJ (1987) Yeast metallothionein and applications in biotechnology. Microbiol Rev 51:351–364

Čertik M, Breierová E, Juršíková P (2005) Effect of cadmium on lipid composition of *Aureobasidium pullulans* grown with added extracellular polysaccharides. Int Biodeter Biodegr 55:195–202

Chae HZ, Chung SJ, Rhee SG (1994) Thioredoxin dependent peroxide reductase from yeast. J Biol Chem 269:27670–27678

Chandran SP, Chaudhary M, Pasricha R, Ahmad A, Sastry M (2006) Synthesis of gold nanotriangles and silver nanoparticles using *Aloe vera* plant extract. Biotechnol Prog 22:577–583

Cherian G, Chan H (1993) Biological roles and medical implications. In: Suzuki KT, Imura N, Kimura M (eds) Metallothionein III. Birkhäuser, Basel

Choi JH, Lou W, Vancura A (1998) A novel membranebound glutathione S-transferase functions in the stationary phase of the yeast *Saccharomyces cerevisiae*. J Biol Chem 273:29915–29922

Clausen CA, Green F (2003) Oxalic acid overproduction by copper-tolerant brown rot basidiomycetes on southern yellow pine treated with copper-based preservatives. Int Biodeter Biodegr 51:139–144

Clemens S, Kim EJ, Neumann D, Schroeder JI (1999) Tolrance to toxic metals by a gene family of phytochelatin synthase from plants and yeast. EMBO J 18:3325–3333

Cobbett CS (2000) Phytochelatins and their roles in heavy metal detoxification. Plant Physiol 123:825–832

Cornejo P, Meier S, Borie G, Rillig MC, Borie F (2008) Glomalin-related soil protein in a Mediterranean ecosystem affected by a copper smelter and its contribution to Cu and Zn sequestration. Sci Total Environ 406:154–160

Culotta VC, Yang M, O'Halloran TV (2006) Activation of superoxide dismutases: putting the metal to the pedal. Biochim Biophys Acta 1763:747–758

Cunningham DP, Lundie LL (1993) Precipitation of cadmium by *Clostridium thermoaceticum*. Appl Environ Microbiol 59:7–14

Dameron CT, Winge DR (1990) Peptide mediated formation of quantum semiconductors. Trends Biotechnol 8:3–6

Dameron CT, Reese RN, Mehra RK, Kortan AR, Carroll PJ, Steigerwald ML, Brus LE, Winge DR (1989) Biosynthesis of cadmium sulphide quantum semiconductor crystallites. Nature 338:596–597

Daniel M-C, Astruc D (2004) Gold nanoparticles: assembly, supramolecular chemistry, quantum-size-related properties, and applications toward biology, catalysis, and nanotechnology. Chem Rev 104:293–346

Dean RT, Fu S, Stocker R, Davies MJ (1997) Biochemistry and pathology of radical-mediated protein oxidation. Biochem J 324:1–18

Deplanche K, Macaskie LE (2008) Biorecovery of gold by *Escherichia coli* and *Desulfovibrio desulfuricans*. Biotechnol Bioeng 99:1055–1064

Deplanche K, Woods RD, Mikheenko IP, Sockett RE, Macaskie LE (2008) Manufacture of stable palladium and gold nanoparticles on native and genetically engineered flagella scaffolds. Biotechnol Bioeng 101:873–880

Du L, Jiang H, Liu X, Wang E (2007) Biosynthesis of gold nanoparticles assisted by *Escherichia coli* DH5α and its application on direct electrochemistry of hemoglobin. Electrochem Commun 9:1165–1170

Durán N, Marcato PD, Alves OL, De Souza GIH, Esposito E (2005) Mechanistic aspects of biosynthesis of silver nanoparticles by several *Fusarium oxysporum* strains. J Nanobiotechnol 3:8. doi:10.1186/1477-3155-3-8

Fayaz AM, Balaji K, Girilal M, Kalaichelvan PT, Venkatesan R (2009) Mycobased synthesis of silver nanoparticles and their incorporation into sodium alginate films for vegetable and fruit preservation. J Agric Food Chem 57:6246–6252

Flores CY, Miñán AG, Grillo CA, Salvarezza RC, Vericat C, Schilardi PL (2013) Citrate-capped silver nanoparticles showing good bactericidal effect against both planktonic and sessile bacteria and a low cytotoxicity to osteoblastic cells. ACS Appl Mater Interfaces 5:3149–3159

Fortin D, Beveridge TJ (2000) From biology to biotechnology and medical applications. In: Aeuerien E (ed). Biomineralization. Wiley-VCH, Weinheim

Gadda G, Fitzpatrick PF (1998) Biochemical and physical characterization of the active FAD-containing form of nitroalkane oxidase from Fusarium oxysporum. Biochem 37:6154–6164

Gan Z (1991) Yeast thioredoxin genes. J Biol Chem 266:1692–1696

Garrido EO, Grant CM (2002) Role of thioredoxins in the response of *Saccharomyces cerevisiae* to oxidative stress induced by hydroperoxides. Mol Microbiol 43:993–1003

Gharieb MM, Kierans M, Gadd GM (1999) Transformation and tolerance of tellurite by filamentous fungi: accumulation, reduction, and volatilization. Mycolog Res 103:299–305

Ghodake G, Lim SR, Lee DS (2013) Casein hydrolytic peptides mediated green synthesis of antibacterial silver nanoparticles. Col Surf B Biointerfaces 108:147–151

González-Chávez M, Carrillo-Gonzalez R, Wright SF, Nichols KA (2004) The role of glomalin, a protein produced by arbuscular mycorrhizal fungi, in sequestering potentially toxic elements. Environ Pollut 130:317–323

González-Guerrero,M., AzcónAguilar C, Mooney M, Valderas A, MacDiarmid CW, Eide DJ, Ferrol N (2005) Characterization of a Glomus intraradices gene encoding a putative Zn transporter of the cation diffusion facilitator family. Fungal Genet Biol 42:130–140

González-Guerrero M, Benabdellah K, Ferrol N, Azcón-Aguilar C (2009) Mechanisms underlying heavy metal tolerance in arbuscular mycorrhizas. In: Azcón-Aguilar C, Barea JM, Gianinazzi S, Gianinazzi-Pearson V (eds) Mycorrhizas-functional processes and ecological impacts. Springer, Berlin

González-Guerrero M, Benabdellah K, Valderas A, Azcón-Aguilar C, Ferrol N (2010) *GintABC1*encodes a putative ABC transporter of the MRP subfamily induced by Cu, Cd, and oxidative stress in *Glomus intraradices*. Mycorrhiza 20:137–146

Grant CM, Collinson LP, Roe J-H, Dawes IW (1996a) Yeast glutathione reductase is required for protection against oxidative stress and is a target gene for yAP-1 transcriptional regulation. Mol Microbiol 21:171–179

Grant CM, MacIver FH, Dawes IW (1996b) Glutathione is an essential metabolite required for resistance to oxidative stress in the yeast *Saccharomyces cerevisiae*. Curr Genet 29:511–515

Grant CM, MacIver FH, Dawes IW (1996c) Stationary phase induction of GLR1 expression is mediated by the yAP-1 transcriptional protein in *Saccharomyces cerevisiae*. Mol Microbiol 22:739–774

Grill E, Winnacker EL, Zenk MH (1985) Phytochelatins: the principal heavy metal complexing pepetides of higher plants. Science 230:674–676

Grill E, Loffler S, Winnacker EL, Zenk MH (1989) Phytochelatins, the heavy- metal-binding peptides of plants, are synthesised from glutathione by a specific gamma-glutamylcystein dipeptidyl transpeptidase (phytochelatin synthase). Proc Natl Acad Sci USA 86:6838–6842

Guo J, Dai X, Xu W, Ma M (2008) Overexpressing *GSH1* and *AsPCS1* simultaneously increases the tolerance and accumulation of cadmium and arsenic in *Arabidopsis thaliana*. Chemosphere 72:1020–1026

Ha SB, Smith AP, Howden R, Dietrich WM, Bugg S, O'Connell MJ, Goldsbrough PB, Cobbett CS (1999) Phytochelatin synthase genes from arabidopsis and the yeast *Schizosaccharomyces pombe*. Plant Cell 11:1153–1164

Halliwell B (2006) Reactive species and antioxidants – redox biology is a fundamental theme of aerobic life. Plant Physiol 141:312–322

Hegedűs N, Emri T, Szilágyi J, Zs K, Nagy I, Penninckx MJ, Pócsi I (2007) Effect of heavy metals on the GSH status in different ectomycorrhizal *Paxillus involutus* strains. World J Microbiol Biotechnol 23:1339–1343

Hiltunen JK, Mursula AM, Rottensteiner H, Wierenga RK, Kastaniotis AJ (2003) The biochemistry of peroxisomal beta-oxidation in the yeast *Saccharomyces cerevisiae*. FEMS Microbiol Rev 27:35–64

Holmgren A (1989) Thioredoxin and glutaredoxin systems. J Biol Chem 264:13963–13966

Huang J, Li Q, Sun D, Lu Y, Su Y, Yang X, Wang H, Wang Y, Shao W, He N, Hong J, Chen C (2007) Biosynthesis of silver and gold nanoparticles by novel sundried *Cinnamomum camphora* leaf. Nanotechnol 18:105104–105115

Inoue Y, Matsuda T, Sugiyama K-I, Izawa S, Kimura A (1999) Genetic analysis of glutathione peroxidase in oxidative stress response of *Saccharomyces cerevisiae*. J Biol Chem 274:27002–27009

Izawa S, Maeda K, Sugiyama K-I, Mano J, Inoue Y (1999) Thioredoxin derficiency causes the constitutive activation of Yap1, an AP-1-like transcription factor in *Saccharomyces cerevisiae*. J Biol Chem 274:28459–28465

Jang HH, Lee KO, Chi YH, Jung BG, Park SK (2004) Two enzymes in one; two yeast peroxiredoxins display oxidative stress-dependent switching from a peroxidase to a molecular chaperone function. Cell 117:625–635

Jarosz-Wilkołazka A, Gadd GM (2003) Oxalate production by wood-rotting fungi growing in toxic metal-amended medium. Chemosphere 52:541–547

Jarosz-Wilkołazka A, Graz M, Braha B, Menge S, Schlosser D, Krauss GJ (2006) Species-specific Cd-stress response in the white rot basidiomycetes *Abortiporus biennis* and *Cerrena unicolor*. Biometals 19:39–49

Jha AK, Prasad K (2010a) Biosynthesis of metal and oxide nanoparticles using *Lactobacilli* from yoghurt and probiotic spore tablets. Biotechnol J 5:285–291

Jha AK, Prasad K (2010b) Synthesis of $BaTiO_3$ nanoparticles: A new sustainable green approach. Integrated Ferroelectr 117:49–54

Jha AK, Prasad K (2010c) Understanding biosynthesis of metallic/oxide nanoparticles: a biochemical perspective. In: Kumar SA, Thiagarajan S, Wang S-F (eds) Biocompatible nanomaterials synthesis, characterization and applications. NOVA Sci. Publication, USA

Jha AK, Prasad K, Kulkarni AR (2008a) Synthesis of nickel nanoparticles: bioreduction method. Nanosci Nanotechnol Ind J 2:26–29

Jha AK, Prasad K, Kulkarni AR (2008b) Yeast mediated synthesis of silver nanoparticles. Int J Nanosci Nanotechnol 4:17–21

Jha AK, Prasad K, Kumar V, Prasad K (2009a) Biosynthesis of silver nanoparticles using *Eclipta* leaf. Biotechnol Prog 25:1476–1479

Jha AK, Prasad K, Prasad K (2009b) A green low-cost biosynthesis of Sb_2O_3 nanoparticles. Biochem Eng J 43:303–306

Jha AK, Prasad K, Prasad K (2009c) Biosynthesis of Sb_2O_3 nanoparticles: a low cost green approach. Biotechnol J 4:1582–1585

Jha AK, Prasad K, Kulkarni AR (2010) Synthesis of Gd_2O_3 nanoparticles using *Lactobacillus* sp.: a novel green approach. Int J Green Nanotechnol: Phys Chem 2:P31–P38

Jia L, Zhang Q, Li Q, Song H (2009) The biosynthesis of palladium nanoparticles by antioxidants in *Gardenia jasminoides* Ellis: long lifetime nanocatalysts forp-nitrotoluene hydrogenation. Nanotechnol 20:385601

Kagi JHR (1993) Biological roles and medical implications. In: Suzuki KT, Imura N, Kimura M (eds) Metallothionein III. Birkhäuser, Basel

Kagi JHR, Schaffer A (1988) Biochemistry of metallothionein. Biochem 27:8509–8515

Kalishwaralal K, Deepak V, Ramkumarpandian S, Nellaiah H, Sangiliyandi G (2008) Extracellular biosynthesis of silver nanoparticles by the culture supernatant of *Bacillus licheniformis*. Mater Lett 62:4411–4413

Karbasian M, Atyabi SM, Siadat SD, Momen SB, Norouzian D (2008) Optimizing nano-silver formation by *Fusarium oxysporum* PTCC 5115 employing response surface methodology. Am J Agric Biol Sci 3:433–437

Kathiresan K, Manivannan S, Nabeel MA, Dhivya B (2009) Studies on silver nanoparticles synthesized by a marine fungus, *Penicillium fellutanum* isolated from coastal mangrove sediment. Col Surf B Biointerfaces 71:133–137

Kondo N, Isobe M, Imai K, Goto T, Murasugi A, Hayashi Y (1983) Structure of cadystin, the unit-peptide of cadmium-binding peptides induced in a fission yeast, *schizosaccharomyces pombe*. Tetrahedron Lett 24:925–928

Kowshik M, Deshmukh N, Vogel W, Urban J, Kulkarni SK, Paknikar KM (2002) Microbial synthesis of semiconductor CdS nanoparticles, their characterization, and their use in the fabrication of an ideal diode. Biotechnol Bioeng 78:583–588

Kowshik M, Ashtaputre S, Kharrazi S, Vogel W, Urban J, Kulkarni SK, Paknikar KM (2003) Extracellular synthesis of silver nanoparticles by a silver-tolerant yeast strain MKY3. Nanotechnol 14:95

Krolikowska A, Kudelski A, Michota A, Bukowska J (2003) SERS studies on the structure of thioglycolic acid monolayers on silver and gold. Surf Sci 532:227–232

Kuge S, Jones N (1994) YAP1 dependent activation of TRX2 is essential for the response of Saccharomyces cerevisiae to oxidative stress by hydroperoxides. EMBO J 13:655–664

Kumar SA, Peter Y-A, Nadeau JL (2008) Facile biosynthesis, separation and conjugation of gold nanoparticles to doxorubicin. Nanotechnol 19:495101

Labrenz M, Druschel GK, Thomsen-Ebert T, Gilbert B, Welch SA, Kemner KM, Logan GA, Summons RE, Stasio GD, Bond PL, Lai B, Kelly SD, Banfield JF (2000) Formation of sphalerite (ZnS) deposits in natural biofilms of sulfate-reducing bacteria. Science 290:1744–1747

Le DT, Lee BC, Marino SM, Zhang Y, Fomenko DE (2009) Functional analysis of free methionine-R-sulfoxide reductase from *Saccharomyces cerevisiae*. J Biol Chem 284:4354–4364

Lewinska A, Bartosz G (2007) Protection of yeast lacking the Ure2 protein against the toxicity of heavy metals and hydroperoxides by antioxidants. Free Radic Res 41:580–590

Limon-Pacheco J, Gonsebatt ME (2009) The role of antioxidants and antioxidant-related enzymes in protective responses to environmentally induced oxidative stress. Mutat Res 674:137–147

Liu XF, Culotta VC (1999) Post-translation control of Nramp metal transport in yeast. Role of metal ions and the *BSD2* gene. J Biol Chem 274:4863–4868

Liu XF, Supek F, Nelson N, Culotta VC (1997) Negative control of heavy metal uptake by the *Saccharomyces cerevisiae BSD2* gene. J Biol Chem 272:11763–11769

Liu L, Yang J, Xie J, Luo Z, Jiang J, Yang Y (2013) The potent antimicrobial properties of cell penetrating peptide-conjugated silver nanoparticles with excellent selectivity for Gram-positive bacteria over erythrocytes. Nanoscale 5:3834–3840

Liu L, Liu T, Tade M, Wang S, Lib X, Liu S (2014) Less is more, greener microbial synthesis of silver nanoparticles. Enzyme Microbial Technol 67:53–58

López-Barea J, Bárcena JA, Bocanegra JA, Florindo J, García-Alfonso C (1990) Structure, mechanism, functions, and regulatory properties of glutathione reductase. In: Vina J (ed) Glutathione: metabolism and physiological functions. CRC, Boca Raton

Luikenhuis S, Dawes IW, Grant CM (1997) The yeast *Saccharomyces cerevisiae* contains two glutaredoxin genes that are required for protection against reactive oxygen species. Mol Biol Cell 9:1081–1091

Haq Manzoor ul, Rathod V, Singh D, Singh AK, Ninganagouda S, Hiremath J (2015) Dried mushroom *Agaricus bisporus* mediated synthesis of silver nanoparticles from Bandipora District (Jammu and Kashmir) and their efficacy against methicillin resistant *Staphylococcus aureus* (MRSA) strains. Nanosci Nanotechnol An Int J 5:1–8

Mehra RK, Mulchandani P (1995) Glutathione-mediated transfer of Cu(I) into phytochelatins. Biochem J 307:697–705

Mehra RK, Winge DR (1991) Metal ion resistance in fungi: molecular mechanisms and their regulated expression. J Cell Biochem 45:30–40

Mehra RK, Tarbet EB, Gray WR, Winge DR (1988) Metal-specific synthesis of two metallothioneins and gamma-glutamyl peptides in *Candida glabrata*. Proc Natl Acad Sci USA 85:8815–8819

Mehra RK, Mulchandani P, Hunter TC (1994) Role of CdS quantum crystallites in cadmium resistance in *Candida glabrata*. Biochem Biophys Res Commun 200:1193–1200

Mendoza-Cózatl D, Loza-Tavera H, Hernández-Navarro A, Moreno-Sánchez R (2005) Sulfur assimilation and glutathione metabolism under cadmium stress in yeast, protists and plants. FEMS Microbiol Rev 29:653–671

Michiels C, Raes M, Toussaint O, Remacle J (1994) Importance of Se-glutathione peroxidase, catalase, and Cu/ Zn-SOD for cell survival against oxidative stress. Free Radic Biol Med 17:235–248

Mokhtari N, Daneshpajouh S, Seyedbagheri S, Atashdehghan R, Abdi K, Sarkar S, Minaian S, Shahverdi HR, Shahverdi AR (2009) Biological synthesis of very small silver nanoparticles by culture supernatant of *Klebsiella pneumonia*: the effects of visible-light irradiation and the liquid mixing process. Mater Res Bull 44:1415–1421
Morano KA, Grant CM, Moye-Rowley WS (2012) The response to heat shock and oxidative stress in *Saccharomyces cerevisiae*. Genetics 190:1157–1195
Morgan BA, Veal EA (2007) Functions of typical 2-Cys peroxiredoxins in yeast. Subcell Biochem 44:253–265
Mukherjee P, Ahmad A, Mandal D, Senapati S, Sainkar SR, Khan MI, Parishcha R, Ajaykumar PV, Alam M, Kumar R, Sastry M (2001a) Fungus-mediated synthesis of silver nanoparticles and their immobilization in the mycelial matrix: a novel biological approach to nanoparticle synthesis. Nano Lett 1:515–519
Mukherjee P, Ahmad A, Mandal D, Senapati S, Sainkar SR, Khan MI, Ramani R, Parishcha R, Ajaykumar PV, Alam M, Sastry M, Kumar R (2001b) Bioreduction of $AuCl_4$-ions by the fungus, *Verticillium* sp. and surface trapping of the gold nanoparticles formed. Angew Chem Int Ed Engl 40:3585–3588
Mukherjee P, Senapati S, Mandal D, Ahmad A, Khan MI, Kumar R, Sastry M (2002) Extracellular synthesis of gold nanoparticles by the fungus *Fusarium oxysporum*. Chembiochem 3:461–463
Murasugi A, Wada C, Hayashi Y (1983) Occurrence of acid-labile sulfide in cadmium-binding peptide 1 from fission yeast. J Biochem Tokyo 93:661–664
Nagy Z, Montigny C, Leverrier P, Yeh S, Goffeau A, Garrigos M, Falson P (2006) Role of the yeast ABC transporter Yor1p in cadmium detoxification. Biochimie 88:1665–1671
Nair B, Pradeep T (2002) Coalescence of nanoclusters and formation of submicron crystallites assisted by *Lactobacillus* strains. Cryst Growth Design 2:293–298
Narayanan KB, Sakthivel N (2010) Biological synthesis of metal nanoparticles by microbes. Adv Colloid Interface Sci 156:1–13
Okazaki S, Tachibana T, Naganuma A, Mano N, Kuge S (2007) Multistep disulfide bond formation in Yap1 is required for sensing and transduction of H_2O_2 stress signal. Mol Cell 27:675–688
Ortiz DF, Ruscitti T, McCue KF, Ow DW (1995) Transport of metal-binding peptides by HMT1, a fission yeast ABC-type vacuolar membrane protein. J Biol Chem 270:4721–4728
Paraszkiewicz K, Długoński J (2009) Effect of nickel, copper, and zinc on emulsifier production and saturation of cellular fatty acids in the filamentous fungus *Curvularia lunata*. Int Biodeter Biodegr 63:100–105
Paraszkiewicz K, Frycie A, Słaba M, Długoński J (2007) Enhancement of emulsifier production by *Curvularia lunata* in cadmium, zinc and lead presence. Biometals 20:797–805
Paraszkiewicz K, Bernat P, Naliwajski M, Długoński J (2010) Lipid peroxidation in the fungus *Curvularia lunata* exposed to nickel. Arch Microbiol 192:135–141
Park SG, Cha M-K, Jeong W, Kim I-H (2000) Distinct physiological functions of thiol peroxidase isoenzymes in *Saccharomyces cerevisiae*. J Biol Chem 275:5723–5732
Pedrajas JR, Kosmidou E, Miranda-Vizuete A, Gustafsson J-A, Wright APH (1999) Identification and functional characterization of a novel mitochondrial thioredoxin system in *Saccharomyces cerevisiae*. J Biol Chem 274:6366–6373
Perego P, Howell SB (1997) Molecular mechanisms controlling toxic metal ions in yeast. Toxicol Appl Pharmacol 147:312–318
Perrone GG, Grant CM, Dawes IW (2005) Genetic and environmental factors influencing GSH homeostasis in *Saccharomyces cerevisiae*. Mol Biol Cell 16:218–230
Philip D (2009a) Honey mediated green synthesis of gold nanoparticles. Spectrochim Acta Part A 73:650–653
Philip D (2009b) Biosynthesis of Au, Ag and Au-Ag nanoparticles using edible mushroom extract. Spectrochim Acta Part A 73:374–381
Pingali KC, Rockstraw DA, Deng S (2005) Silver nanoparticles from ultrasonic spray pyrolysis of aqueous silver nitrate. Aerosol Sci Technol 39:1010–1014

Pócsi I, Prade RA, Penninckx MJ (2004) GSH, altruistic metabolite in fungi. Adv Microb Physiol 49:1–76
Poljšak B, Gazdag Z, Jenko-Brinovec Š, Fujs Š, Pesti M, Bélagyi J, Plesničar S, Raspor P (2005) Pro-oxidative vs antioxidative properties of ascorbic acid in chromium(VI)-induced damage: an in vivo and in vitro approach. J Appl Toxicol 25:535–548
Prasad R (2014) Synthesis of silver nanoparticles in photosynthetic plants. J Nanoparticles 963961, doi:10.1155/2014/963961
Prasad K, Jha AK (2009) ZnO nanoparticles: synthesis and adsorption study. Nat Sci 1:129–135
Prasad K, Jha AK, Kulkarni AR (2007) *Lactobacillus* assisted synthesis of titanium nanoparticles. Nanoscale Res Lett 2:248–250
Prasad K, Jha AK, Prasad K, Kulkarni AR (2010) Can microbes mediate nano-transformation? Ind J Phys 84:1355–1360
Prasad R, Kumar V, Prasad KS (2014) Nanotechnology in sustainable agriculture: present concerns and future aspects. Afr J Biotechnol 13:705–713
Prasad R, Pandey R, Barman I (2015) Engineering tailored nanoparticles with microbes: *quo vadis*. WIREs Nanomed Nanobiotechnol. doi:10.1002/wnan.1363
Prévéral S, Gayet L, Moldes C, Hoffmann J, Mounicou S, Gruet A, Reynaud F, Lobinski R, Verbavatz JM, Vavasseur A, Forestier C (2009) A common highly conserved cadmium detoxification mechanism from bacteria to humans: heavy metal tolerance conferred by the ATP-binding cassette (ABC) transporter SpHMT1 requires GSH but not metal-chelating phytochelatin peptides. J Biol Chem 284:4936–4943
Rand JD, Grant CM (2006) The thioredoxin system protects ribosomes against stress-induced aggregation. Mol Biol Cell 17:387–401
Rauser WE (1995) Phytochelatins and related peptides. Structure, biosynthesis, and function. Plant Physiol 109:1141–1149
Rauser WE (1999) Structure and function of metal chelators produced by plants: the case for organic acids, amino acids, phytin, and metallothioneins. Cell Biochem Biophys 31:19–48
Ray S, Sarkar S, Kundu S (2011) Extracellular biosynthesis of silver nanoparticles using the mycorrhhizal mushroom *Tricholoma crassum* (BERK.) SACC: its antimicrobial activity against pathogenic bacteria and fungus, including multidrug resistant plant and human bacteria. Digest J Nanomater Biostruct 6:1289–1299
Reese RN, Winge DR (1988) Sulfide stabilization of the cadmium-gamma-glutamyl peptide complex of Schizosaccharomyces pombe. J Biol Chem 263:12832–12835
Roveri A, Maiorino M, Ursini F (1994) Enzymatic and immunological measurements of soluble and membrane-bound phospholipid-hydroperoxide glutathione peroxidases. Methods Enzymol 233:202–212
Saifuddin N, Wong CW, Yasumira AAN (2009) Rapid biosynthesis of silver nanoparticles using culture supernatant of bacteria with microwave irradiation. E-J Chem 6:61–70
Salt DE, Wagner GJ (1993) Cadmium transport across tonoplast of vesicles from oat roots. Evidence for a Cd^{2+}/H^+ antiport activity. J Biol Chem 268:12297–12302
Sanghi R, Verma P (2009) Biomimetic synthesis and characterisation of protein capped silver nanoparticles. Bioresource Technol 100:501–504
Sastry M, Ahmad A, Khan MI, Kumar R (2003) Biosynthesis of metal nanoparticles using fungi and actinomycete. Curr Sci India 85:162–170
Schafer FQ, Buettner GR (2001) Redox environment of the cell as viewed through the redox state of the glutathione disulfide/ glutathione couple. Free Radic Biol Med 30:1191–1212
Schlosser D, Höfer C (2002) Laccase-catalyzed oxidation of Mn^{2+} in the presence of natural Mn^{3+} chelators as a novel source of extracellular H_2O_2 production and its impact on manganese peroxidase. Appl Environ Microbiol 68:3514–3521
Schmid G (1992) Large clusters and colloids. Metals in the embryonic state. Chem Rev 92:1709–1727
Senapati S, Ahmad A, Khan MI, Sastry M, Kumar R (2005) Extracellular biosynthesis of bimetallic Au-Ag alloy nanoparticles. Small 1:517–520

Shaligram NS, Bule M, Bhambure R, Singhal RS, Singh SK, Szakacs G, Pandey A (2009) Biosynthesis of silver nanoparticles using aqueous extract from the compactin producing fungal strain. Process Biochem 44:939–943

Shankar SS, Ahmad A, Pasricha R, Sastry M (2003) Bioreduction of chloroaurate ions by geranium leaves and its endophytic fungus yields gold nanoparticles of different shapes. J Mater Chem 13:1822–1826

Shanti SS, Karl JD (2006) The significance of amino acids and amino acid derived molecules in plant responses and adaptation to heavy metal stress. J Exp Botany 57:711–726

Sharma KG, Mason DL, Liu G, Rea PA, Bachhawat AK, Michaelis S (2002) Localization, regulation, and substrate transport properties of Bpt1p, a *Saccharomyces cerevisiae* MRP-type ABC transporter. Eukaryotic Cell 1:391–400

Sheehan D, Meade G, Foley VM, Dowd CA (2001) Structure, function and evolution of glutathione transferases: implications for classification of non-mammalian members of an ancient enzyme superfamily. Biochem J 360:1–16

Shi ZZ, Osei-Frimpong J, Kala G, Kala SV, Barrios RJ (2000) Glutathione synthesis is essential for mouse development but not for cell growth in culture. Proc Natl Acad Sci USA 97:5101–5106

Shin HS, Yang HJ, Kim SB, Lee SS (2004) Mechanism of growth of colloidal silver nanoparticles stabilized by polyvinyl pyrrolidone in γ-irradiated silver nitratesolution. J Colloid Interface Sci 274:89–94

Song WY, Sohn EJ, Martinoia E, Lee YJ, Yang YY, Jasinski M, Forestier C, Hwang I, Lee Y (2003) Engineering tolerance and accumulation of lead and cadmium in transgenic plants. Nat Biotechnol 21:914–919

Song JY, Kwon EY, Kim BS (2010) Biological synthesis of platinum nanoparticles using *Diopyros kaki* leaf extract. Bioprocess Biosyst Eng 33:159–164

Southam G, Beveridge TJ (1994) The in vitro formation of placer gold by bacteria. Geochim Cosmochim Acta 58:4527–4530

Srikanth CV, Vats P, Bourbouloux A, Delrot S, Bachhawat AK (2005) Multiple cis-regulatory elements and the yeast sulphur regulatory network are required for the regulation of the yeast GSH transporter, Hgt1p. Curr Genet 47:345–358

Stadtman ER, Levine RL (2003) Free radical-mediated oxidation of free amino acids and amino acid residues in proteins. Amino Acids 25:207–218

Stadtman ER, Moskovitz J, Levine RL (2003) Oxidation of methionine residues of proteins: biological consequences. Antioxid Redox Signal 5:577–582

Stephen DW, Jamieson DJ (1997) Amino acid-dependent regulation of the *Saccharomyces cerevisiae GSH1* gene by hydrogen peroxide. Mol Microbiol 23:203–210

Suman V, Prasad R, Jain VK, Varma A (2010) Role of nanomaterials in symbiotic fungus growth enhancement. Curr Sci 99:1189–1191

Tamás MJ, Labarre J, Toledano MB, Wysocki R (2005) Mechanisms of toxic metal tolerance in yeast. In: Tamás MJ, Martinoia E (eds) Molecular biology of metal homeostasis and detoxification: from microbes to man. Springer, Heidelberg

Tanvir S, Oudet F, Pulvin S, Anderson WA (2012) Coenzyme based synthesis of silver nanocrystals. Enzyme Microb Technol 51:231–236

Tripathi AK, Harsh NSK, Gupta N (2007) Fungal treatment of industrial effluents: a mini-review. Life Sci J 4:78–81

Trotter EW, Grant CM (2005) Overlapping roles of the cytoplasmic and mitochondrial redoc regulatory systems in the yeast *Saccharomyces cerevisiae*. Eukaryot Cell 4:392–400

Ulla AJ, Patrick AWV, Ulla SL, Roger DF (2000) Organic acids produced by mycorrhizal *Pinus sylvestris* exposed to elevated aluminium and heavy metal concentrations. New Phytologist 146:557–567

Valtchev V, Tosheva L (2013) Porous nanosized particles: preparation, properties, and applications. Chem Rev 113:6734–6760

Vatamaniuk OK, Mari S, Lu YP, Rea PA (1999) At PCS1, a Phytochelatin synthase from *Arabidopsis*: isolation and in vitro reconstitution. Proc Natl Acad Sci USA 96:7110–7115
Verbavatz JM, Vavasseur A, Forestier C (2009) A common highly conserved cadmium detoxification mechanism from bacteria to humans heavy metal tolerance conferred by the ATP-binding cassette (ABC) transporter sphmt1 requires glutathione but not metal-chelating phytochelatin peptides. J Biol Chem 284:4936–4943
Vesentini D, Dickinson DJ, Murphy RJ (2006) Fungicides affect the production of fungal extracellular mucilaginous material (ECMM) and the peripheral growth unit (PGU) in two woodrotting basidiomycetes. Mycol Res 110:1207–1213
Wang X, Zuo J, Keil P, Grundmeier G (2007) Comparing the growth of PVD silver nanoparticles on ultra thin fluorocarbon plasma polymer films and self-assembled fluoroalkyl silane monolayers. Nanotechnol 18:265303
Wemmie JA, Szczypka MS, Thiele DJ, Moye-Rowley WS (1994) Cadmium tolerance mediated by the yeast AP-1 protein requires the presence of an ATP-binding cassette transporter-encoding gene, *YCF1*. J Biol Chem 269:32592–32597
Westwater J, McLaren NF, Dormer UH, Jamieson DJ (2002) The adaptive response of *Saccharomyces cerevisiae* to mercury exposure. Yeast 19:233–239
Winkler BS, Orselli SM, Rex TS (1994) The redox couple between glutathione and ascorbic acid: a chemical and physiological perspective. Free Radic Biol Med 17:333–349
Wong CM, Siu KL, Jin DY (2004) Peroxiredoxin-null yeast cells are hypersensitive to oxidative stress and are genomically unstable. J Biol Chem 22:23207–23213
Wood ZA, Schroder E, Harris JR, Poole LB (2003) Structure, mechanism and regulation of peroxiredoxins. Trends Biochem Sc 28:32–40
Wright SF, Franke-Snyder M, Morton JB, Upadhyaya A (1996) Time-course study and partial characterization of a protein on hyphae of arbuscular mycorrhizal fungi during active colonization of roots. Plant Soil 181:193–203
Wu AL, Moye-Rowley WS (1994) *GSH1*, which encodes γ-glutamylcysteine synthetase, is a target gene for yAP-1 transcriptional regulation. Mol Cell Biol 14:5832–5839
Wysocki R, Tamás MJ (2010) How *Saccharomyces cerevisiae* copes with toxic metals and metalloids. FEMS Microbiol Rev 34:925–951
Xie H, Lee JY, Wang DIC, Ting YP (2007) Silver nanoplates: from biological to biomimetic synthesis. ACS Nano 1:429–439
Yompakdee C, Bun-ya M, Shikata K, Ogawa N, Harashima S, Oshima Y (1996) A putative new membrane protein, Pho86p, in the inorganic phosphate uptake system of *Saccharomyces cerevisiae*. Gene 171:41–47
Zenk MH (1996) Heavy metal detoxification in higher plants-a review. Gene 179:21–30

Chapter 2
Innovation of Strategies and Challenges for Fungal Nanobiotechnology

Necdet Saglam, Ozfer Yesilada, Ahmet Cabuk, Mesut Sam, Semran Saglam, Sedef Ilk, Ezgi Emul, Pınar Aytar Celik, and Ekrem Gurel

Abstract Nanotechnology involves the study and use of materials under the 100 nm scale, exploiting the different physiochemical properties exhibited by these materials at the nanoscale level. Microorganisms are the best model and role of action for the nano/biotechnological applications. This technology has become increasingly important for the biotechnology and the related sectors. Promising applications have been already employed in the areas of drug delivery systems using bioactive nanoencapsulation, biosensors to detect and quantify pathogens, chemical and organic compounds, alteration of food compositions, and high-performance sensors and film to preserve fruits and vegetables. Moreover, the taste of food and food safety can be improved by new nano-materials from the microbiological sources. The huge benefits from this technology have led to increases in the market investments in nanoscience and nanoproducts in several areas.

N. Saglam (✉) • E. Emul
Division of Nanotechnology and Nanomedicine, Graduate School of Science and Engineering, Hacettepe University, Ankara, Turkey
e-mail: saglam@hacettepe.edu.tr

O. Yesilada
Department of Biology, Faculty of Science and Letters, Inonu University, Malatya, Turkey

A. Cabuk • P.A. Celik
Department of Biotechnology, Graduate School of Natural and Applied Science, Osmangazi University, Eskisehir, Turkey

M. Sam
Department of Biology, Faculty of Science and Letters, Aksaray University, Aksaray, Turkey

S. Saglam
Department of Atomic and Molecular Physics, Faculty of Science, Gazi University, Ankara, Turkey

S. Ilk
Faculty of Agricultural Sciences and Technologies, Nigde University, Nigde, Turkey

E. Gurel
Department of Biology, Faculty of Science and Letters, Abant Izzet Baysal University, Bolu, Turkey

R. Prasad (ed.), *Advances and Applications Through Fungal Nanobiotechnology*, Fungal Biology, DOI 10.1007/978-3-319-42990-8_2

Fungi are the common source of industrial enzymes by cause of their excellent capacity for extracellular protein production. These industrial enzymes are applied in pulp and paper chemical and biomedical products, food, starch, textile, drinks, baking, leather, detergents and animal feed. For industrial application, immobilization of enzymes has advantages due to their improvement in the stability and storage ability because of reuse, easy separation of enzymes from the reaction mixture, a possible increase in pH and thermal stability and low product cost. The reusability and the cost of immobilized enzymes display a great advantage comparing to those of free enzymes. Using nanoscale structures for immobilization is preferred due to an increase in the functional surface area to maximize enzyme loading and reducing diffusion limitations. In addition, the physical characteristics of nanostructure such as enhanced diffusion, thermal stability, irradiation resistance and support mobility can impact catalytic activity of immobilized enzymes. This chapter deals with the strategies, challenges, applications and benefits of fungal nanobiotechnology in different areas and, also, antifungal activity of nanoparticles from the microbial sources.

Fungal nanobiotechnology based agro-industries and environmental spheres created the enormous range of possible applications of fungi. The successful and promising studies in these areas have provided a better understanding of fungi in nanobiotechnological disciplines. The utilization of fungi in the environmental biotechnology is a more recent development with many advantages related to bioremediation, treatment of industrial wastes and biotransformation of specific compounds. The objective of this chapter is to summarize recent developments in fungal nanobiotechnology and fungal synthesis of nanoparticles.

The manufacture and use of dyes are widespread industries. The utilization of these pigments is an integral part of almost all manufacturing processes. Wastewaters are produced during the synthesis and use of dyes. Decolorization of water is a significant and a critical part of wastewater treatment processes. Furthermore, metal contaminated industrial wastewater treatment is also acknowledged as one of the bionanotechnological issues. Microorganisms, especially fungi, are possible and strong candidates for heavy metal removal from wastewaters due to its binding ability to a toxic metal or metal ion. Thus, nanobiotechnological aspects of fungal studies in wastewater treatment applications are explained in a separate section in this chapter.

Nanoparticles can be used in various areas such as medicine, biosensors, environmental treatment and so on. These could be produced by conventional chemical and physical methods although conventional methods have some disadvantages. Therefore, a relatively simple, economical and nonhazardous (i.e., eco-friendly) method must be used in order to synthesize various nanoparticles. Biotechnological methods have several advantages over conventional ones. Nanoparticles can be synthesized by using various organisms such as fungi and bacteria. Here, some of the fungi used in the synthesis of nanoparticles are reviewed. The mechanism of bionanoparticle synthesis and biological activity of these nanoparticles are also discussed.

1 Introduction

Nanotechnology is known as the art of the creation and modification of materials in the general size up to 100 nm and this science is greatly affected by various disciplines especially physics, chemistry and biology (Fig. 2.1). This technology provides great opportunities by changing the significant properties of materials when they are in nano size. Because, when the dimension of a material is getting closer to nanometer, quantum physics becomes to play an important role instead of conventional laws of physics. This situation brings different and unique properties to the material. In other words, nanoscience helps us understand these new behaviours that emerge in nanometer size due to quantum theories. Thus, nanotechnology lets us to design and synthesize nanostructures and provides us to use these extraordinary properties in order to generate new material forms and new application areas such as medicine, environment, green technological methods, pharmacology, electronics etc. (Rai et al. 2009; Suman et al. 2010; Gupta et al. 2012; Aziz et al. 2015). Nanobiotechnology is a new branch of nanotechnology, combining biological principles with physical and chemical procedures to generate nanoscale particles with specific functions and structures.

Nowadays, metal nanoparticles like silver, gold and platinum have great popularity because of their many advantageous properties in several fields of application. Furthermore, metals and metal oxides can be used as antimicrobial agents. During the studies with antimicrobial nanoparticles, the developing resistance of microorganisms against nanoparticles was not evident (Mühling et al. 2009).

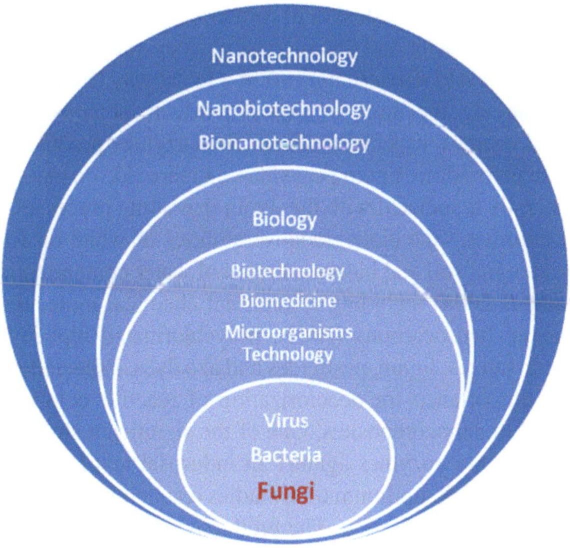

Fig. 2.1 Nanotechnology and its related areas

Silver nanoparticles are one of the most researched and popular ones among the other metal nanomaterials. Its usage in wound healing applications and treatments of some infections makes them very preferable and attainable particularly in biomedical field. Even if they are widely utilized in many areas, these nanoparticles have many side effects especially against environment because of their toxicity (Rai et al. 2009; Gupta et al. 2012). Therefore, new eco-friendly approaches have been improved for green biosynthesis of nanoparticles. Fungal nanotechnology has emerged due to these requirements and studies. In addition, energy shortage and malnutrition problems have occurred with increasing population. Environmental pollution is also getting a threatening issue for human life and, recently, nanotechnology has appeared as a promising tool to solve such environmental- and health-related issues.

Nanobio/Bionanotechnology is playing a significant role for solving current difficulties, many scientists specifying that the studies of microorganisms would provide precious contributions. Fungal nanotechnology can be defined as the fabrication of nanoparticles with green synthesis by fungi, which bring significant advantages to the nanomaterials. Because of the unsatisfactory researches that include both the activity of microorganisms and nanotechnology for solving the problems that mentioned above is the main reason for focusing recently more on fungal approaches in nanobiotechnology.

Fungi have a number of advantages for green synthesis of nanoparticles compared with other organisms, particularly for their easy isolation and ability of extracellular enzyme secretion (Singh et al. 2014; Prasad et al. 2015). Also, many of the proteins that secreted by fungi can transform metal ions rapidly with non-hazardous processes. Therefore, there is an increasing interest in using fungi for these processes, and they may have a significant potential to provide a quick and safe process for production of metallic nanoparticles (Rai et al. 2009). For example, an endophytic fungus, *Penicillium* sp., isolated from the leaves of turmeric (*Curcuma longa*) was studied for producing silver nanoparticles and this successful study was used as a weapon against *Staphylococcus aureus* and *Escherichia coli* in a facial way (Singh et al. 2014).

Enzymes are among the most important products obtained for human needs through microbial sources, especially fungi. The enzymes produced from white rot fungi (manganese peroxidase, lignin peroxidase, superoxide dismutase, ligninase and laccase etc.) generally associated with the lignin degrading processes. There is a popular interest in improving the biodegradative abilities of white rot fungi for treating contaminated water and soil and bioremediation of water contamination with colored substances (Bumpus 2004) including recalcitrant chemical molecules, for example, polycyclic aromatic hydrocarbons (PAH), polychlorinated biphenyls and dioxins, phenols and its derivatives, lignin, pesticides and azo dyes. Especially, laccase enzyme has an industrial importance for decolorization of reactive coloring substances by being immobilized to nanocomposites. One of the significant studies was to use free and immobilized laccase enzymes against an industrial textile dye, Reactive Red 5 (Ilk et al. 2016). As a result, optimum time studies resulted in 120 min for free laccase and 90 min for immobilized laccase. The productivity of free laccase was measured as 33 % while immobilized laccase was 65 %. Also, during the reapplication of laccase complex to the nanostructure for 10 times over a period of 3 days; it was found that 77.3 % of the initial productivity was maintained. In another similar study, Cetin et al.

(unpublished data/submitted paper) focused on *Phanerochaete chrysosporium* loaded monolithic composite cryogel columns for the removal of mercury (Hg^{2+}). The columns showed a great efficiency about biosorption of Hg^{2+}.

Based on these studies and aspects, because of these products are really hazardous, their removal from clean sources with conventional methods is not sufficient nowadays, so fungal nanotechnology may contribute to these refining studies. These effective, economical, eco-friendly and quick nanobiotechnological methods can provide a new perspective to these researches.

In brief, nanotechnology brings new aspects to biological applications and provides many advantages unlike the conventional methods. Bionano/Nanobiotechnological studies include the scientific researches with living organisms; bacteria, yeasts, fungi etc. and when nanotechnological studies are observed from fungal aspects, these new approaches will contribute a lot to the conventional nanotechnological methods (Prasad et al. 2015). The significant roles of fungi in green technology and its biological activities in many areas make these unique organisms to provide a variety of bio-engineering applications. Fungal nanotechnology also creates more effective and eco-friendly synthesis methods for metal nanoparticles and the other toxic and hazardous methods are about to become invalid nowadays. Additionally, nano-sized fibres created with electrospinning method are getting popular in encapsulation of microorganisms and in microbial biosensors; some microorganisms being able to detect comprehensive chemical substances. Dar and Soytong (2014) indicated the ability of electrospinning method for the encapsulation of anti-fungal compounds. For example, they encapsulated the active compounds from *Chaetomium* species that are widely known as an effective anti-fungal agent. Also, El-Newehy et al. (2012) tested polyvinyl alcohol (PVA) and polyethylene oxide (PEO) nanofibers against pathogenic fungi such as *Penicillium notatum*, *Aspergillus niger* and *Aspergillus flavus*. The results showed that these nanofibers are able to exhibit zone of inhibitions.

As mentioned above, because of the insufficient studies related to environmental/green nanotechnological methods with using living organisms and the increasing importance of nanotechnology in industrial applications, fungal nanotechnology is emphasized within this chapter by utilising both general nanotechnology description and its fungal applications.

2 Biosynthesis of Metallic Nanoparticles by Fungi

2.1 *Filamentous Fungi*

Several products obtained from the filamentous fungi are produced at commercial scale, including organic acids, antibiotics and enzymes. Moreover, fungi and their secondary metabolites are important biological agents for industrial applications such as biodegradation, bioremediation, bioaugmentation and biotransformation (Çabuk et al. 2013; Grimm et al. 2005). Currently, fungi are used for the fabrication of nanoparticles.

Progresses in the field of nanobiotechnology have resulted in the production of different metal nanoparticles, which have found a decent ground for several applications. Production of nanoparticles by fungi has some practical advantages. Fungi that are typically not exposed to large concentrations of metal(s) can be transformed to a form that can tolerate high concentrations of metal ions owing to their inherent ability to yield higher concentrations of proteins which aids in the reduction of metal ions to less toxic forms (Sastry et al. 2003; Prasad et al. 2015).

Additionally, fungal mycelial mesh can withstand flow pressure and agitation and other conditions in bioreactors or other chambers compared to bacteria. These are fastidious to grow and easy to handle and produce. The extracellular secretions of reductive proteins are more common and can be easily handled in downstream processing. Also, since the nanoparticles precipitated outside the cell is devoid of unnecessary cellular components, it can be directly used in various applications (Narayanan and Sakthivel 2010). Therefore, several fungi have been utilized for many biotechnological processes on the industrial scale; the utilization of waste mycelium would be promising for feasible, eco-friendly and cost effective biosynthesis of nanoparticles. Nowadays, fungi are excellent candidates for the synthesis of various metals and metal sulfides nanoparticles (Sastry et al. 2003; Sadowski et al. 2008).

Metallic nanoparticles have potential applications in various fields, such as electronics, cosmetics, coatings, packaging and medical applications (Prasad et al. 2014). Nanoparticles can be induced to amalgamate into a solid at relatively lower temperatures, often without melting, leading to the improvement of easy-to-create coatings for electronics applications. Nanoparticles possess a wavelength below the critical wavelength of light, which renders them transparent, a property that makes them very useful for applications in cosmetics and packaging. Nanoparticles can be employed as an efficient tool to explore the finest processes in various biotechnologies including biomedical sciences (Hutten et al. 2004; Safarik and Safarikova 2002). Apart from this, nanoparticles play an essential role in drug delivery, diagnostics, imaging, sensing, gene delivery and tissue engineering (Morones et al. 2005; Arruebo et al. 2007; Prasad 2014; Prasad et al. 2014).

Verticillium sp. and *Fusarium oxysporum* are known to produce nanoparticles when challenged with aqueous solutions of metal ions either intracellularly or extracellularly (Mukherjee et al. 2001a; Gericke and Pinches 2006). *Fusarium*, *Aspergillus*, and *Penicillium* have capable potential for the extracellular production of different metal nanoparticles. The biosynthesis of nanoparticles by various strains of filamentous fungi is presented in Table 2.1 and some examples of nanoparticles are described below.

2.2 Gold Nanoparticles

In the study of Castro-Longoria et al. (2011), the filamentous fungus *Neurospora crassa* was screened and found to be successful for the production of mono and bimetallic Au/Ag nanoparticles. Scanning electron microscopy (SEM), energy

Table 2.1 The list of nanoparticles synthesized from different fungi

Nanoparticle type	Fungus name	Size of the nanoparticle (nm)	Production location	References
Gold	*Aspergillus oryzae*	10–60	Extracellular	Binupriya et al. (2010)
	Verticillium luteoalbum	10	Extracellular	Gericke and Pinches (2006)
	Trichothecium sp.	5–200	Extracellular	Ahmad et al. (2005)
	Collitotrichium sp.	20–40	Extracellular	Shankar et al. (2003)
	Fusarium oxysporium	20–40	Extracellular	Mukherjee et al. (2002)
	Collitotrichum sp.	20–40	Extracellular	Shankar et al. (2003)
	Verticillium Luteoalbum	<10–100	Intracellular	Gericke and Pinches (2006)
	Verticillium sp.	2–20	Intracellular	Mukherjee et al. (2001a), Mukherjee et al. (2001b)
Silver	*Rhizopus nigricans*	35–48	Extracellular	Ravindra and Rajasab (2014)
	Trichoderma sp.	5–40	Extracellular	Fayaz et al. (2010)
	Alternaria alternata	20–60	Extracellular	Gajbhiye et al. (2009)
	Phoma glomerata	60–80	Extracellular	Birla et al. (2009)
	Penicillium fellutanum	5–25	Extracellular	Kathiresan et al. (2009)
	Penicillium brevicompactum	58.35 ± 17.88	Extracellular	Shaligram et al. (2009)
	Fusarium solani	16.23	Extracellular	Ingle et al. (2009)
	Humicola sp.	5–25	Extracellular	Syed et al. (2013)
	Cladosporium cladosporioides	10–100	Extracellular	Balaji et al. (2009)
	Aspergillus fumigates	5–25	Extracellular	Gade et al. (2008); Prabhu et al. (2009)
	Volvariella volvacea	Ag and Au-Ag 15 and 20–150 nm	Extracellular	Daizy (2009)
	Fusarium semitectum	10–60	Extracellular	Basavaraja et al. (2008)
	Aspergillus niger	20 nm/3–30 nm	Extracellular	Gade et al. (2008); Jaidev and Narasimha (2010)
	Aspergillus flavus	8.92	Extracellular	Vigneshwaran et al. (2007)
	Penicillium purpurogenum	5–200	Extracellular	Nayak et al. (2011)
	Penicillium atramentosum	5–25	Extracellular	Sarsar et al. (2015)

(continued)

Table 2.1 (continued)

Nanoparticle type	Fungus name	Size of the nanoparticle (nm)	Production location	References
	Fusarium oxysporium	5–50 nm/8–14 nm Au-Ag	Intracellular/ extracellular	Senapati et al. (2004); Senapati et al. (2005)
	Aspergillus clavatus	550–650	Extracellular	Saravanan and Nanda (2010)
	Neurospora crassa	11 nm for silver and 32 nm for gold	Extracellular	Castro-Longoria et al. (2011)
	Trichoderma asperellum	13–18	Extracellular	Mukherjee et al. (2008)
Silver	*Tricoderma viride*	2–4	Extracellular	Mohammed-Fayaz et al. (2009)
	Trichoderma reesei	5–50	Extracellular	Vahabi et al. (2011), Mansoori (2010)
	Alternaria alternate	20–60	Extracellular	Gajbhiye et al. (2009)
	Fusarium oxysporum	30	Extracellular	Duran et al. (2007)
Zirconia	*Fusarium oxysporium*	3–11	Extracellular	Bansal et al. (2005)
Zinc oxide	*Aspergillus terreus*	54.8–82.6	Extracellular	Baskar et al. (2013)
Cadmium	*Fusarium oxysporium*	9–15/5–20	Extracellular	Ahmad et al. (2002), Kumar et al. (2007)
Silica	*Fusarium oxysporium*	5–15	Extracellular	Bansal et al. (2005)
Platinum	*F. oxysporum* f. sp. *lycopersici*	10–100	Intracellular/ Extracellular	Riddin et al. (2006)
Titanium	*Fusarium oxysporium*	6–13	Extracellular	Bansal et al. (2005)
Magnetite	*Fusarium oxysporium*	20–50	Extracellular	Bharde et al. (2006)

dispersive X-ray spectroscopy (EDS), and transmission electron microscopy (TEM) confirmed the biosynthesis of nanoparticles by *N. crassa*. The size of nanoparticles was found to be average diameter of 32 nm for gold when the fungus was exposed to the aqueous solutions of $HAuCl_4$. The results obtained indicate that *N. crassa* can be a potential "nanofactory" for the synthesis of metallic nanoparticles.

Vala (2015) produced gold nanoparticles using a marine-derived fungal isolate, *Aspergillus sydowii*. The mode of biosynthesis (extracellular/intracellular) depended on supplied gold ion concentration. Higher concentrations supported synthesis of smaller particles. The particles biosynthesized at 3 mM gold chloride were found to be spherical and nearly monodisperse in nature. The particles were found to be in the size range of 8.7–15.6 nm with a mean diameter of 10 nm (Vala 2015). Roy and co-workers studied the extracellular biosynthesis of gold nanoparticles (GNPs) using the fungal species *Aspergillus foetidus*. X-ray diffraction (XRD) results revealed distinctive formation of face centered cubic crystalline GNPs. The spherical and polydispersed GNPs in the range of 10–40 nm were observed by TEM analysis (Roy et al. 2016).

Gopinath and Arumugam (2014) investigated the extracellular synthesis of gold nanoparticles from *Fusarium solani* culture filtrate. Synthesized gold nanoparticles were characterized by UV–VIS, FTIR, XRD, AFM, and TEM analysis. TEM results revealed that the gold nanoparticles were highly stable in the diameter range between 20 and 50 nm. Mukherjee et al. (2002) also studied *Fusarium oxysporum* strain for producing gold nanoparticles using green chemistry approach. The same researchers utilized *Trichoderma asperellum* for biosynthetic route to nanocrystalline silver particles (Mukherjee et al. 2008).

2.3 Silver Nanoparticles

Microorganisms may play a significant role in toxic metal remediation through reduction of metal ions. Studies demonstrated that silver ions might be reduced extracellularly using *Fusarium oxysporum* to generate stable gold or silver nanoparticles in water. These nanoparticles synthesized by some microbes can be incorporated with several kinds of materials such as cloths. These cloths with silver nanoparticles are sterile and can be useful in hospitals to prevent or to minimize infection with pathogenic bacteria such as *S. aureus*. Duran et al. (2007) investigated the extracellular production of silver nanoparticles by *F. oxysporum* and its antimicrobial effect when incorporated with cotton fabrics against *S. aureus*. In conclusion, it was demonstrated the application of biological synthesis to silver nanoparticles production and its incorporation in cloths, providing them sterile properties. Similar results were also obtained by Ahmad et al. (2002). Basavaraja et al. (2008) produced highly stable and crystalline silver nanoparticles (10–60 nm) in solution by treating the filtrate of the fungus *F. semitectum* with the aqueous silver nitrate solution.

In another study, the biosorption of silver in the form of nanoparticles by the fungus *Aspergillus flavus* was demonstrated. These nanoparticles were found to be stable in aqueous for more than three months, which could be attributed to surface binding of stabilizing materials secreted by the fungus. The average size of the nanoparticles was determined to be 8.92 ± 1.61 nm. Vigneshwaran et al. (2007) concluded that process of nanoparticle production is eco-friendly as it is free from any solvent or toxic chemicals. Gade et al. (2008) also performed biosynthesis of silver nanoparticles by *Aspergillus niger* isolated from soil. The nanoparticles characterized by TEM exhibited spherical silver nanoparticles with a diameter of around 20 nm. The silver nanoparticles showed remarkable antibacterial activity against *Staphylococcus aureus* and *Escherichia coli*. The reduction of the silver ions might have occurred by a nitrate-dependent reductase enzyme and a shuttle quinone extracellular process. Potential of fungal-mediated biosynthesis of silver nanoparticles may be important for the development of effective antibacterial agents showing resistance to drugs available in the market. Another *Aspergillus* strain, *A. fumigatus*, was also used for extracellular biosynthesis of silver nanoparticles, synthesis process of which was quite fast and silver nanoparticles formed within minutes of silver ion coming in contact with the cell filtrate. TEM micrograph showed formation of well-dispersed silver nanoparticles in the range of 5–25 nm (Bhainsa and D'souza 2006).

Vahabi et al. (2011) reported the extracellular biosynthesis of silver nanoparticles (AgNPs) by using a fungus named *Trichoderma reesei*. In the biosynthesis of AgNPs by this fungus, the fungus mycelium is exposed to the silver nitrate solution. That prompts the fungus to produce enzymes and metabolites for its own survival. In this process the toxic Ag^+ ions are reduced to the non-toxic metallic AgNPs through the catalytic effect of the extracellular enzyme and metabolites of the fungus. They suggested that this process is an excellent candidate for industrial scale production of silver nanoparticles

2.4 Other Nanoparticles

Bansal et al. (2005) showed that *Fusarium oxysporum* secreted proteins capable of hydrolyzing aqueous zirconia ions extracellularly at room temperature. Particularly gratifying is the fact that the fungus is capable of hydrolyzing tough metal halide precursors under acidic conditions. To use as biosorbent agent, the combination of SiO_2-nanoparticles (N-Si) with *Penicillium funiculosum* for the formation of (N-Si-Pen) was investigated as a solid sorbent phase. This biosynthesized nanoparticle was utilized to adsorb Pb(II). The maximum capacity value was 1266.7 $\mu mol\ g^{-1}$ for N-Si-Pen combined particle, at pH 5. Sorption equilibrium was established in about 20 min.

2.5 White Rot Fungi

White rot fungi are able to degrade a wide variety of environmentally pollutants such as organopollutants and xenobiotics by means of laccase enzyme (Birhanli and Yesilada 2010; Birhanli et al. 2013; Yesilada et al. 2014). Due to their many well-known advantages, various researchers are interested in the development of white rot fungi technology for biodegradation of pollutants (Cihangir and Saglam 1999), decolorization of textile dyes (Yesilada et al. 2003) and biodesulphurization of coal (Aytar et al. 2008).

There is a great interest in metallic nanoparticles because of their various applications in several areas such as medicine, biosensors preparation, water purification, food packaging and cosmetic industry (Li et al. 2011; Das and Thiagarajan 2012).

Various metallic nanoparticles including silver, iron, gold, cadmium, selenium and copper could be synthesized by physical and chemical methods. However, these methods are not eco-friendly as well as having some disadvantages. These nanoparticles must be produced via environment-friendly and safe methods due to their possible uses, especially in biomedical applications. Therefore, there is need to produce this metallic nanoparticles via safe approaches. A biological method, which is the green way of synthesizing, may be an alternative and eco-friendly method for producing metallic nanoparticles. There have been serious attempts to develop biological processes for nanoparticle production. The main advantage of biological methods is that they are safe and eco-friendly. In this green way, biological systems such as fungi, bacteria, plants and also enzymes can be used to produce these nanoparticles. Due to the pathogenic properties of various fungi, the use of safe fungi is important. The solution could be white rot fungi. These fungi are non-pathogenic and many of them have also medicinal properties. There is limited study on the production of nanoparticles by white rot fungi. The extract, mycelia or culture filtrate of white rot fungi can be used to synthesize metallic nanoparticles with various sizes via bioreduction process (Table 2.2).

Table 2.2 Some nanoparticles (AgNPs) produced by white rot fungi and their applications

Fungus	NP	Shape	Size (nm)	Application	References
P. sajor caju	AgNP	Spherical and well distributed without aggregation	40	Decolorization	Nithya and Ragunathan (2011)
P. sajor-caju	AgNP	Spherical	5–50	Antibacterial	Nithya and Ragunathan (2009)
P. florida	AgNP	Spherical	20 ± 5	Antibacterial	Bhat et al. (2011)
P. ostreatus	AgNP	–	100	Antibacterial	Mirunalini et al. (2012)
P. ostreatus	AgNP	Polydispersed with spherical	4–15	Anticandidal, Anticancer	Yehia and Al-Sheikh (2014)
G.neo-japonicum Imazeki	AgNP	Spherical and well dispersed	5–8	Anticancer	Gurunathan et al. (2013)
Pleurotus djamor var. *roseus*	AgNP	Spherical	90–370	Anticancer	Raman et al. (2015)
Pleurotus cornucopiae var. *citrinopileatus*	AgNP	Spherical	20–30	Anticandidal	Owaid et al. (2015)

(continued)

Table 2.2 (continued)

Fungus	NP	Shape	Size (nm)	Application	References
T.trogii	AgNP	–	<10	–	Unpublished data
Pleurotus sp.	FeNP		–	–	Mazumdar and Haloi (2011)
Stereum hirsutum	CuNP and copper oxide	Monodispersed and spherical	5–20	–	Cuevas et al. (2015)
Lentinula edodes	AuNP	Spherical	5–50	–	Vetchinkina et al. (2013)
Ganoderma sp.	AuNP	Monodispersed and spherical	20	Biocompatibility	Gurunathan et al. (2014)
Pycnoporus sanguineus	AuNP	Various shapes	Several to several hundred nm	Biodegradation	Shi et al. (2015)
P. ostreatus	CdS	Spherical	4–5	–	Borovaya et al. (2015)

3 Biosynthesis of Silver Nanoparticles

Silver nanoparticles (AgNPs) have various applications including therapeutic ones. Therefore, their production by biotechnological approach is very important. There are some studies on AgNP production potential of white rot fungi. Vigneshwaran et al. (2006) used the mycelial mat of *Phanerochaete chrysoporium* to synthesize the stable AgNPs. The workers stated that the extracellular proteins of this fungus keep the NPs stable by minimizing the aggregation with capping. They reported the peak maximum at 470 nm instead of 413 nm which might be due to a decrease in particle size.

The same group also produced uniform AgNPs, which remained stable for more than 6 months by spent substrate (stalks) of edible *Pleurotus sajor caju* (Gade et al. 2008). The proteins secreted by this fungus on solid substrate functioned as capping agent and thus stabilized the NPs. AgNPs showed the surface plasmon absorption band at 436 nm. They also had an excellent antibacterial activity (Vigneshwaran et al. 2007). Nithya and Ragunathan (2011) produced spherical shaped AgNPs with an average size of 40 nm by the biomass of *P. sajor caju* and used these *P. sajor caju* NPs to decolorize congo red dye. They emphasized that these *P. sajor caju* NPs could be used for treatment of dyes. The biosynthesis of spherical shaped AgNPs, with an antibacterial activity against Gram positive and Gram negative bacteria, using *P. sajor caju* culture filtrate has also been reported by Nithya and Ragunathan (2009).

It was shown that stable AgNPs could be synthesized by culture filtrate or mycelium of *Coriolus versicolor* MUCL (Sanghi and Verma 2009a). The reduction capacity of filtrate (third day) was faster than the mycelium (fourth day) at pH 5.5–6.0. Under alkaline conditions (pH 10) the reaction was much faster and only 60 min was enough for fast reduction of silver ions. They attributed this effect to the increase in the reducing power of the responsible proteins at alkaline conditions. The fungal proteins acted as reducing and capping agents in NPs production process and spherical shaped NPs were obtained. The pH of the solution was reported to be an important factor affecting the morphology of the nanoparticle obtained.

Mycosynthesis of AgNPs was also achieved using the *Pleurotus florida* mushroom extract as reducing agent (Bhat et al. 2011). The spherical shaped NPs formed by the photo-irradiation technique using the extract as bioreductant were in a size range of 20 ± 5 nm. They showed antimicrobial activity against various pathogenic microorganisms. The workers claimed that their approach was biologically safe and eco-friendly due to the biosynthetic nature of this process.

Various extracts of *Pleurotus ostreatus* are also able to synthesize AgNPs. Mirunalini et al. (2012) described the biosynthesis of AgNPs by *P. ostreatus* mushroom extract. The authors stated that the extract containing proteins could act as the reducing and capping agents, and the NPs with antimicrobial activity against *Staphylococcus aureus* could be obtained. Both Devika et al. (2012) and Yehia and Al-Sheikh (2014) demonstrated that *P. ostreatus* extract reduced silver nitrate solutions and produced AgNPs with the diameter of about 50 and 450 nm, respectively. These NPs were stable due to capping by proteins. The AgNPs exhibited antimicrobial activity against various pathogenic microorganisms. Devika et al. (2012) also verified that the antimicrobial activity of AgNPs increased when the mixture of AgNP and antibiotic were used. Yehia and Al-Sheikh (2014) reported the dose dependent antiproliferative effect of biologically synthesized AgNPs against human breast carcinoma cells (MCF-7)

Mycosynthesis of AgNPs by various extracts of *Ganoderma lucidum* was also described by Mirunalini et al. (2012), Karwa et al. (2011) and Paul et al. (2005). Mirunalini et al. (2012) and Paul et al. (2005) used the extract of mushroom form directly, while Karwa et al. (2011) preferred the extract of mycelia. The bioreduction of silver ions was achieved by *Ganoderma lucidum* mushroom and silver ions were completely reduced to AgNPs with an average size of 50 nm after 48 h (Mirunalini et al. 2012). These NPs were found to have antibacterial activity against *S. aureus*. Paul et al. (2005) also reported the fast bioreduction activity of the extract of *G. lucidum* mushroom, with the presence of large amounts of polyphenols in the sample. The authors incorporated the AgNPs in cotton fabrics and their bacteriostatic activity against various bacteria was also tested. The fabrics incorporated with AgNPs showed a high antibacterial activity against various pathogens. Karwa et al. (2011) also described the biosynthesis of stable AgNPs with the average size of 45 nm by mycelial extract of *G. lucidum*. FT-IR analysis confirmed that large particles may be due to the protein capping of the particles and the authors concluded that the extracellular enzymes were responsible for the reduction. The obtained NPs showed antibacterial activity against *S. aureus* and *E. coli*. The combined effect of

NPs and tetracycline antibiotic was also studied. The results demonstrated a higher antibacterial activity of the mixture than the tetracycline alone. This indicates the synergistic effect of NPs and antibiotic.

Gurunathan et al. (2013) described the green synthesis AgNPs by the extract of *Ganoderma neo-japonicum* Imazeki KUM61076. This NPs solution had long-term stability which might be due to the proteins in the extract as capping agent. TEM and DSL analysis showed that most of the AgNPs are spherical in shape and relatively uniform and monodispersed with an average size of 5 nm. The study demonstrated the cytotoxic effect of biologically synthesized AgNPs against MDA-MB-231 human breast cancer cells. It was observed that cell death and membrane leakage were dose-dependent.

Chan and Don (2012) showed that *Schizophyllum commune* and *Pycnoporus sanguineus* could be used for biological synthesis of AgNPs. They used directly mycelia or culture supernatant of these white rot fungi for testing their reduction effect. The mycelia or culture supernatant produced AgNPs with different sizes. The obtained AgNPs were detected as an effective antimicrobial agent against various bacteria and fungi. The authors speculated that AgNPs produced extracellularly as well as by culture supernatant have better antimicrobial activity when compared to AgNPs synthesized intracellularly.

More recently, Raman et al. (2015) and Owaid et al. (2015) described the mycosynthesis of AgNPs by the extract of *Pleurotus djamor* var. *roseus* basidocarps and *P. cornucopiae* var *citrinopileatus*. When the extract of *P. djamor* var. *roseus* basidocarps exposed to silver ions, the spherical AgNPs with an average size ranging from 5-50 nm could be obtained. Spectral and electron microscopy results revealed that the extract serves as a reducing agent in the biosynthesis of AgNPs. The cytotoxic effects of these nanoparticles against human prostate carcinoma cells (PC3) were also described in this study. These AgNPs significantly inhibited the cell viability and induced cell death in dose-dependent manner. The extract of *P. cornucopiae* var *citrinopileatus* was also able to synthesis the spherical shaped AgNPs with an average size ranging from 20 to 30 nm. The authors speculated that ions were reduced by the active molecules like polysaccharides and proteins present in this extract. The nanoparticles exhibited antifungal activity against *Candida albicans*, *C. glabrata*, *C. krusei* and *C. pseudotropicalis* especially at a concentration of 60 μg/well.

Yesilada (unpublished data) observed that it was possible to synthesis AgNPs by the culture filtrate of *P. ostreatus*, *T. trogii* 200800 and *T. versicolor* 200801. When *T. trogii* filtrate incubated with $AgNO_3$ solution, a change in the color of the solution was observed and AgNPs with the diameter below 10 nm were obtained (Table 2.2).

4 Biosynthesis of Other Metallic Nanoparticles

White rot fungi can also be used to synthesize different nanoparticles such as iron, copper, gold and cadmium sulphide nanoparticles. Iron nanoparticles have been successfully synthesized by the mycelia of *Pleurotus* sp. (Mazumdar and Haloi

2011). No color change was observed in medium or biomass. The authors suggested that some biochemical changes may occur in the medium due to the oxidation of ferrous ions.

Cuevas et al. (2015) described the use of the extract of *Stereum hirsutum* for preparing copper and copper oxide nanoparticles. In their study, three copper salts ($CuCl_2$, $CuSO_4$ and $Cu(NO_3)_2$) were used and the effect of various pH on reduction activity of this extract was investigated. $CuCl_2$ (5 mM) gave the highest nanoparticle formation at alkaline conditions. The extracellular protein in this extract may be responsible for nanoparticle formation and stabilization. The obtained nanoparticles (5–20 nm) are spherical in shape, monodispersed and they were embedded in a biopolymer. The XRD analysis also revealed the polycarbohydrate nature associated with stability of the nanoparticles.

Sanghi et al. (2011) compared the utilization of the mycelium and culture filtrate of *P. chrysosporium* for biosynthesis of gold nanoparticles (AuNPs). The rate and shape of particle formation was temperature dependent. Both the mycelium and culture filtrate were ineffective in biosynthesis of AuNPs at room temperature. However, when the temperature increased to 37 °C nanoparticle formation started within 3 min. During the mycelia studies the nanoparticles could only form on the surface of the mycelia and not in the solution (intracellular synthesis); the characteristic absorbance band of 525 nm being detected in the culture filtrate studies (extracellular synthesis). It was possible to obtain stable nanoparticles in the form of spheres and with the diameter of 10–100 nm. Mycelium age influenced the rate and extent of the nanoparticle formation due to the amount of protein secreted. The results of this study demonstrated that the rate of particle formation and size of particles might be controlled by temperature, gold ion concentration and exposure time to $HAuCl_4$. It was also speculated that laccase and lignin peroxidase enzymes are responsible for the extracelullarly and the intracellularly bioreduction of gold ions, respectively.

The AuNP biosynthesis activity of *Lentinula edodes* was investigated during the growth under agitated culture conditions. The growing mycelial cells of this fungus produced and accumulated the spherical nanoparticles with 5–50 nm in diameters. The authors emphasized that elemental gold was reduced and accumulated either on the surface or inside the mycelia (Vetchinkina et al. 2013).

Gurunathan et al. (2014) reported the extract of *Ganoderma* sp. as a good reducing and stabilizing agent. The proteins in the extract help the biosynthesis of AuNPs. The obtained monodispersed AuNPs with an average size of 20 nm were detected as nontoxic and biocompatible against MDA-MB-231 human breast cancer cells. The authors speculated that the biocompatible AuNPs could be used in catalysis, sensors, electronics and biomedical applications, especially for cancer therapy.

Another study on AuNPs biosynthesis is the bioreduction of gold ions by intracellular protein extract of *Pycnoporus sanguineus* (Shi et al. 2015). The AuNPs with high catalytic activity could be obtained using this extract. The amount of extract and also gold ion concentration influenced the nanoparticle production and particle size. The solution pH and bioreduction rate were also important for particle size distribution and characteristic of nanoparticles. The authors also tested the catalytic activity of the obtained nanoparticles in 4-nitroaniline (4-NP) degradation and reported the complete degradation of 4-NP in 6 min.

Sanghi and Verma (2009b) described a continuous and extracellular formation of cadmium sulphide nanoparticle (CdS) by immobilized *Coriolus versicolor* in a column reactor. The protein was reported as a capping agent. TEM images showed that the embedded nanoparticles in the fungal matrix were well dispersed spherical nanoparticles with uniform size (about 5-9 nm). *P. chrysosporium* was able to synthesize various nanoparticles as stated above. Chen et al. (2014) demonstrated that *P. chrysosporium* synthesized uniform and sphere shaped fluorogenic CdS nanoparticles with the average size of 2.56 nm. The rate of production was dependent on pH level. They also stated the role of proteins and amino acids in the formation of the CdS particles. The mycelium of *P. ostreatus* can also be used for the production of luminescent CdS nanoparticles. It was reported that spherical shaped CdS nanoparticles with a particle size of 4–5 nm could be obtained by the mycelium of this fungus (Borovaya et al. 2015).

The fungal-mediated eco-friendly synthesis approach of nanoparticles has many advantages, for example, the ease with which the process can be scaled up, economic viability, downstream processing, and simpler handling of the biomass and possibility of easily covering large surface areas by appropriate growth of the mycelia (Prasad et al. 2015). A number of filamentous fungi have been successfully used for extracellular biosynthesis of silver and gold nanoparticles and others. Generally, analytical techniques, such as ultraviolet-visible spectroscopy, X-ray powder diffraction, and transmission electron microscopy and zeta potential measurements were applied to characterize the morphology of nanoparticles.

5 Biosynthesis of Metallic Nanoparticles by Laccase

Fungi secrete high amount of enzymes in their growth medium. Due to the protein nature of the enzymes, this is an advantage for the biogenic formation of NPs. There are some studies on possible bioreduction role of laccase from fungi (Table 2.3). Faramarzia and Forootanfara (2011) reported that purified laccase enzyme from the ascomycete *Paraconiothyrium variabile* is responsible from production of AuNPs. The gold nanoparticles obtained at 70 °C after 20 min of incubation were in the size

Table 2.3 Nanoparticles (NPs) biosynthesized by laccase enzyme

Source of laccase	NP	Shape	Size (nm)	References
Paraconiothyrium variabile	AuNP	Well dispersed	71–266	Faramarzia and Forootanfara (2011)
P. ostreatus	AuNP	Mono dispersed	22–39	El-Batal et al. (2015)
Trametes versicolor	Ag@AgClNP	Spherical	<100	Jose et al. (2013)

range of 71–266 nm and this was the best temperature for this nanoparticle production activity. It was concluded that the reductive groups of the enzyme might be responsible from AuNPs. El-Batal et al. (2015) also described the AuNP production ability of the partially purified laccase enzyme from solid state culture of *P. ostreatus*. The AuNPs were highly mono dispersed nanoparticles with the size range of 22–39 nm. This production activity was found to be dependent on temperature, radiation and substrate concentration. The authors emphasized that laccase performed the reduction as a protein and not as an active enzyme. Durán et al. (2014) described silver and silver chloride nanoparticles production by semi-purified laccase from *Trametes versicolor*. It was also stated that the main pathway for reduction was the interaction of silver ions with T1 site of laccase enzyme, which the sulfhydryl group is the reducing agent.

6 Future Prospects

Starting with the last century, microorganisms (especially fungi) have been widely used for medical treatments and disease prevention efforts. For that purpose, several primary and secondary metabolites (antibiotics, immune suppressor subtances, enzymes, biosurfactants and organic acids) have been produced at large scales. Considering that we know about a more or less 5 % only of the fungi available in nature, it is highly plausible that there will be an increase in the scientific and technological interest in fungal nanotechnology and the related fields. Since nanoparticles produced by chemical synthesis are highly toxic, a greener alternative (production of nanoparticles via biological pathways) seems to have already gained a strong interest in the scientific community (Jose et al. 2013; Devika et al. 2012; Popescu et al. 2010). On the other hand, functional nanofibrous scaffolds produced by electrospinning have important potential in nanobiotechnology, such as nanomembranes for environmental applications, heavy metals removal from waste water treatment, tissue engineering, enzyme immobilization and drug delivery for biomedical/nanomedicine applications. Thus, all these nanomaterials have a sustainable, biocompatible, biodegradable, antimicrobial and non-toxic of great relevance in nanotechnology (Spasova et al. 2011; Prasad et al. 2015).

References

Adrio JL, Demain AL (2013) Fungal biotechnology. Int Microbiol 6:191–193

Ahmad A, Mukherjee P, Mandal D, Senapati S, Khan M, Kumar R, Sastry M (2002) Enzyme mediated extracellular synthesis of CdS nanoparticles by the fungus, *Fusarium oxysporum*. J Am Chem Soc 124:12108–12109

Ahmad A, Senapati S, Khan MI, Kumar R, Sastry M (2005) Extra-/intracellular biosynthesis of monodisperse gold nanoparticles by a an alkolotolerant fungus, *Trichothecium* sp. J Biomed Nanotechnol 1:47–53

Arruebo M, Fernandez-Pacheco R, Ibarra MR, Santamaria J (2007) Magnetic nanoparticles for drug delivery. Nano Today 2:22–32

Aytar P, Şam M, Çabuk A (2008) Microbial desulphurization of Turkish lignites. Energy Fuels 22:1196–1199

Aziz N, Faraz M, Pandey R, Sakir M, Fatma T, Varma A, Barman I, Prasad R (2015) Facile algae-derived route to biogenic silver nanoparticles: synthesis, antibacterial and photocatalytic properties. Langmuir 31:11605–11612

Balaji DS, Basavaraja SB, Mahesh D, Prabhakar BK, Ventkataraman A (2009) Extracellular biosynthesis of functionalized silver nanoparticles by strains of *Cladosporium cladosporioides* fungus. Colloid Surf B 68:88–92

Bansal V, Rautaray D, Bharde A, Ahire K, Sanyal A, Ahmad A, Sastry M (2005) Fungus-mediated biosynthesis of silica and titania particles. J Mater Chem 15:2583–2589

Basavaraja SS, Balaji SD, Lagashetty AK, Rajasab AH, Venkataraman A (2008) Extracellular biosynthesis of silver nanoparticles using the fungus *Fusarium semitectum*. Mater Res Bull 43:1164–1170

Baskar G, Chandhuru J, Sheraz Fahad K, Praveen AS (2013) Mycological synthesis, characterization and antifungal activity of zinc oxide nanoparticles. Asian J Pharm Tech 3:142–146

Bhainsa KC, D'Souza SF (2006) Extracellular biosynthesis of silver nanoparticles using the fungus *Aspergillus fumigates*. Colloid Surf B 47:160–164

Bharde A, Rautaray D, Bansal V, Ahmad A, Sarkar I, Yusuf SM, Sanyal M, Sastry M (2006) Extracellular biosynthesis of magnetite using fungi. Small 2(1):135–141

Bhat R, Deshpande R, Sharanabasava VG, Do SH, Venkataraman A (2011) Photo-irradiated biosynthesis of silver nanoparticles using edible mushroom *Pleurotus florida* and their antibacterial activity studies. Bioinorg Chem Appl. doi:10.1155/2011/650979

Binupriya AR, Sathishkumar M, Vijayaraghavan K, Yun SI (2010) Bioreduction of trivalent aurum to nano-crystalline gold particles by active and inactive cells and cell-free extract of *Aspergillus oryzae* var. *viridis*. J Hazard Mater 177:539–545

Birhanli E, Yesilada O (2010) Enhanced production of laccase in repeated-batch cultures of *Funalia trogii* and *Trametes versicolor*. Biochem Eng J 52:33–37

Birhanli E, Erdoğan S, Yesilada O, Onal Y (2013) Laccase production by newly isolated white rot fungus *Funlaia trogii:* effect of immobilization matrix on laccase production. Biochem Eng J 71:134–139

Birla SS, Tiwari VV, Gade AK, Ingle AP, Yadav AP, Rai MK (2009) Fabrication of silver nanoparticles by *Phoma glomerata* and its combined effect against *Escherichia coli*, *Pseudomonas aeruginosa* and *Staphylococcus aureus*. Lett Appl Microbiol 48:173–179

Borovaya M, Pirko Y, Krupodorova T, Naumenko A, Blume Y, Yemets A (2015) Biosynthesis of cadmium sulphide quantum dots by using *Pleurotus ostreatus* (Jacq.) P. Kumm. Biotechnol Biotec Eq 29:1156–1163

Bumpus J (2004) Biodegradation of azo dyes by fungi. In: Dilip KA (ed) Fungal biotechnology in agricultural, food, and environmental applications, 21st edn. Basel, New York, pp 457–469

Çabuk A, Aytar P, Gedikli S, Özel YK, Kocabıyık E (2013) Biosorption of acidic textile dyestuffs from aqueous solution by *Paecilomyces* sp. isolated from acidic mine drainage. Environ Sci Pollut Res 20:4540–4550

Castro-Longoria E, Vilchis-Nestor AR, Avalos-Borja M (2011) Biosynthesis of silver, gold and bimetallic nanoparticles using the filamentous fungus *Neurospora crassa*. Colloid Surf B 83:42–48

Chan YS, Don MM (2012) Characterization of Ag nanoparticles produced by white-rot fungi and its in vitro antimicrobial activities. Int Arabic J Antimicrobial Agents 2:1–8

Chen G, Yi B, Zeng G, Niu Q, Yan M, Chen A, Du J, Huang J, Zhang Q (2014) Facile green extracellular biosynthesis of CdS quantum dots by white rot fungus *Phanerochaete chrysosporium*. Colloid Surf B 117:199–205

Cihangir N, Saglam N (1999) Removal of cadmium by *Pleurotus sajor-caju* basidiomycetes. Acta Biotech 19:171–177

Cuevas R, Durán N, Diez MC, Tortella G, Rubilar O (2015) Extracellular biosynthesis of copper and copper oxide nanoparticles by *Stereum hirsutum*, a native white rot fungus from Chilean forests. J Nanomater. doi:10.1155/2015/789089
Daizy P (2009) Biosynthesis of Au, Ag and Au–Ag nanoparticles using edible mushroom extract. Spectrochim Acta A 73:374–381
Dar J, Soytong K (2014) Construction and characterization of copolymer nanomaterials loaded with bioactive compounds from *Chaetomium* species. J Agr Technol 10:823–831
Das K, Thiagarajan P (2012) Mycobiosynthesis of metal nanoparticles. Int J Nanotech Nanosci 1:1–10
Devika R, Elumalai S, Manikandan E, Eswaramoorthy D (2012) Biosynthesis of silver nanoparticles using the Fungus *Pleurotus ostreatus* and their antibacterial activity. Open Access Scientific Reports. doi:10.4172/scientificreports.557
Duran N, Marcato PD, De Souza GIH, Alves OL, Esposito E (2007) Antibacterial effect of silver nanoparticles produced by fungal process on textile fabrics and their effluent treatment. J Biomed Nanotechnol 3:203–208
Durán N, Cuevas R, Cordi L, Rubilar O, Diez MC (2014) Biogenic silver nanoparticles associated with silver chloride nanoparticles (Ag@AgCl) produced by laccase from *Trametes versicolor*. Springer Plus 3:1–7
El-Batal AI, El Kenawy NM, Yassin AS, Amin MA (2015) Laccase production by *Pleurotus ostreatus* and its application in synthesis of gold nanoparticles. Biotechnol Reports 5:31–39
El-Newehy MH, Al-Deyab SS, Kenawy E, Abdel-Megeed A (2012) Fabrication of electrospun antimicrobial nanofibers containing metronidazole using nanospider technology. Fiber Polym 13:709–717
Faramarzia MA, Forootanfara H (2011) Biosynthesis and characterization of gold nanoparticles produced by laccase from *Paraconiothyrium variabile*. Colloid Surf B 87:23–27
Fayaz M, Balaji K, Girilal M, Yadav R, Kalaichelvan PT, Venketesan R (2010) Biogenic synthesis of silver nanoparticles and their synergistic effect with antibiotics: a study against gram-positive and gram-negative bacteria. Nanomedicine 6:103–109
Gade AK, Bonde P, Ingle AP, Marcato PD, Duran N, Rai MKJ (2008) Exploitation of *Aspergillus niger* for synthesis of silver nanoparticles. J Biobased Mater Bioenergy 2:243–248
Gajbhiye M, Kesharwani J, Ingle A, Gade A, Rai M (2009) Fungus-mediated synthesis of silver nanoparticles and their activity against pathogenic fungi in combination with fluconazole. Nanomed Nanotechnol 5:382–386
Gericke M, Pinches A (2006) Biological synthesis of metal nanoparticles. Hydrometallurgy 83:132–140
Gopinath K, Arumugam A (2014) Extracellular mycosynthesis of gold nanoparticles using *Fusarium solani*. App Nanosci 4:657–662
Grimm LH, Kelly S, Krull R, Hempel DC (2005) Morphology and productivity of filamentous fungi. App Microbiol Biotechnol 69:375–384
Gupta S, Sharma K, Sharma R (2012) Myconanotechnology and application of nanoparticles in biology. Recent Res Sci Technol 4:36–38
Gurunathan S, Raman J, Abd Malek SN, John PA, Vikineswary S (2013) Green synthesis of silver nanoparticles using *Ganoderma neo-japonicum* Imazeki: a potential cytotoxic agent against breast cancer cells. Int J Nanomed 8:4399–4413
Gurunathan S, Han J, Park JH, Kim YK (2014) A green chemistry approach for synthesizing biocompatible gold nanoparticles. Nanoscale Res Lett 9:1–11
Hutten A, Sudfeld D, Ennen I, Reiss G, Hachmann W, Heinzmann U, Wojczykowski K, Jutzi P, Saikaly W, Thomas G (2004) New magnetic nanoparticles for biotechnology. J Biotechnol 112:47–63
Ilk S, Demircan D, Saglam S, Saglam N, Rzayev ZMO (2016) Immobilization of laccase onto a porous nanocomposite: application for textile dye degradation. Turkish J Chem 40:262–276.
Ingle A, Rai MK, Gade A, Bawaskar M (2009) *Fusarium solani*: a novel biological agent for the extracellular synthesis of silver nanoparticles. J Nanoparticle Res 11:2079–2085

Jaidev LR, Narasimha G (2010) Fungal mediated biosynthesis of silver nanoparticles, characterization and antimicrobial activity. Colloid Surf B 81:430–433

Karwa A, Gaikwad S, Rai M (2011) Mycosynthesis of silver nanoparticles using lingzhi or reishi medicinal mushroom, *Ganoderma lucidum* (W.Curt.:Fr.) P. Karst and their role as antimicrobials and antibiotic activity enhancers. Int J Med Mushrooms 13:483–491

Kathiresan K, Manivannan S, Nabeel MA, Dhivya B (2009) Studies on silver nanoparticles synthesized by a marine fungus, *Penicillium fellutanum* isolated from coastal mangrove sediment. Colloids Surf B 71:133–137

Kumar SA, Ayoobul AA, Absar A, Khan MI (2007) Extracellular biosynthesis of CdSe quantum dots by the fungus, *Fusarium oxysporium*. J Biomed Nanotechnol 3:190–194

Li X, Xu H, Chen ZS, Chen G (2011) Biosynthesis of nanoparticles by microorganisms and their applications. J Nanomater. doi:10.1155/2011/270974

Mansoori GA (2010) Synthesis of nanoparticles by fungi. US patent. US 2010/0,055,199 A1

Mazumdar H, Haloi N (2011) A study on biosynthesis of iron nanoparticles by *Pleurotus* sp. J Microbiol Biotechnol Res 1:39–49

Mirunalini S, Arulmozhi V, Deeppalakshmi K, Krishnaveni M (2012) Intracellular biosynthesis and antibacterial activity of silver nanoparticles using edible mushrooms. Notulae Scientia Biologicae 4:55–61

Mohammed-Fayaz A, Balaji K, Kalaichelvana PT, Venkatesan R (2009) Fungal based synthesis of silver nanoparticles—an effect of temperature on the size of particles. Colloid Surf B 74:123–126

Morones JR, Elechiguerra JL, Camacho A, Holt K, Kouri JB, Ramfrez JT, Yacaman MJ (2005) The bactericidal effect of silver nanoparticles. Nanotechnol 16:2346–2353

Mühling M, Bradford A, Readman A, Somerfield PJ, Handy RD (2009) An investigation into the effects of silver nanoparticles on antibiotic resistance of naturally occurring bacteria in an estuarine sediment. Mar Environ Res 68:278–283

Mukherjee P, Ahmad A, Mandal D, Senapati S, Sainkar SR, Khan MI, Parischa R, Ajayakumar PV, Alam M, Kumar R, Sastry M (2001a) Fungus mediated synthesis of silver nanoparticles and their immobilization in the mycelial matrix: a novel biological approach to nanoparticle synthesis. Nano Lett 1:515–519

Mukherjee P, Ahmad A, Mandal D, Senapati S, Sainkar SR, Khan MI, Ramani R, Parischa R, Ajayakumar PV, Alam M, Sastry M, Kumar R (2001b) Bioreduction of AuCl4– ions by the fungus, *Verticillium* sp. and surface trapping of the gold nanoparticles formed. Angew Chem Int 40:3585–3588

Mukherjee P, Senapati S, Mandal D, Ahmad A, Khan MI, Kumar R, Sastry M (2002) Extracellular synthesis of gold nanoparticles by the fungus *Fusarium oxysporum*. Chembiochem 3:461–463

Mukherjee P, Roy M, Mandal B, Dey G, Mukherjee P, Ghatak J (2008) Green synthesis of highly stabilized nanocrystalline silver particles by a non-pathogenic and agriculturally important fungus *T. asperellum*. Nanotechnol 19:103–110

Narayanan KB, Sakthivel N (2010) Biological synthesis of metal nanoparticles by microbes. Adv Coll Int Sci 156:1–13

Nayak RR, Pradhan N, Behera KM, Pradhan S, Mishra S, Sukla LB, Mishra KB (2011) Green synthesis of silver nanoparticle by *Penicillium purpurogenum* NPMF: the process and optimization. J Natopart Res 13:3129–3137

Nithya R, Ragunathan R (2009) Synthesis of silver nanoparticle using *Pleurotus sajor caju* and its antimicrobial study. Digest J Nanomater Biostruct 4:623–629

Nithya R, Ragunathan R (2011) Decolorization of the dye Congo red by *Pleurotus sajor caju* Silver nanoparticle. Int Conf Food Eng Biotechnol 9:12–15

Owaid MN, Raman J, Lakshmanan H, Al-Saeedi SS, Sabaratnam V, Abed IA (2015) Mycosynthesis of silver nanoparticles by *Pleurotus cornucopiae* var. *citrinopileatus* and its inhibitory effects against *Candida* sp. Mater Lett 153:186–190

Paul S, Sasikumar CS, Singh AR (2005) Fabrication of silver nanoparticles synthesized from *Ganoderma lucidium* into the cotton fabrica and its antimicrobial property. Int J Pharma Sci 7:53–56

Popescu M, Velea A, Lorinzi (2010) Biogenic production of nanoparticles. Digest J Nanomater Biosufactans 5:1035–1040

Prabhu N, Revathi N, Darsana R, Sruthi M, Chinnaswamy P, Innocent DJP (2009) Antibacterial activities of silver nanoparticles synthesized bu *Aspergillus fumigatus*. Icfai Univ J Biotechnol 3:50–55

Prasad R (2014) Synthesis of silver nanoparticles in photosynthetic plants. J Nanoparticles. doi:10.1155/2014/963961

Prasad R, Kumar V, Prasad KS (2014) Nanotechnology in sustainable agriculture: present concerns and future aspects. Afr J Biotechnol 13(6):705–713

Prasad R, Pandey R, Barman I (2015) Engineering tailored nanoparticles with microbes: quo vadis. WIREs Nanomed Nanobiotechnol. doi:10.1002/wnan.1363

Rai M, Yadav A, Bridge PD, Gade A (2009) Myconanotechnology: a new and emerging science. In: Rai M, Bridge PD (eds) Applied mycology. Wallingford, UK, pp 258–267

Raman JG, Reddy R, Lakshmanan H, Selvaraj H, Gajendran B, Nanjian R, Chinnasamy A, Sabaratnam V (2015) Mycosynthesis and characterization of silver nanoparticles from *Pleurotus djamor* var. *roseus* and their in vitro cytotoxicity effect on PC3 cells. Pro Biochem 50:140–147

Ravindra BK, Rajasab AH (2014) A comparative study on biosynthesis of silver nanoparticles using four different fungal species. Int J Pharm Sci 6:372–376

Riddin TL, Gericke M, Whiteley CG (2006) Analysis of the inter and extracellular formation of platinum nanoparticles by *Fusarium oxysporum* f. sp *lycopersici* using response surface methodology. Nanotechnology 17:3482–3489

Roy S, Kumar Das T, Prasad Maiti G, Basu U (2016) Microbial biosynthesis of nontoxic gold nanoparticles. Mat Sci Eng B 203:41–51

Sadowski Z, Maliszewska IH, Grochowalska B, Polowczyk I, Kozlecki T (2008) Synthesis of silver nanoparticles using microorganisms. Mater Sci Poland 26:419–424

Safarik I, Safarikova M (2002) Magnetic nanoparticles and biosciences. In: Hofmann IH, Rahman Z, Schubert U (eds) Nanostructured materials. Springer, Vienna, pp 1–23

Sanghi R, Verma P (2009a) Biomimetic synthesis and characterization of protein capped silver nanoparticles. Biores Technol 100:501–504

Sanghi R, Verma P (2009b) A facile green extracellular biosynthesis of CdS nanoparticles by immobilized fungus. Chem Eng J 155:886–891

Sanghi R, Verma P, Puri S (2011) Enzymatic formation of gold nanoparticles using *Phanerochaete chrysosporium*. Adv Chem Eng Sci 1:154–162

Saravanan M, Nanda A (2010) Extracellular synthesis of silver bionanoparticles from *Aspergillus clavatus* and its antimicrobial activity against MRSA and MRSE. Colloid Surf B 77:214–218

Sarsar V, Selwal MK, Selwal KK (2015) Biofabrication, characterization and antibacterial efficacy of extracellular silver nanoparticles using novel fungal strain of *Penicillium atramentosum* KM. J Saudi Chem Soc 19:682–688

Sastry M, Ahmad A, Khaan MI, Kumar R (2003) Biosytnhesis of metal nanoparticles using fungi and actinomycetes. Curr Sci 85:162–170

Senapati S, Mandal D, Ahmad A, Khan MI, Sastry M, Kumar R (2004) Fungus mediated synthesis of silver nanoparticles: a novel biological approach. Ind J Phys A 78:101–105

Senapati S, Ahmad A, Khan MI, Sastry M, Kumar R (2005) Extracellular biosynthesis of bimetallic Au-Ag alloy nanoparticle. Small 1(5):517–520

Shaligram NS, Bule M, Bhambure R, Singhal RS, Singh SK, Szakacs G et al (2009) Biosynthesis of silver nanoparticles using the aqueous extract from the compaction producing fungal strain. Process Biochem 44:939–943

Shankar SS, Ahmad A, Pasrichaa R, Sastry M (2003) Bioreduction of chloroaurate ions by geranium leaves and its endophytic fungus yields gold nanoparticles of different shapes. J Mater Chem 13:1822–1826

Shi C, Zhu N, Cao Y, Wu P (2015) Biosynthesis of gold nanoparticles assisted by the intracellular protein extract of *Pycnoporus sanguineus* and its catalysis in degradation of 4-nitroaniline. Nanoscale Res Lett 10(147):1–8

Singh D, Rathod V, Ninganagouda S, Hiremath J, Singh AK, Mathew J (2014) Optimization and characterization of silver nanoparticle by endophytic fungi *Penicillium* sp. isolated from *Curcuma longa* (turmeric) and application studies against MDR *E.coli* and *S. aureus*. Bioinorg Chem Appl 2014, Article ID 408021

Spasova M, Monolova N, Naydenov M, Kuzmonova J, Rashkov I (2011) Electrospun biohybrid materials for plant biocontrol containing chitosan and *Trichoderma viride* spores. J Bioact Compat Pol 26:148–155

Suman PR, Jain VK, Varma A (2010) Role of nanomaterials in symbiotic fungus growth enhancement. Curr Sci 99:1189–1191

Syed A, Saraswati S, Kundu GC, Ahmad A (2013) Biological synthesis of silver nanoparticles using the fungus *Humicola* sp. and evaluation of their cytoxicity using normal and cancer cell lines. Spectrochim Acta A Mol Biomol Spect 114:144–147

Vahabi K, Mansoori GA, Karimi S (2011) Biosynthesis of silver nanoparticles by fungus *Trichoderma reesei*. Insciences J 1:65–79

Vala A (2015) Exploration on green synthesis of gold nanoparticles by a marine-derived fungus *Aspergillus sydowii*. Environ Prog Sus Energy 34:194–197

Vetchinkina EP, Burov AM, Ageeva MV, Dykman LA, Nikitin VE (2013) Biological synthesis of gold nanoparticles by the xylotrophic basidiomycete *Lentinula edodes*. App Biochem Microbiol 49:406–411

Vigneshwaran N, Kathe AA, Varadarajan PV, Nachane RP, Balasubramanya RH (2006) Biomimetics of silver nanoparticles by white rot fungus, *Phanerochaete chrysosporium*. Colloid Surf B 53:55–59

Vigneshwaran N, Kathe AA, Varadarajan PV, Nachane RP, Balasubramany RH (2007) Silver-protein (core-shell) nanoparticle production using spent mushroom substrate. Langmuir 23:7113–7117

Yehia RS, Al-Sheikh H (2014) Biosynthesis and characterization of silver nanoparticles produced by *Pleurotus ostreatus* and their anticandidal and anticancer activities. World J Microbiol Biotechnol 30:2797–2803

Yesilada O, Asma D, Cing S (2003) Decolorization of textile dyes by fungal pellets. Process Biochem 38:933–938

Yesilada O, Birhanli E, Ozmen N, Ercan S (2014) Highly stable laccase from repeated-batch culture of *Funalia trogii* ATCC 200800. App Biochem Microbiol 50:55–61

Chapter 3
Marine-Derived Fungi: Potential Candidates for Fungal Nanobiotechnology

Anjana K. Vala, Hiral B. Trivedi, and Bharti P. Dave

Abstract Due to unique properties, gold and silver nanoparticles (GNPs and AgNPs, respectively) have wide applications in diverse fields like biomedicine, catalysis, imaging and photonics, solar energy conversion and nanoelectronics etc. and hence, are in great demand. Available physicochemical synthesis protocols generally face limitations like high cost, polluting nature, and also have restricted use in clinical and pharma applications. In order to overcome these limitations, biosynthesis of nanoparticles could be a promising alternative. While it has been suggested that initiatives should be taken for exploitation of marine microbial resources in the area of nanobiotechnology, despite the unique traits of marine-derived fungi, they are comparatively less explored for biosynthesis of GNPs and AgNPs. Though a few, available reports suggest marine-derived fungi as promising candidates for such purpose. Recent reports on observation of the laser speckle pattern and weak localization of light by AgNPs and GNPs biosynthesized by marine-derived fungi assert their novel application potentialities.

1 Introduction

Nanotechnology is considered as cutting edge technology today. Though considered as one of the recent branches of modern science, nanotechnology is in practice since ancient times. Nano level preparations of gold (swarna bhasma) was widely used in ancient Indian medical system (Ayurveda) (Rimal Isac et al. 2013). In ninth century silver and gold nanoparticles (AgNPs and GNPs, respectively) were used for generating glittering effect on pots (Prathna et al. 2010).

Michael Faraday (1857) had provided the first scientific description of properties of nanoparticles in paper entitled "Experimental relations of gold (and other metals) to light". After more than 100 years from this, the first talk on nanotechnology was

A.K. Vala (✉) • H.B. Trivedi • B.P. Dave (✉)
Department of Life Sciences, Maharaja Krishnakumarsinhji
Bhavnagar University, Bhavnagar 364 002, India
e-mail: akv@mkbhavuni.edu.in; bpd8256@gmail.com

R. Prasad (ed.), *Advances and Applications Through Fungal Nanobiotechnology*, Fungal Biology, DOI 10.1007/978-3-319-42990-8_3

given by Richard Feynman in 1959. But the conceptual revolution was started in 1981, when Eric Drexler published first paper on nanotechnology (Drexler 1981; Prathna et al. 2010). Nanomaterials/nanoparticles are one of the integral components of nanotechnology.

2 Classification of Nanoparticles

Nanoparticles can be classified in to two broad groups: (a) organic nanoparticles e.g. carbon nanotubes and (b) inorganic nanoparticles e.g. magnetic nanoparticles, noble metal nanoparticles (gold and silver nanoparticles) and semiconductor nanoparticles (cadmium sulfide, titanium oxide, zinc oxide).

3 Strategies Employed for Synthesizing Nanoparticles

Unique properties of metal nanoparticles are distinctly different from their bulk counter parts. Due to increasing demand of nanomaterials, developing versatile protocols for synthesizing nanoparticles with controlled properties is of utmost interest.

Nanoparticles are synthesized basically by two approaches: (1) Top down approach and (2) Bottom up approach. Figure 3.1 shows schematic representation of synthesis of nanoparticles by these approaches.

Conventionally, nanoparticles have been synthesized by physical and chemical methods. Some of the frequently used physical and chemical methods for nanoparticles synthesis include:

- Sol–gel technique
- Solvothermal synthesis
- Chemical reduction
- Laser ablation
- Inert gas condensation

In general, the traditionally used physical and chemical protocols are:

- cost intensive and not environment-friendly
- In addition, use of certain chemicals during synthesis of nanoparticles limit their use in clinical and pharma applications also (Narayanan and Sakthivel 2010; Prakasham et al. 2012; Vala 2015a; Nachiyar et al. 2015)

These restrictions call for development of cleaner and safer protocols for synthesis of nanoparticles.

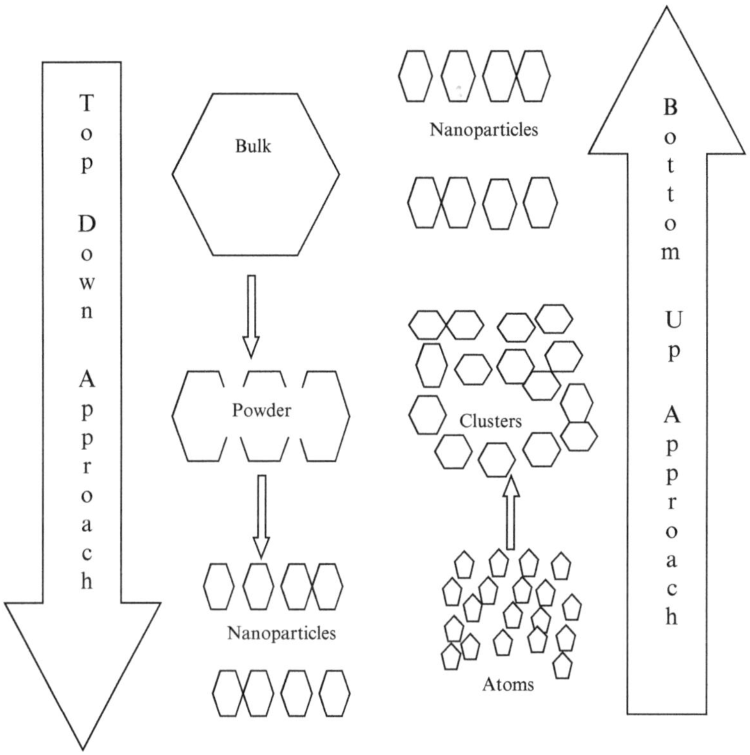

Fig. 3.1 Schematic representation of synthesis of nanoparticles (adopted from http://www.gitam.edu/eresource/nano/NANOTECHNOLOGY/role_of_bottomup_and_topdown_a.htm)

3.1 Biological Approach for the Synthesis of Nanoparticles

In the search for less toxic and cleaner methods for synthesizing nanomaterials, recent developments in the biosynthesis of nanoparticles have highlighted the important role of microorganisms (Prasad et al. 2015; Aziz et al. 2015). Microorganisms offer potential leads for producing nanoparticles through a biological route (Gwynne 2013).

A number of biological systems including microbes and plants have been examined for synthesizing nanoparticles in general and AgNPs and GNPs in particular (Mukherjee et al. 2001a, b; Prasad 2014; Rauwel et al. 2015). However, compared to their terrestrial counterparts, marine microbes have been overlooked in spite of the fact that researches on nanobiotechnology in future may progressively depend more on marine microbes having the capability to grow under extreme conditions (Chandramohan 2004; Kathiresan et al. 2009; Vala et al. 2012). It has been recommended that initiatives should be taken for exploiting these important resources for nanobiotechnological research (Chandramohan 2004). Harnessing fungi is even more beneficial, because of their unique traits. Rai et al. (2009) proposed the term "Myconanotechnology" to indicate synthesis of nanoparticles using fungi.

Advantages of using fungi include: ease in handling the culture, high wall binding capacity and intracellular metal uptake ability, their capability to resist conditions like agitation and high flow pressure in bioreactors, ability to secrete large amount of enzymes extracellularly and higher yield of nanoparticles (Prasad et al. 2015). As the nanoparticles synthesized outside the cell are devoid of cellular components, can be directly used for various applications. Economic viability is also one of the most important advantages of using fungal cultures for nanoparticle synthesis (Ingle et al. 2008; Sanghi and Verma 2009; Narayanan and Sakthivel 2010; Hemath et al. 2010; Jain et al. 2011; Chan and Mashitah 2012; Soni and Prakash 2012; Alani et al. 2012; Honary et al. 2013; Vala 2014a; Vala et al. 2014; Yadav et al. 2015). One serious concern involved with biomedical applications of biosynthesized nanoparticles is generation of immune response due to the capping proteins present on the biosynthesized nanoparticles. Glycosylation of the proteins reduce or even suppress the immune response of the host (Rudd et al. 2001). Glycosylation is more prominent in fungi than in bacteria, hence, nanoparticles capped and stabilized by fungal proteins are quite less likely to produce immune response in the host (Li and d'Anjou 2009; Jain et al. 2012; Kitching et al. 2014). Hence, fungus-mediated nanoparticles can prove better from biomedical application point of view.

3.1.1 Fungi from Marine Habitats

Duriers and Montagene (1846–1850) were first to describe fungi in marine habitats around the middle of nineteenth century in France (Verma 2011). Marine fungi may be divided into majority of Ascomycota, few Basidiomycota and anamorphic fungi. Marine fungi can be grouped in to temperate, tropical, subtropical and cosmopolitan species according to their biogeochemical distribution. Assessment of taxonomic diversity of marine fungi can be done by direct observations of sporulating structures, culturing and metagenomics.

Fungi from marine habitats have been classified as (1) obligate marine fungi—that grow and sporulate exclusively in a marine or estuarine (brakish water) habitat and (2) facultative marine fungi, which have freshwater or terrestrial origin and have ability to grow and possibly sporulate in marine environment (Kohlmeyer and Kohlmeyer 1979; Kohlmeyer and Volksmann- Kohlmeyer 2003; Li and Wang 2009). To classify more generally the term 'marine-derived fungi' is used (Christophersen et al. 1999; Osterhage 2001; Bonugli-Santos et al. 2015). Marine fungal strains have been found to inhabit nearly all possible marine habitats like inorganic matter, marine plants, marine vertebrates and invertebrates and marine microbial communities.

Fungal isolates with distinguishable metal removal efficiency have been screened out from marine habitats (Vala et al. 2004; Vala 2010; Vala and Sutariya 2012) however, very less information is available on utilization of marine-derived fungi for biosynthesis of metal nanoparticles (Vala 2014a).

3.2 *Marine Environment: A Novel Gateway for Green Nano Technology*

Recently, Singh et al. (2015) has reviewed various marine sources for synthesizing nanoparticles. The present chapter focuses on marine-derived fungi as biofactories for synthesizing AgNPs and GNPs.

AgNPs and GNPs are synthesized intracellularly, extracellulary and even surface synthesis is possible. Common fundamental mechanism involves reduction of Ag^{+} and Au^{3+} to form AgNPs and GNPs, respectively. Though many reports show role of certain enzymes or molecules in biosynthesis, complete mechanistic aspects are yet to be established (Shedbalkar et al. 2014). Figure 3.2 shows generalized mechanisms of biosynthesis of AgNPs and GNPs.

3.2.1 Marine-Derived Fungi in Synthesizing AgNPs

While first reports on fungus-mediated AgNPs and GNPs biosynthesis appeared in 2001 using fungus *Verticillium* isolated from *Taxus* plant (Mukherjee et al. 2001a, b; Rauwel et al. 2015), studies on biosynthesis of nanoparticles using marine-derived fungi started in 2009, when Kathiresan et al. (2009) reported biosynthesis of silver nanoparticles by *Penicillium fellutanum* isolated from coastal mangrove sediment. Figure 3.3 shows time line for marine-derived fungal nanotechnology.

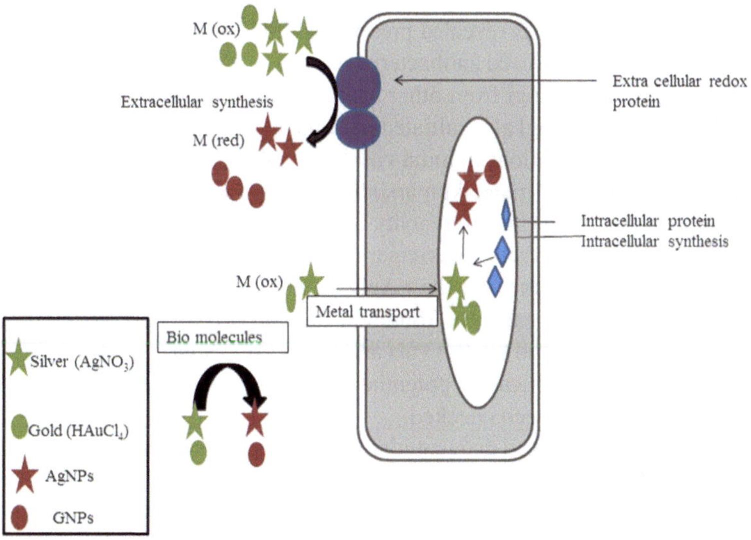

Fig. 3.2 General mechanisms of microbial biosynthesis of AgNPs and GNPs (Adopted from Das et al. 2012, ACS Nano 6:6165–6173; Shedbalkar et al. 2014, Advances in Colloid and Interface Science 209: 40–48)

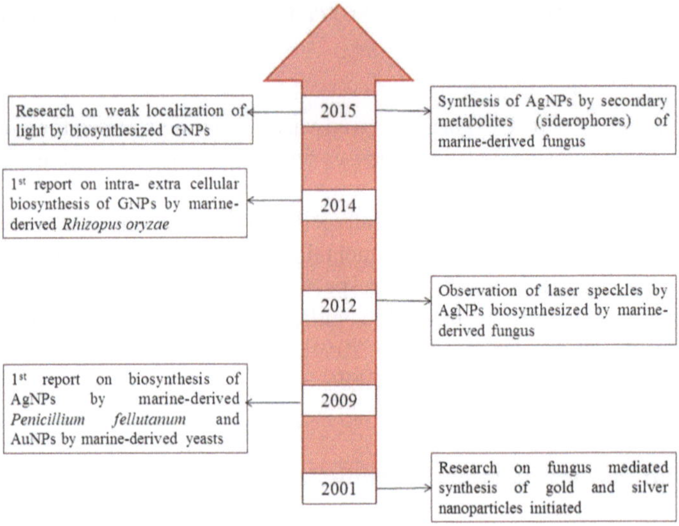

Fig. 3.3 Time Line for marine-derived fungal nanotechnology

Kathiresan et al. (2009) observed that maximum biosynthesis of AgNP was achieved upon treating the culture filtrate with 1.0 mM $AgNO_3$, pH 6.0, maintaining at 0.3 % NaCl and incubating at 5°C for 24 h. Further, Kathiresan et al. (2010) reported synthesis of extracellular AgNPs by *Aspergillus niger* isolated from coastal mangrove sediment of southeast India. The particles were spherical ranging in size from 5 to 35 nm. SDS-PAGE analysis revealed presence of 70 KDa protein in the filtrate. The biosynthesized AgNPs showed antibacterial as well as antifungal potentials.

As in the case with fungi from other sources, biosynthesis of AgNPs and GNPs using marine-derived fungi also initiated in India. Mainly pioneering work has been carried out at three institutions in India viz. Annamalai University (AgNPs, filamentous fungi), Maharaja Krishnakumarsinhji Bhavnagar University (AgNPs and GNPs, filamentous fungi) and University of Pune (AgNPs and GNPs, yeast).

Marine-derived fungi from Bhavnagar Coast, Gulf of Khambhat, West Coast of India, have been investigated for their AgNPs and GNPs biosynthesis potentials. It was observed that the fungal biota from these habitats are very promising source for biosynthesizing AgNPs and GNPs (Vala and Shah 2012; Vala et al. 2012, 2014; Vala 2014a, b, 2015a, b). Application potentialities of these biosynthesized particles in various fields have also been checked.

Vala et al (2012) observed extracellular biosynthesis of spherical AgNPs with size range of 5–26 nm, within 24 h by a marine-derived *Aspergillus niger*. Laser optical speckles were observed for these biosynthesized AgNPs (Fig. 3.4). The authors claim this to be the first ever report on laser speckle pattern of the biosynthesized AgNPs. Speckle phenomena have utility in fields like optical image processing, metrology, bio-optics and photonics (Ulyanov 1998; Gatti et al. 2008; Vala et al. 2012).

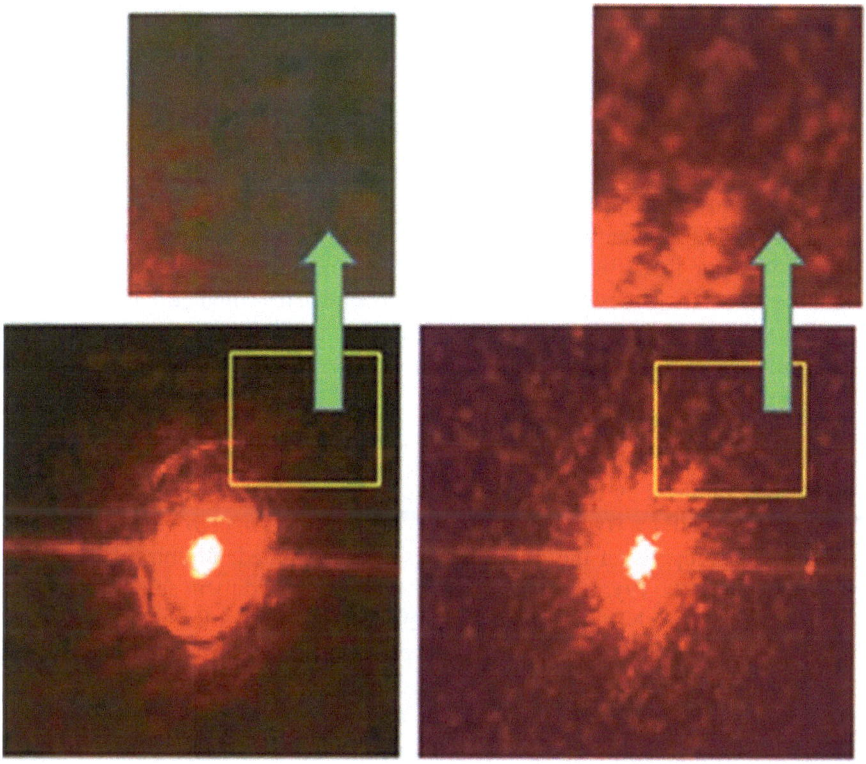

Fig. 3.4 Shows the typical light transmission experiment images. (**a**) Laser light transmission through the first sample. No scattering or speckle is observed. The enlarge image is also clear. (**b**) Transmission pattern for Sample 2. This shows the clear speckle pattern. This speckle pattern is because of the presence of silver nanoparticles synthesised using *A. niger*. The enlarged image shows typical laser speckle pattern because of the light matter interaction, especially with metal nanoparticles. [Reproduced by permission of the Institution of Engineering & Technology (Vala AK, Chudasama B, Patel RJ. Green synthesis of silver nanoparticles using marine-derived fungus Aspergillus niger. 2012, Micro & Nano Letters 7(8):859–862)]

Apart from its fundamental importance, the longitudinal coherence of speckled light has relevance in areas including speckle holography, speckle interferometry and speckle photography as well as in novel optical imaging techniques like dynamic speckle illumination microscopy, and thermal ghost imaging (Gatti et al. 2008). Further studies on biosynthesized AgNPs in this line demands attention.

By tuning the physico-chemical parameters, better biosynthesis output could be achieved by the same organism. Vala and Shah (2012) reported rapid synthesis of AgNPs through biological route using marine-derived *Aspergillus niger*, where biosynthesis could be achieved in 3 min extracellularly. It has been reported that microbe mediated biosynthesis of nanoparticles is time consuming taking generally

24–120 h (Bai et al. 2011). However, biosynthesis using marine-derived fungi could easily be achieved within minutes at alkaline pH. Alkaline pH is more conducive for efficiency of enzymes involved in biosynthesis of AgNPs (Sanghi and Verma 2009). Vala and Shah (2012) examined antimicrobial potentials of AgNPs biosynthesized at different pH against four test bacteria viz. *Bacillus megaterium*, *Proteus vulgaris*, *Staphylococcus aureus* and *Shigella sonnei*. Combined effect of Gentamicin and biosynthesized AgNPs on test bacteria was also examined. AgNPs biosynthesized at pH 10 exerted highest antimicrobial activity. Variation in particle size could be possible reason for this (Brunner et al. 2006; Vala and Shah 2012). The observed smaller size of biosynthesized particles with increase in pH could be due to increased dynamics of the ions and formation of more nucleation regions because of the availability of ^{-}OH ions. There is an increase in conversion of Ag^{+} to Ag^{0} followed by increase in the kinetics of the deposition of the silver atoms (Deepak et al. 2011; Vala and Shah 2012).

Antimicrobial activities of biosynthesized AgNPs using marine-derived fungi have been examined by other workers as well. Bhimba et al. (2011) reported biosynthesis of AgNP by *Hypocrea lixii* MV1 isolated from mangrove sediment soil. Antibacterial property of the AgNP has also been reported. Bhimba et al. (2015) reported *Aspergillus oryzae* isolated from the mangrove sediment soil to biosynthesize extracellular AgNPs in ranging 6-37 nm in size. Sathiya Rathna et al. (2013) observed extracellular biosynthesis of silver nanoparticles by endophytic fungus *Aspergillus terreus* associated with mangrove leaves of *Rhizophora annamalayanna* and reported its anti-dermatophytic activity against *Trichophyton rubrum, Epidermophyton floccosum and Trichophyton mentagrophytes*. Nayak and Anitha (2014) reported extracellular synthesis of AgNPs by *Aspergillus sydowii* isolated from sand dunes. The antimicrobial efficacy of these AgNPs in combinantion with ampicillin and vancomycin was found to be encouraging against Gram positive and Gram negative bacteria viz. *Bacillus cereus, Eschericha coli, Klebsiella pneumoniae, Pseudomonas aeruginosa, Staphylococcus aureus* and *Shigella* sp.

Recently, Anand et al. (2015) studied biosynthesis of silver nanoparticles by using cell free filtrates of five marine sediment fungi viz. *Aspergillus flavus* SP-3, *Trichoderma gamsii* SP-4, *Talaromyces flavus* SP-5, and *Aspergillus oryzae* SP-6. They found that all test filtrates could synthesize the AgNPs. The AgNPs synthesized were found to be in the size range of 20–60 nm in all the fungal isolates. They reported that *T. gamsii* SP-4 showed an enhanced antimicrobial activity, anti-oxidant activity and a dose dependent cytotoxic activity against HEp2 cell lines.

Vala et al (2014) checked AgNPs biosynthesis potential of a marine-derived *Aspergillus flavus*. They observed that the test isolate could synthesize AgNPs intracellularly at different concentrations (0.25, 0.5 and 1.0 mM) of silver nitrate within 24 h.When the test fungus was exposed to 1 mM $AgNO_3$ in the pH range 3-10, extracellular biosynthesis of AgNPs was observed as revealed by dark brownish coloration of the solution. The alkaline pH had a significant impact not only on the mode of synthesis but also on the time required for biosynthesis of AgNPs. At pH 10, formation of AgNPs was observed just within 10 min, followed by 30 min and 1 h at pH 9 and 8, respectively. Hence, by altering the parameters two criteria important from large scale production point of view could be fulfilled (i) recovery of biosynthesized particles and (ii) less time required for biosynthesis. Analytical techniques like XRD, UV–vis spectrophotometry and transmission electron microscopy were employed for characterization of AgNPs and it was observed that the particles were spherical in the size range of 2–22 nm.

Recently, two marine-derived fungi *Aspergillus candidus* and *A. sydowii*, were studied for AgNPs biosynthesis. Both the test isolates exhibited ability to biosynthesize silver nanoparticles intracellularly. Upon challenging with 1 mM silver nitrate at different pH (3–10), both the test fungi showed rapid extracellular AgNP biosynthesis (within 20 min) at alkaline pH (Vala 2015b). Plate 3.1 shows biosynthesis of AgNPs by *A. candidus*. Marine-derived strain of *Rhizopus oryzae* was observed to biosynthesize AgNPs intracellularly (Vala, data unpublished).

One of the objectives of explorations on marine and related environments for marine-derived fungi mediated biosynthesis of nanoparticles is to examine if as synthesized nanoparticles exhibit novelty in properties. A breakthrough observation has been made recently (Rodrigues et al. 2013). They examined mangrove

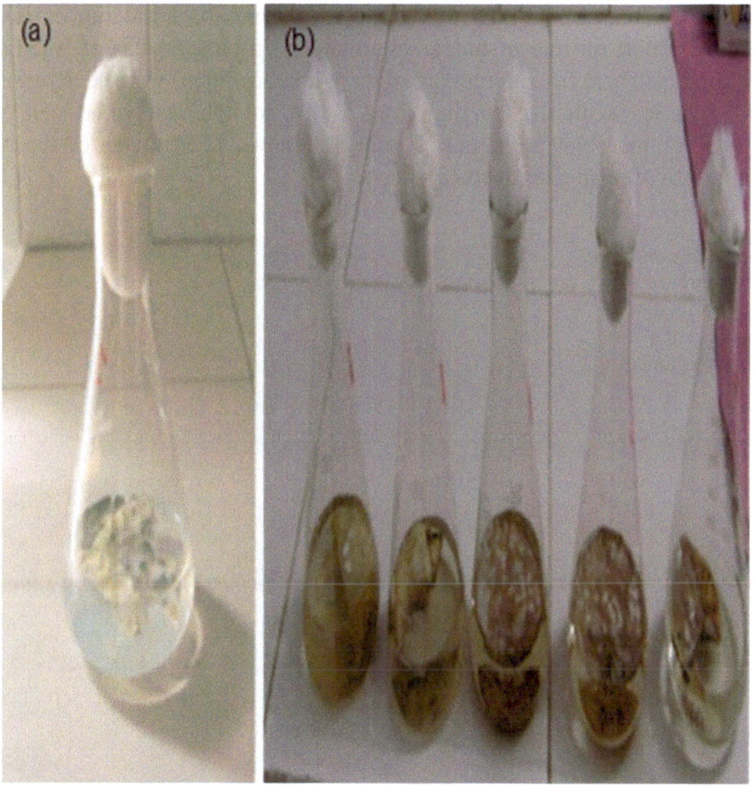

Plate 3.1 *Aspergillus candidus* (**a**) before biosynthesis and (**b**) after biosynthesis of silver nanoparticles at different concentrations (0.25, 0.5, 1, 2, 3 mM) of silver nitrate from *right* to *left*

associated *Aspergillus tubingensis* and *Bionectria ochroleuca* for AgNPs biosynthesis and antimicrobial properties of as synthesized AgNPs. This is the first report on AgNPs biosynthesis potential of manglicolous fungi from a Brazilian mangrove. Interestingly, *A. tubingensis* synthesized AgNPs with positive surface potential. Again, this is the first report about any known fungus synthsizing AgNPs with a positive zeta potential. Hence, unique property of AgNPs synthesized by a marine-derived fungus has been observed. The authors claim that the data open the prospect of obtaining biosynthesized AgNPs with positive zeta potential and new applications.

Animals from marine habitats have also been explored as source of marine-derived fungi for biosynthesizing nanoparticles. A sponge-associated strain of *Aspergillus terreus* MP1 has been reported to produce extracellular AgNPs with antimicrobial capabilities (Meenupriya et al. 2011).

Secondary metabolites from marine-derived fungi have recently been harnessed for production of AgNPs. Bioinspired AgNPs synthesis has been achieved using purified siderophore of a manglicolous fungi for the first time (Trivedi et al. 2015). Siderophores are low molecular weight iron chelating compounds produced by microorganisms. So far only one report exists on siderophore mediated synthesis of nanoparticles (Bharde 2007). The siderophore was from a bacterial source. However, there is no report on role of fungal siderophores in biosynthesis of AgNPs. The AgNPs obtained here (by siderophore of manglicolous fungi) were nearly monodispersed, spherical with size ranging from 5 to 8 nm. The as synthesized AgNPs exhibited antimicrobial properties. Figure 3.5 shows TEM image of the AgNPs synthesized by fungal siderophores.

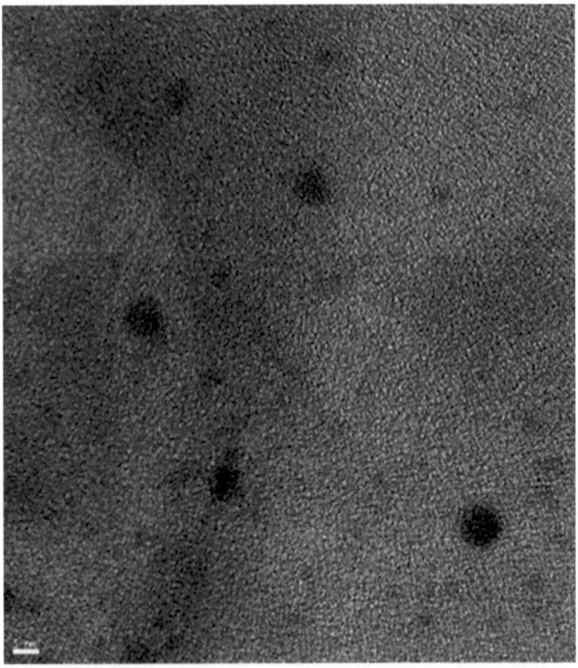

Fig. 3.5 TEM image of AgNPs synthesized by partially purified siderophores of a manglicolous fungus

Not only filamentous marine-derived fungi but marine-derived yeasts have also been examined for biosynthesis of AgNPs. Manivannan et al. (2010) tested twelve species of mangrove sediment-derived yeasts for synthesis of silver nanoparticles. *Pichia capsulata* was found to be the most efficient producer of AgNPs followed by *Saccharomyces cerevisiae, Rhodotorula minuta, Pichia salicaria* and *Debaromyces hansenii*. Synthesis of spherical particles (5–25 nm) was suggested to be mediated by an NADH-dependent protein similar to nitrate reductase.

NP biosynthesis work has been carried out using strains of *Yarrowia lipolytica*. Recently, Apte et al. (2013a) reported cell associated biosynthesis of AgNPs by a psychrotrophic marine yeast strain *Yarrowia lipolytica* (NCYC 789). Further, they extracted melanin from the yeast cells and harnessed it for AgNPs biosynthesis. They proposed possible mechanism for AgNPs biosynthesis. AgNPs biosynthesized using melanin exhibited antibiofilm properties.

Apte et al. (2013b) isolated pigment melanin from *Yarrowia lipolytica* NCIM 3590 cells. Melanin mediated synthesis of fairly mono-disperse AgNP and AuNP with average size 7 and 20 nm, respectively was achieved. They proposed mechanism involved in the synthesis of nanoparticles by L-DOPA induced-melanin derived from *Y. lipolytica* and demonstrated an application of the as synthesized AgNPs as paint-additives with anti-fungal properties.

Recently, Prasad et al. (2015) highlighted importance of careful strain selection in biosynthesis of nanoparticles. Marine-derived fungal strains so far examined prove promising in this connection. Compared to other sources of fungi, marine-derived fungi seem to be more potent in synthesizing nanoparticles. The statement can be comprehended/justified by the examples of pioneering work on fungus mediated synthesis of nanoparticles carried out NCL, Pune and that carried out using marine-derived fungi at the MK Bhavnagar University. The success rate for the former was 1 % (Mukherjee et al. 2001a, b, 2002; Du et al. 2011) while that for the latter was 100 %. Each marine-derived strain examined in our laboratory was able to biosynthesize nanoparticles (Vala et al. 2012; Vala 2014a, b, 2015a, b). Conditions in marine environment are dynamic and influenced continuously by industrial effluents and other pollutants, organisms inhabiting marine environment have developed the capability to adapt rapidly to these changes by synthesizing a range of Biomolecules to combat the stress of polluting load including metal species. Mechanisms developed (like biosorption, bioprecipitation, extracellular sequestration, and/or chelation) to adapt to presence of metals may be ultimately leading to nanoparticle synthesis (Zinjarde and Pant 2002; Haferburg and Kothe 2007; Pawar et al. 2012; Mohite et al. 2015). This could be the reason for the success rate or marine-derived fungi for synthesizing nanoparticles.

3.2.2 Marine-Derived Fungi in Synthesizing GNPs

Less than 30 fungal species have been explored so far for GNPs biosynthesis (Kitching et al. 2014). Six of these species are marine-derived (Agnihotri et al. 2009; Apte et al. 2013b, c; Vala 2014a, b; 2015a).

Biosynthesis of GNPs by marine-derived fungi also like that of AgNPs started in 2009, however, unlike for AgNPs, the study initiated using marine yeasts (Agnihotri et al. 2009). Marine yeasts are able to produce variety of bioproducts (Zhen-Ming et al. 2010).

Agnihotri et al. (2009) reported cell wall associated biosynthesis of GNP by tropical marine yeast *Yarrowia lipolytica* NCIM 3589. Pimprikar et al. (2009) examined influence of biomass and gold salt concentration on GNP synthesis by tropical marine yeast *Yarrowia lipolytica* NCIM 3589. They reported that a range of nanoparticles and nanoplates can be synthesized by varying the cell numbers and the gold salt concentration. Possible involvement of amide, carboxyl and hydroxyl groups on the cell surfaces in GNPs synthesis has also been suggested.

Apte et al. (2013b) identified melanin as a factor responsible for GNPs synthesis by *Yarrowia lipolytica*. Apte et al. (2013c) isolated pigment melanin from *Yarrowia lipolytica* NCIM 3590 cells. They reported synthesis of fairly mono-disperse AgNPs as well as GNPs with average size 7 and 20 nm, respectively using melanin.

First report on biosynthesis of GNPs by filamentous marine-derived fungi appeared recently (Vala 2014a). Marine-derived *Rhizopus oryzae* has been reported to biosynthesize GNPs both, extra and intracellularly. The author claims this as first ever report on extracellular biosynthesis of GNPs by *R. oryzae* under static condition. Extracellular biosynthesis is advantageous especially for downstream processing. The as synthesized particles were analysed using UV-vis spectrophotometry, XRD and transmission electron microscopy. The GNPs were found to be spherical ranging in size from 28 to 52 nm. Sheikhloo and Salouti (2012) has reported Intra-extra biosynthesis of GNPs by *Rhizopus oryza* under shaking condition at pH 2. Das et al. (2012) has observed GNPs synthesis on cell wall as well as in cytoplasmic region of *Rhizopus oryzae*. Plate 3.2 shows biosynthesis of GNP by *R. Oryzae*. Figure 3.6 shows FTIR spectra of the GNPs biosynthesized by marine-derived *R. Oryzae*.

Vala (2014b) examined three marine-derived fungal isolates viz. *Aspergillus candidus*, *Aspergillus flavus* and *Aspergillus niger* for their capability to biosynthesize GNPs. When the test isolates were challenged with different concentrations of gold (III) chloride, mostly extracellular biosynthesis of spherical GNPs was achieved. In certain cases, mode of synthesis (extra/intracellular) was observed to be dependent on supplied gold concentrations.

A marine-derived *Aspergillus sydowii* (GU004536.1) was examined for biosynthesizing GNPs (Vala 2015a). The test fungus exhibited potential to synthesize GNPs at different gold chloride concentrations. Interestingly, here also, mode of synthesis was governed by supplied gold concentrations. Spherical, nearly monodisperse particles were obtained at 3 mM gold chloride. The particles were in the size range of 8.7–15.6 nm with a mean diameter of 10 nm.

"Trick of the light" in Nature by Neil Savage (2013) throws light on applications of gold nano particles. Properties of light can be manipulated with the help of invisibly small gold particles (Savage 2013). Recently, weak localization of light has been observed for the first time in gold nanofluids synthesized using marine-derived fungus *Aspergillus niger* (Dave et al. 2015). Weak localization of light results due

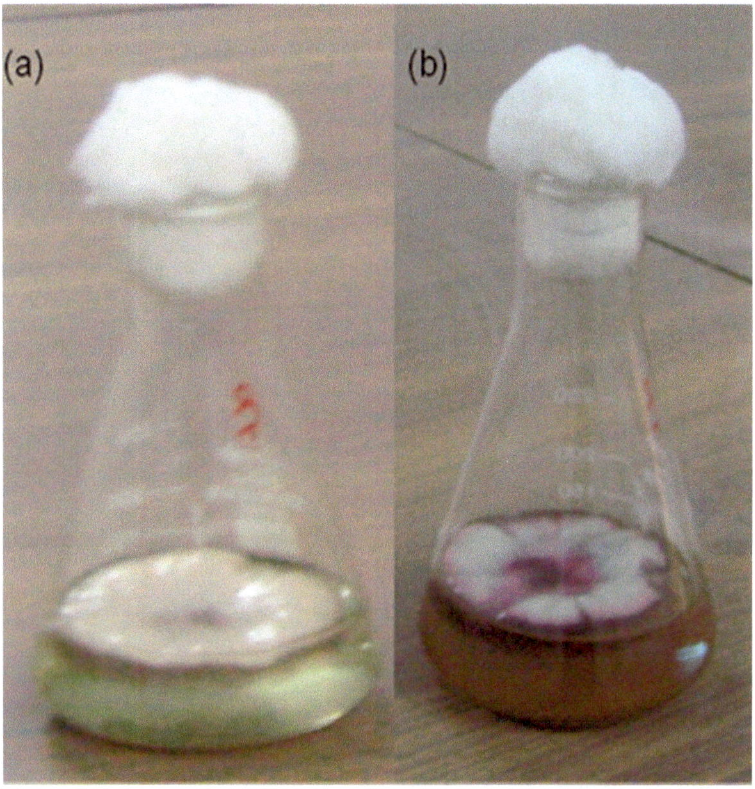

Plate 3.2 *Rhizopus oryzae* after (**a**) 1 h and (**b**) 48 h of exposure to 2 mM gold chloride

to coherent backscattering of waves by a disordered scattering medium (Dave et al. 2015). The experiments were carried out with 15 and 35 nm spherical GNPs synthesized using marine-derived fungus *Aspergillus niger* to measure enhancement of intensity in the backscattered direction i.e. the coherent backscattering cone for the particles, the experimental data corresponded well to the classical theoretical cone shape. The localization parameter kl* was observed to be $1 < kl^* < 5$.

The GNPs synthesized by marine-derived fungi exhibited remarkable stability, they were well dispersed even after one year of their synthesis, which is quite higher than reported by Das et al. (2012) for GNPs biosynthesized with protein extract of a *Rhizopus oryzae.* Detailed study on the noteworthy stability of GNPs could be very appealing. As per Das et al. (2012) electrostatic repulsion arising as a result of the negative charge of the conjugate proteins is the main stabilization factor. By altering the physicochemical characteristics GNPs of different size could be synthesized. Plate 3.3 shows GNPs biosynthesized at different pH.

Bimetallic NP synthesis is receiving much attention due to various applications in medicines, catalysis and sensors (Reith et al. 2009; Shedbalkar et al. 2014).

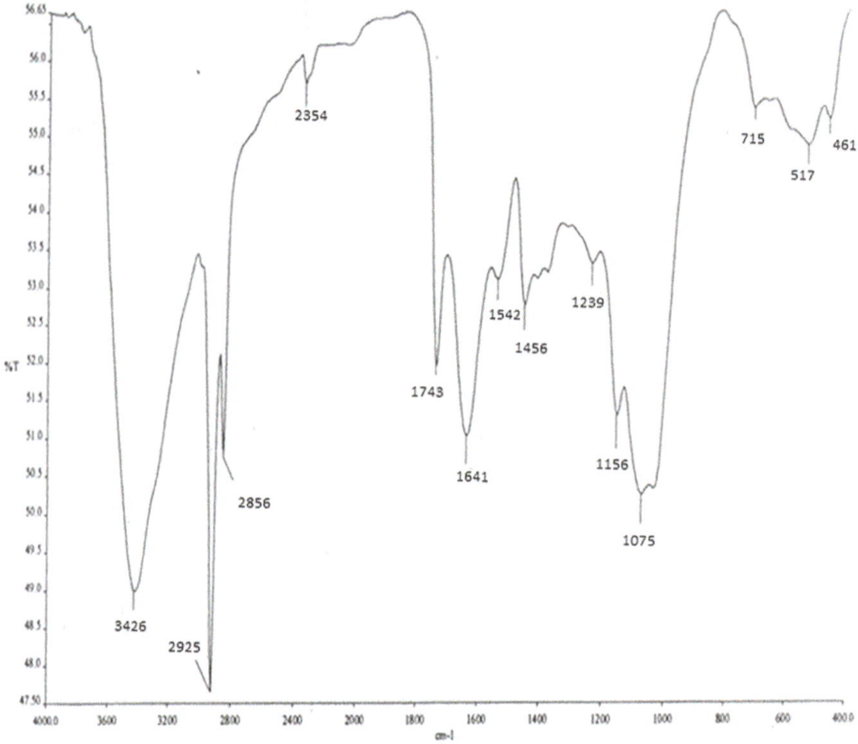

Fig. 3.6 IR spectra of GNPs biosynthesized by marine-derived *R. oryzae*

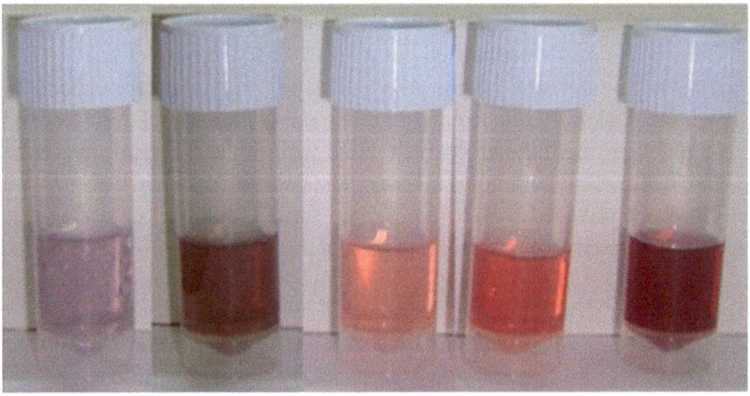

Plate 3.3 GNPs biosynthesized at different pH by a marine-derived fungus

However, very few reports exist on the microbe-mediated synthesis of gold nanoalloys (Shedbalkar et al. 2014). In our laboratory we examined bimetallic Au–Ag synthesis using above mentioned marine-derived fungi, among them only *Rhizopus oryzae* could synthesize bimetallic Au-AgNP (Data unpublished).

GNPs conjugated with various drugs, antibiotics, antibodies and other biomolecules also have diverse applications. Microbe-mediated synthesis of conjugate particles would be more eco-friendly and would open new avenues of research in nanomedicine (Shedbalkar et al. 2014), however, they have not yet been reported. Synthesis of such nanostructures using marine-derived fungi demands attention.

4 Applications of Nanoparticles

Due to their unique properties nanoparticles have wide spread applications in diverse fields including biomedicine, environment and agriculture. Noble metal nanoparticles like AgNPs and GNPs have important role in the field of biomedicine.

4.1 Silver Nanoparticles (AgNPs)

World's first nano silver-based wound dressing was developed by Dr. Robert Burrell in 1995 (Burrell et al. 1995). AgNPs can be used as an excellent candidate for anti-inflammatory agent in various therapies, also are used in artificial joint replacements in bone cements (Alt et al. 2004). AgNPs contain plasmonic nature and this characteristic could be employed to destroy target cells or unwanted cell (Loo et al. 2005). Nanosilver based biosensor has the ability to bio sense large number of protein that is not easily detected by normal biosensors and this unique characteristic helps to detect abnormalities and diseases including cancer (Zhou et al. 2011). AgNPs can be used as disinfectant as its anti-microbial property is already documented (Brady et al. 2003; Prasad 2014; Prasad et al. 2014).

4.1.1 Antibacterial Effect

Figure 3.7 summarizes some common mechanisms for antibacterial effects exerted by AgNPs. It has also been reported that the antibacterial activity of AgNPs also depends on surface modifications (Kvitek et al. 2008).

The AgNPs have shown broad spectrum bactericidal activity against both, Gram positive and Gram negative including some highly pathogenic strains (Jones and Hoek 2010). Antibacterial activities of AgNPs (in the range of 1–100 nm) against different types of Gram positive and Gram negative bacteria were tested by Morones et al. (2005).

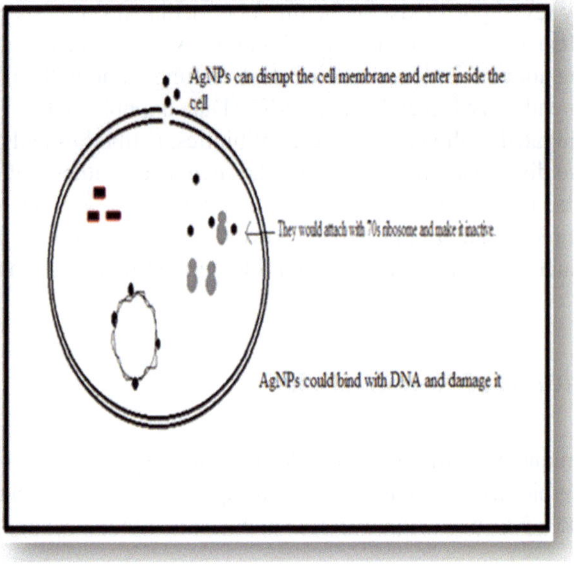

Fig. 3.7 Antibacterial mechanism of AgNPs (Adopted from Chaloupka et al. 2010, Trends Biotechnol 28(11):580–588)

Recently, remarkable synergistic effect of AgNPs on activities of antibiotics tetracycline and kanamycin has been reported by Khurana et al. (2016). Such studies would be helpful in combating issues related to antibiotic resistance in microbes.

4.1.2 Antifungal Effect

Some fungi can work as critical pathogens especially during fungal infections like nosocomial infection. 44 strains of six fungal species from clinical isolates and ATCC strains of *Trichophyton mentagrophytes* and *Candida albanicans* were exploited to test the antifungal activity of AgNPs and encouraging results were observed (Kim et al. 2008). Plastic catheters coated with AgNPs were investigated for their anti-fungal activity and almost complete inhibition of *C. albicans* was observed (Roe et al. 2008).

4.1.3 Antiviral Effect

Antiviral activity of AgNPs is comparatively less investigated. However, available reports suggest promising results.

Elechiguerra et al. (2005) revealed the interaction between AgNPs and HIV-1. This first report suggests that AgNPs interact with HIV-1 through glycol-protein knob as a result virus could not bind with host. This mechanism was explained by Lara et al. (2010). Different sizes of AgNPs (10, 50 and 800 nm) were exploited for the inhibition of hepatitis B virus (HBV) in, in vitro conditions (Lu et al. 2008).

Recently antiviral effect of AgNPs was investigated against H1N1 influenza A virus in vitro (Xiang et al. 2011). Biologically synthesized AgNPs have been demonstrated to exert dose-dependent antiviral activity against herpes simplex virus type 1 and 2 (HSV 1and 2) and human parainfluenza virus type 3 (HPIV-3) Gaikwad et al. (2013).

AgNPs have also been employed as water disinfectant (Bhattacharya and Mukherjee 2008). Besides AgNPs have other important applications including single electron transistors, fuel cells, nanocomputers, insect pest management and agriculture (Zhang et al. 2011; Singh et al. 2015).

4.2 Gold Nanoparticles (GNPs)

GNPs have applications especially in diagnosis and treatment of cancer, AIDS, tuberculosis and other diseases (Yadav et al. 2015). GNPs have been reported to be more promising than Au(I) and Au(III) based antirheumatic drugs (Griem and Gleichmann 1996; Mirabelli et al. 1985; Shukla et al. 2005). Ashiq et al. (2013) reported application of GNPs in breast cancer therapy using the technique laser induced coulomb explosion. Madhusudhan et al. (2014) reported conjugates of dauxorubicin and GNPs to be more effective than dauxorubicin alone.

GNPs have been demonstrated to be effective in photothermal therapy (Hwang et al. 2014) and radiotherapy (McLaughlin et al. 2013). Biosynthesized GNPs have been studied for detection of liver cancer (Chauhan et al. 2011). GNPs also play important role in treatment of AIDS (Berry et al. 2007; Bowman et al. 2008). Detection of mycobacterial infection is possible using GNP based biosensors (Duman et al. 2009; Thiruppathiraja et al. 2011).

GNPs are used for vaccine delivery. GNP-based DNA vaccines are more efficient than the conventional vaccines (Yadav et al. 2015). Besides therapeutics and drug delivery, GNPs also have high technology applications like electronic conduc tors, catalysts, organic photovoltaics and sensory probes (Daniel and Astruc 2004; Huang et al. 2003). As stated earlier in the text, weak localization of light has been observed in biosynthesized GNPs (Dave et al. 2015).

5 Conclusion and Future Challenges

Marine-derived fungi can be regarded as novel biofactories for synthesizing nanoparticles with diverse applications. Issues generally involved with biosynthesis of nanoparticles also are tackled by marine-derived fungi mediated biosynthesis. However, marine-derived fungal nanotechnology is still in infancy and core investigation is needed especially to elucidate the mechanism involved in nanoparticles biosynthesis. Extensive research is also required for commercially viable large-scale production of nanoparticles.

Acknowledgement AKV acknowledges Council of Scientific and Industrial Research, New Delhi, for sanctioning and sponsoring the work under Scientists' Pool Scheme. Thanks are due to the Department of Science and Technology, Government of India, New Delhi, for financial support to AKV under Women Scientist Scheme (WOS-A). AKV also acknowledges Prof. S P Bhatnagar, Head, Department of Physics, Maharaja Krishnakumarsinhji Bhavnagar University for support and encouragement.

References

Agnihotri M, Joshi S, Kumar AR, Zinjarde S, Kulkarni S (2009) Biosynthesis of gold nanoparticles by the tropical marine yeast *Yarrowia lipolytica* NCIM 3589. Mater Lett 63:1231–1234

Alani F, Moo-Young M, Anderson W (2012) Biosynthesis of silver nanoparticles by a new strain of *Streptomyces* sp. compared with *Aspergillus fumigatus*. World J Microbiol Biotechnol 28:1081–1086

Alt V, Bechert T, Steinrücke P, Wagener M, Seidel P, Dingeldein E, Scheddin D, Domann E, Schnettler R (2004) Nanoparticulate silver. A new antimicrobial substance for bone cement. Orthopade 33:885–892

Anand BG, Navin CK, Thomas S, Prakash C, Kumar S (2015) Biosynthesis of silver nano-particles by marine sediment fungi for a dose dependent cytotoxicity against HEp2 cell lines. Biocat Agricult Biotechnol 4:150–157

Apte M, Sambre D, Gaikawad S, Joshi S, Ashok Bankar A, Ravi AK, Zinjarde S (2013a) Psychrotrophic yeast *Yarrowia lipolytica* NCYC 789 mediates the synthesis of antimicrobial silver nanoparticles via cell-associated melanin. AMB Exp 3:32

Apte M, Girme G, Nair R, Bankar A, Kumar AR, Zinjarde S (2013b) Melanin mediated synthesis of gold nanoparticles by *Yarrowia lipolytica*. Mater Lett 95:149–152

Apte M, Girme G, Bankar A, Kumar AR, Zinjarde S (2013c) 3,4-Dihydroxy-L-phenylalanine-derived melanin from *Yarrowia lipolytica* mediates the synthesis of silver and gold nanostructures. J Nanobiotechnol 11:2

Ashiq MGB, Saeed MA, Tahir BA et al (2013) Breast cancer therapy by laser-induced Coulomb explosion of gold nanoparticles. Chin J Cancer Res 25:756–761

Aziz N, Faraz M, Pandey R, Sakir M, Fatma T, Varma A, Barman I, Prasad R (2015) Facile algae-derived route to biogenic silver nanoparticles: synthesis, antibacterial and photocatalytic properties. Langmuir 31:11605–11612. doi:10.1021/acs.langmuir.5b03081

Bai H, Yang B, Chai C, Yang G, Jia W, Yi W (2011) Green synthesis of silver nanoparticles using *Rhodobacter sphaeroides*. World J Microbiol Biotechnol Vol 27:2723–2728

Berry CC, de la Fuente JM, Mullin M et al (2007) Nuclear localization of HIV-1 Tat functionalized gold nanoparticles. IEEE Trans Nanobiosci 6:262–269

Bharde A (2007) Microbial synthesis of metal oxide, metal sulfide and metal nano particles. Ph.D. Thesis. University of Pune, India

Bhattacharya R, Mukherjee P (2008) Biological properties of naked nanoparticles. Adv Drug Del Rev 60:1289–1306

Bhimba BV, Gurung S, Nandhini US (2015) Silver nanoparticles synthesized from marine fungi *Aspergillus oryzae*. Int J Chem Tech Res 7(01):68–72

Bhimba BV, Nath N, Sinha P (2011) Characterization and antibacterial analysis of silver nanoparticles synthesized by the marine fungi *Hypocrea lixii* MV1 isolated from mangrove sediment soil. J Pharm Res 4(2):477–479

Bonugli-Santos RC, dos Santos Vasconcelos MR, Passarini MRZ, Vieira GAL, Lopes VCP, Mainardi PH, dos Santos JA, de Azevedo Duarte L, Otero IVR, da Silva Yoshida AM, Feitosa VA, Pessoa A Jr, Sette LD (2015) Marine-derived fungi: diversity of enzymes and biotechnological applications. Front Microbiol 6:269 doi:10.3389/fmicb.2015.00269

Bowman MC, Ballard TE, Ackerson CJ et al (2008) Inhibition of HIV fusion with multivalent gold nanoparticles. J Am Chem Soc 130:6896–6897

Brady MJ, Lissay CM, Yurkovetskiy AV, Sarwan SP (2003) Persistent silver disinfectant for environmental control of pathogenic bacteria. Am J Infect Control 31:208–214

Brunner TJ, Wick P, Manser P, Spohn P, Grass RN, Limbach LK, Bruinink A, Stark WJ (2006) In vitro cytotoxicity of oxide nanoparticles: comparison to asbestos, silica, and effects of particle solubility. Environ Sci Technol 40(14):4374–4381

Burrell RE, McIntosh CL, Morris LR (1995) Process of activating anti-microbial materials. Patent 5454886, 3

Chaloupka K, Malam Y, Seifalian AM (2010) Nanosilver as a new generation of nanoproduct in biomedical applications. Trends Biotechnol 28(11):580–588

Chan YS, Mashitah MD (2012) Instantaneous biosynthesis of silver nanoparticles by selected macro-fungi. Aust J Basic App Sci 6:222–226

Chandramohan D (2004) Marine microbiology: challenges and future directions. In: Ramaiah N (ed) Marine microbiology: facets & opportunities. National Institute of Oceanography: Goa, pp 7–13

Chauhan A, Zubair S, Tufail S, Sherwani A, Sajid M, Raman SC, Azam A, Owais M (2011) Fungusmediated biological synthesis of gold nanoparticles: potential in detection of liver cancer. Int J Nanomedicine 6:2305–2319

Christophersen CO, Crescente JC, Frisvad L, Gram J, Nielsen PH, Nielsen RL (1999) Antibacterial activity of marine-derived fungi. Mycopathologia 143:135–138

Daniel MC, Astruc D (2004) Gold nanoparticles: assembly, supramolecular chemistry, quantum-size-related properties, and applications toward biology, catalysis, and nanotechnology. Chem Rev 104(1):293–346

Das SK, Liang J, Schmidt M, Laffir F, Marsili E (2012) Biomineralization mechanism of gold by zygomycete fungi *Rhizopus oryzae*. ACS Nano 6:6165–6173

Dave V, Vala AK, Patel RJ (2015) Observation of weak localization of light in gold nanofluids synthesized using the marine derived fungus *Aspergillus niger*. RSC Adv 5:16780–16784

Deepak K, Kalishwaralal S. Pandian RK, Gurunathan S (2011) An insight into the bacterial biogenesis of silver nanoparticles, industrial production and scale-up. In: Rai M, Duran N (eds) Metal nanoparticles in microbiology. Springer-Verlag, Berlin/Heidelberg, pp 17–35

Drexler KE (1981) Molecular engineering: an approach to the development of general capabilities for molecular manipulation. Proc Natl Acad Sci USA 78:5275–5278

Du L, Xian L, Feng JX (2011) Rapid extra-/intracellular biosynthesis of gold nanoparticles by the fungus *Penicillium* sp. J Nanopart Res 13:921–930

Duman M, Caglayan Demirel G, Piskin E (2009) Detection of *Mycobacterium tuberculosis* complex using surface Plasmon resonance based sensors carrying self-assembled nano-overlayers of probe oligonucleotide. Sensor Lett 7:535–542

Elechiguerra JL, Burt JL, Morones JR, Camacho-Bragado A, Gao X, Lara HH, Yacaman MJ (2005) Interaction of silver nanoparticles with HIV-1. J Nanobiotechnol 29:3–6

Faraday M (1857) Experimental relations of gold (and other metals) to light. Phil Trans Roy Soc Lond 147:145–181

Gaikwad S, Birla S, Ingle A, Gade A, Marcato P, Rai M, Duran N (2013) Screening of different *Fusarium* species to select potential species for the synthesis of silver nanoparticles. J Braz Chem Soc 24(12):1974–1982

Gatti A, Magatti D, Ferri F (2008) Three-dimensional coherence of light speckles: theory. Phys Rev A 78:063806

Griem P, Gleichmann E (1996) Gold antirheumatic drug: desired and adverse effects of Au(I) and Au(III) [corrected] on the immune system. Z Rheumatol 55(5):348–358

Gwynne P (2013) There's gold in them bugs. Nature 495:S12–S13
Haferburg G, Kothe E (2007) Microbes and metals: interactions in the environment. J Basic Microbiol 47(6):453–467
Hemath NKS, Kumar G, Karthik LB, Rao HKV (2010) Extracellularbiosynthesis of silver nanoparticles using the filamentous fungus *Penicillium* sp. Arch App Sci Res 2:161–167
Honary S, Barabadi H, Gharaei-Fatahabad E, Naghibi F (2013) Green synthesis of silver nanoparticles induced by the fungus *Penicillium citrinum*. Trop J Pharma Res 12:7–11
Huang D, Liao F, Molesa S, Redinger D, Subramanian V (2003) Plastic-compatible low resistance printable gold nanoparticle conductors for flexible electronics. J Electrochem Soc 150:G412–G417
Hwang S, Nam J, Song J et al (2014) A sub 6 nanometer plasmonic gold nanoparticle for pH-responsive near-infrared photothermal cancer therapy. New J Chem 38:918–922
Ingle A, Gade A, Pierrat S et al (2008) Mycosynthesis of silver nanoparticles using the fungus *Fusarium acuminatum* and its activity against some human pathogenic bacteria. Curr Nanosci 4:141–144
Jain N, Bhargava A, Majumdar S et al (2011) Extracellular biosynthesis and characterization of silver nanoparticles using *Aspergillus flavus* NJP08: a mechanism perspective. Nanoscale 3:635–641
Jain K, Kesharwani P, Gupta U, Jain NK (2012) A review of glycosylated carriers for drug delivery. Biomaterials 33:4166–4186
Jones CM, Hoek EMV (2010) A review of the antibacterial effects of silver nanomaterials and potential implications for human health and the environment. J Nanopart Res 12:1531–1551
Kathiresan K, Manivannan S, Nabeel MA, Dhivya B (2009) Studies on silver nanoparticles synthesized by a marine fungus, *Penicillium fellutanum* isolated from coastal mangrove sediment. Colloids Surf B Biointerfaces 71:133–137
Kathiresan K, Alikunhi NM, Pathmanaban S, Nabikhan A, Kandasamy S (2010) Analysis of antimicrobial silver nanoparticles synthesized by coastal strains of *Escherichia coli* and *Aspergillus niger*. Can J Microbiol 56:1050–1059
Khurana C, Vala AK, Andhariya N, Pandey OP, Chudasama B (2016) Influence of antibiotic adsorption on biocidal activities of silver nanoparticles. IET Nanobiotechnol 10(2):69–74
Kim KJ, Sung WS, Moon SK, Choi JS, Kim JG, Lee DG (2008) Antifungal effect of silver nanoparticles on dermatophytes. J Microbiol Biotechnol 18:1482–1484
Kitching M, Ramani M, Marsili E (2014) Fungal biosynthesis of gold nanoparticles: mechanism and scale up. Microbial Biotechnol. doi:10.1111/1751-7915.12151
Kohlmeyer J, Kohlmeyer E (1979) Marine mycology-the higher fungi. Academic, New York
Kohlmeyer J, Volksmann- Kohlmeyer B (2003) Fungi from coral reefs: a commentary. Mycol Res 107:386–387
Kvitek L, Panacek A, Soukupova J, Kolar M, Vecerova R, Prucek R (2008) Effect of surfactants and polymers on stability and antibacterial activity of silver nanoparticles (NPs). J Phys Chem Part 112:5825–5834
Lara HH, Ayala-Nuñez NV, Ixtepan-Turrent L, Rodriguez-Padilla C (2010) Mode of antiviralaction of silver nanoparticles against HIV-1. J Nanobiotechnol 8:1–10
Li H, d'Anjou M (2009) Pharmacological significance of glycosylation in therapeutic proteins. Curr Opin Biotechnol 20:678–684
Li Q, Wang G (2009) Diversity of fungal isolates from three Hawaiian marine sponges. Microbe Res 164:233–241
Loo C, Lowery A, Halas N, West J, Drezek R (2005) Immunotargeted nanoshellsfor integrated cancer imaging and therapy. Nano Lett 5:709–711
Lu L, Sun RW, Chen R, Hui CK, Ho CM, Luk JM, Lau GK, Che CM (2008) Silver nanoparticles inhibit hepatitis B virus replication. Antivir Ther 13:253–262
Madhusudhan A, Reddy GB, Venkatesham M et al (2014) Efficient pH dependent drug delivery to target cancer cells by gold nanoparticles capped with carboxymethyl chitosan. Int J Mol Sci 15:8216–8234

Manivannan S, Alikunhi NM, Kandasamy K (2010) In vitro synthesis of silver nanoparticle by marine yeasts from coastal mangrove sediment. Adv Sci Lett 3:428–433
McLaughlin MF, Woodward J, Boll RA et al (2013) Gold coated lanthanide phosphate nanoparticles for targeted alpha generator radiotherapy. PLoS One 8, e54531
Meenupriya J, Majumdar A, M. Thangaraj M (2011) Biogenic silver nanoparticles by *Aspergillus terreus* MP1 and its promising antimicrobial activity. J Pharmacy Res 4(6):1648–1650
Mirabelli CK, Johnson RK, Sung C, Faucette KF, Muirhead K, Crooke ST (1985) Evaluation of the in vivo antitumor activity and in vitro cytotoxic properties of auranofin, a coordinated gold compound, in murine tumor models. Cancer Res 45(1):32–9
Mohite P, Apte M, Kumar AR, Zinjarde S (2015) Marine organisms in nanoparticle synthesis. In: Kim S-K (ed) Handbook of marine biotechnology. Springer, Dordrecht/Heidelberg/London/New York, pp 1229–1246
Morones JR, Elechiguerra JL, Camacho A, Holt K, Kouri JB, Ramirez JT, Yacaman MJ (2005) The bactericidal effect of silver nanoparticles. Nanotechnology 6:2346–2353
Mukherjee P, Ahmad A, Mandal D, Senapati S, Sainkar SR, Khan MI, Ramani R, Parischa R, Ajayakumar PV, Alam M, Sastry M, Kumar R (2001a) Bioreduction of $AuCl_4^-$ ions by the fungus, *Verticillium* sp. and surface trapping of the gold nanoparticles formed. Angew Chem Int Ed 40:3585–3588
Mukherjee P, Ahmad A, Mandal D, Senapati S, Sainkar SR, Khan MI, Parishcha R, Ajaykumar PV, Alam M, Kumar R, Sastry M (2001b) Fungus-mediated synthesis of silver nanoparticles and their immobilization in the mycelial matrix: a novel biological approach to nanoparticle synthesis. Nano Lett 1:515–519
Mukherjee P, Senapati S, Mandal D, Ahmad A, Khan MI, Kumar R, Sastry M (2002) Extracellular synthesis of gold nanoparticles by the fungus *Fusarium oxysporum*. Chem Bio Chem 3:461–463
Nachiyar V, Sunkar S, Prakash P, Bavanilatha (2015) Biological synthesis of gold nanoparticles using endophytic fungi. Scholars Research library 7(2):31–38
Narayanan KB, Sakthivel N (2010) Biological synthesis of metal nanoparticles by microbes. Adv Colloid Interface Sci 156:1–13
Nayak BK, Anitha K (2014) Amplified antibiotic potency of two different drugs combined with biosynthesized AgNPs from *Aspergillus sydowii* isolated from sand dunes. Int J Pharm Tech Res 6(6):1751–1755
Osterhage C (2001) Isolation, structure determination and biological activity assessment of secondary metabolites from marine-derived fungi. Technischen Universität Carolo-Wilhelmina, Braunschweig
Pawar V, Shinde A, Kumar AR, Zinjarde S, Gosavi S (2012) Tropical marine microbe mediated synthesis of cadmium nanostructures. Sci Adv Mater 4(1):135–142
Pimprikar PS, Joshi SS, Kumar AR, Zinjarde SS, Kulkarni SK (2009) Influence of biomass and gold salt concentration on nanoparticle synthesis by the tropical marine yeast *Yarrowia lipolytica* NCIM 3589. Colloids Surfaces B 74(1):309–316
Prakasham RS, Buddana SK, Yannam SK, Guntuku GS (2012) Characterization of silver nanoparticles synthesized by using marine isolate *Streptomyces albidoflavus*. J Microbiol Biotechnol 22(5):614–621
Prasad R (2014) Synthesis of silver nanoparticles in photosynthetic plants. J Nanoparticles 963961, doi:10.1155/2014/963961
Prasad R, Kumar V, Prasad KS (2014) Nanotechnology in sustainable agriculture: present concerns and future aspects. Afr J Biotechnol 13(6):705–713
Prasad R, Pandey R, Barman I (2015) Engineering tailored nanoparticles with microbes: quo vadis? WIREs Nanomed Nanobiotechnol. doi:10.1002/wnan.1363
Prathna TC, Mathew L, Chandrasekaran N, Raichur AM, Mukherjee A (2010) Biomimetic synthesis of nanoparticles: Science, Technology and Applicability. In: Mukherjee A (ed) Biomimetics, learning from nature. doi:10.5772/8776
Rai MK, Yadav AP, Gade AK (2009) Silver nanoparticles as a new generation of antimicrobials. Biotechnol Adv 27:76–82

Rauwel P, Siim K, Ferdov S, Rauwel E (2015) A Review on the Green synthesis of silver nanoparticles and their morphologies studied via TEM. Adv Mater Sci Eng. ID 682749

Reith F, Etschmann B, Grosse C, Moors H, Benotmane MA, Monsieurs P, Grass G, Doonan C, Vogt S, Lai B, Martinez-Criado G, George GN, Nies DH, Mergeay M, Pring A, Southam G, Brugger J (2009) Mechanisms of gold biomineralization in the bacterium *Cupriavidus metallidurans*. Proc Natl Acad Sci USA 106:17757–17762

Rimal Isac RS, Sakthivel G, Murthy Ch (2013) Green synthesis of gold and silver nanoparticles using *Averrhoa bilimbi* fruit extract. J Nanotechnol ID 906592, 6 pages, DOI:10.1155/2013/906592

Rodrigues AG, Ping LY, Marcato PD, Alves OL, Silva MCP, Ruiz RC, Melo IS, Tasic L, De Souza AO (2013) Biogenic antimicrobial silver nanoparticles produced by fungi. Appl Microbiol Biotechnol 3:775–782. doi:10.1007/s00253-012-4209-7

Roe D, Karandikar B, Bonn-Savage N, Gibbins B, Roullet JB (2008) Antimicrobial surface functionalization of plastic catheters by silver nanoparticles. J Antimicrob Chemoth 61:869–876

Rudd PM, Elliott T, Cresswell P, Wilson IA, Dwek RA (2001) Glycosylation and the immune system. Science 291:2370–2376

Sanghi R, Verma P (2009) Biomimetic synthesis and characterisation of protein capped silver nanoparticles. Bioresour Technol 100(1):501–504

Sathiya Rathna G, Elavarasi AA, Peninal S, Subramanian J, Mano G, Kalaiselvam M (2013) Extracellular biosynthesis of silver nanoparticles by endophytic fungus *Aspergillus terreus* and its anti-dermatophytic activity. Int J Pharmaceut Biol Arch 4(3):481–487

Savage N (2013) Trick of life. Nature 495:S8

Shedbalkar U, Singh R, Wadhwani S, Gaidhani S, Chopade BA (2014) Microbial synthesis of gold nanoparticles: current status and future prospects. Adv Colloid Interfac 209:40–48

Sheikhloo Z, Salouti M (2012) Intracellular biosynthesis of gold nanoparticles by fungus *Phoma macrostoma*. Synth React Inorg Met Org Nano Met Chem 42:65–67

Shukla R, Bansal V, Chaudhary M, Basu A, Bhonde RR, Sastry M (2005) Biocompatibility of gold nanoparticles and their endocytotic fate inside the cellular compartment: a microscopic overview. Langmuir 21(23):10644–10654

Singh CR, Kathiresan K, Anandhan S (2015) A review on marine based nanoparticles and their potential applications. Afr J Biotechnol 14(18):1525–1532

Soni N, Prakash S (2012) Synthesis of gold nanoparticles by the fungus *Aspergillus niger* and its efficacy against mosquito larvae. Rep Parasitol 2:1–7

Thiruppathiraja C, Kamatchiammal S, Adaikkappan P et al (2011) Specific detection of *Mycobacterium* sp. Genomic DNA using dual labeled gold nanoparticle based electrochemical biosensor. Anal Biochem 417:73–79

Trivedi HB, Vala AK, Dave BP (2015) Siderophore mediated biosynthesis of silver nanoparticles. (in communication)

Ulyanov SS (1998) Speckled speckle statistics with a small number of scatterers: implications for blood flow measurement. J Biomed Optics 3(3):237–245

Vala AK (2010) Tolerance and removal of arsenic by a facultative marine fungus *Aspergillus candidus*. Bioresour Technol 101:2565–2567

Vala AK (2014a) Intra- and extracellular biosynthesis of gold nanoparticles by a marine-derived fungus *Rhizopus oryzae*. Synth React Inorg Met Org Nano Met Chem 44(9):1243–1246

Vala AK (2014b) Investigations on gold nanoparticles biosynthesis potential of marine derived fungi. BioTechnol An Ind J 9(5):206–209

Vala AK (2015a) Exploration on green synthesis of gold nanoparticles by a marine derived fungus *Aspergillus sydowii*. Env Prog Sustain Energy 34(1):194–197

Vala AK (2015b) Rapid biogenesis of silver nanoparticles by two marine-derived fungi: a green chemistry approach. (In communication)

Vala AK, Shah S (2012) Rapid synthesis of silver nanoparticles by a marine-derived fungus *Aspergillus niger* and their antimicrobial potentials. Int J Nanosci Nanotechnol 8(4):197–206

Vala AK, Sutariya V (2012) Trivalent arsenic tolerance and accumulation in two facultative marine fungi. Jundishapur J Microbiol 5:542–545

Vala AK, Anand N, Bhatt PN, Joshi HV (2004) Tolerance and accumulation of hexavalent chromium by two seaweed associated aspergilli. Mar Pollut Bull 48(9):983–985

Vala AK, Chudasama B, Patel RJ (2012) Green synthesis of silver nanoparticles using marine-derived fungus *Aspergillus niger*. Micro Nano Lett 7:859–862

Vala AK, Shah S, Patel R (2014) Biogenesis of silver nanoparticles by marine-derived fungus *Aspergillus flavus* from Bhavnagar Coast, Gulf of Khambhat, India. J Mar Biol Oceanogr 3:1

Verma AK (2011) Potential of marine derived fungi and their enzymes in bioremediation of industrial pollutants. Ph.D. Thesis, National Institute of Oceanography, Dona-Paula

Xiang DX, Chen Q, Pang L, Zheng CL (2011) Inhibitory effects of silver nanoparticles on H1N1 influenza A virus in vitro. J Virol Methods 178:137–142. doi:10.1016/j.jviromet.2011.09.00

Yadav A, Kateryna K, Gabriela K, Nelson D, Avinash P, Ingle RM (2015) Fungi as an efficient mycosystem for the synthesis of metal nanoparticles: progress and key aspects of research. Biotechnol Lett 37:2099–2120

Zhang X, Yan S, Tyagi RD, Surampalli RY (2011) Synthesis of nanoparticles by microorganisms and their application in enhancing microbiological reaction rates. Chemosphere 82:489–494

Zhen-Ming C, Guanglei L, Shoufeng Z, Jing L, Ying P (2010) Marine yeasts as biocontrol agents and producers of bio-products. Appl Microbiol Biotechnol 86:1227–1241

Zhou W, Ma YY, Yang HA, Ding Y, Luo XG (2011) A label-free biosensor based on silver nanoparticles array for clinical detection of serum p53 in head and neck squamous cell carcinoma. Int J Nanomed 6:381–386

Zinjarde SS, Pant A (2002) Hydrocarbon degraders from tropical marine environments. Mar pollut bull 44(2):118–121

Chapter 4
Green Synthesis of Metal Nanoparticles by Fungi: Current Trends and Challenges

Luciano Paulino Silva, Cínthia Caetano Bonatto, and Vera Lúcia Perussi Polez

Abstract The approaches for synthesis of metal nanoparticles (MNPs) through green chemistry methods have become a recent trend of studies that focus on sustainability and innovation. Fungi are among the many groups of living organisms that have been known as useful for the synthesis of MNPs and there are many advantages of their use over other organisms since this group is directly or indirectly dependent of metals to its growth, metabolism and differentiation. They share efficient mechanisms of tolerance to high metal concentrations, being considered an important source of molecules able to transform metal ions into MNPs. The MNPs synthesis by fungi can be intracellular or extracellular and the latter is the most used because fungi secrete high amounts and diversity of enzymes making the process of synthesis sustainable, reliable, versatile and scalable. Indeed, the MNPs synthesis by fungi can use gold, silver, copper, iron, cadmium, nickel and others. However, the mechanisms of MNPs synthesis using fungi are not fully understood. The MNPs synthesis by fungi relies on many factors including biological material (e.g. species and/or strains; cultivation and sample preparation) and reaction conditions (e.g. metal species content and concentration; pH; temperature; and time of incubation) being necessary new strategies to improve the reproducibility of the processes. In the future, MNPs synthesized by fungi and their parts thereof can have unprecedented novel applications to several areas such as medical, agricultural and environmental.

L.P. Silva (✉) • C.C. Bonatto
Embrapa Genetic Resources and Biotechnology, Laboratory of Nanobiotechnology (LNANO), Nanotechnology and Synthetic Biology Group, Brasilia 70770-917, DF, Brazil

Institute of Biological Sciences, University of Brasilia, Brasilia 70910-900, DF, Brazil
e-mail: luciano.paulino@embrapa.br; lucianopaulinosilva@gmail.com

V.L.P. Polez
Embrapa Genetic Resources and Biotechnology, Laboratory of Prospection of Bioactive Compounds (LPCB), Bioactive Compounds Group, Brasilia 70770-917, DF, Brazil

R. Prasad (ed.), *Advances and Applications Through Fungal Nanobiotechnology*, Fungal Biology, DOI 10.1007/978-3-319-42990-8_4

Abbreviations

Ag-MNPs	Silver nanoparticles
Au-MNPs	Gold nanoparticles
CdS	Cadmium sulfide
CdS-NPs	Cadmium sulfide nanoparticles
CFE	Cell free extract
Cu-MNPs	Copper nanoparticles
MNPs	Metal nanoparticles
NiO-MNPs	Nickel oxide nanoparticles
QD	Quantum dots

1 Green Nanotechnology as a Sustainable Approach

Green nanotechnology is the application of green chemistry, green engineering, and sustainability principles to eliminate or at least minimize the use and generation of hazardous substances in the nanotechnology field (Albrecht et al. 2006; Schmidt 2007; Nath and Benerjee 2013). There are an unlimited number of opportunities, challenges and implications of this emerging area (Fleischer and Grunwald 2008; Hutchison 2008; Matus et al. 2011). However, one of the most promising approaches lies in synthesizing, characterizing and evaluating the effects of nanoparticles (NPs) with desired physical, chemical and/or biological properties (Albrecht et al. 2006; Katti et al. 2009; Nune et al. 2009; Suman et al. 2010).

Polymeric, lipid and metal nanoparticles are representative examples of nanosystems developed by green nanotechnologies. Among them, metal nanoparticles (MNPs) are probably the most studied and scientifically researched due to their unique properties that can be attributed to a high surface-to-volume ratio and metal specificities at the nanoscale (Raveendran et al. 2003; Vigneshwaran et al. 2006; Sharma et al. 2009; Thakkar et al. 2010; Iravani 2011; Narayanan and Sakthivel 2011; Philip et al. 2011; Prasad 2014; Bonatto and Silva 2014; Silva et al. 2015). Thus, MNPs may have unique optical, electrical, magnetic and catalytic properties; and exhibit medical, agricultural and environmental applications (Prasad and Swamy 2013; Salvadori et al. 2014a, b; Prasad et al. 2014, 2015; Alghuthaymi et al. 2015; Moghaddam et al. 2015; Bhuyan et al. 2015; Aziz et al. 2015; Yadav et al. 2015).

MNPs are useful platforms for numerous applications including microorganisms' control (Sharma et al. 2009), sensing (Dubas and Pimpan 2008), contrast agents (Liao et al. 2006), catalytic reactions (Liu et al. 2005), therapeutics (Philip 2009), electronics (Fu et al. 2013), photonics (Shah et al. 2012), wastewater treatment (Aziz et al. 2015). A variety of chemical, physical and biological methods could be used for synthesis of MNPs (Duran et al. 2011). However, some of these methods are fraught with challenges and drawbacks such as the role of toxic compounds (chemical methods) and the high-energy demand for production (physical methods). Thus, approaches using living organisms such as microorganisms (Duran et al. 2011) and plants (Silva et al. 2015) constitute an alternative to classical methods and they are called biological (or green) methods of synthesis.

2 Green Synthesis of MNPs

Green synthesis is an emerging and frontier method for synthesizing MNPs. Low-cost, sustainability, non-toxicity and simplicity are common benefits shared by green synthesis methods towards the production of MNPs (Vigneshwaran et al. 2006; Thakkar et al. 2010; Gan et al. 2012; Venkatessham et al. 2014). Green synthesis of MNPs may be performed using prokaryotic or eukaryotic organisms (e.g. bacteria, fungi or plants) or their parts thereof (e.g. extracts or specific tissues), through intra- and/or extracellular mechanisms (Thakkar et al. 2010; Venkatessham et al. 2014; Yadav et al. 2015). MNPs' properties such as size, shape, surface charge, chemical functionality and composition are the major features and measurements which have been explored by researchers using analytical tools (Albanese et al. 2012). The balance of these characteristics, and others, is essential to ensure that the MNPs could be used for potential applications ranging from biomedical, agricultural and environmental areas to commodity materials (Prasad et al. 2014).

Virtually all green synthesis approaches aiming the production of MNPs are based on the utilization of reducing or oxidizing agents that are able to react with metal ions from aqueous or nonaqueous solutions to produce metallic or metal oxide nanoparticles. A myriad of macromolecules and secondary metabolites from biological sources have been related to the oxidation-reduction (redox) reactions which can lead to MNPs formation (Liu et al. 2005; Sivaraman et al. 2009; Yang et al. 2010; Iravani 2011; Park et al. 2011; Bonatto and Silva 2014). Many of these compounds may also serve as coating and stabilizing agents for MNPs. In addition, several compounds in the surface of MNPs can act as bioactive molecules in either physiological or pathological conditions (Brila et al. 2009; Sathishkumar et al. 2009; Park et al. 2011). Among the biological resources useful for the green synthesis of MNPs, fungi are now recognized as important members of this recent trend and the choice of researchers worldwide (Fig. 4.1).

2.1 *Green Synthesis of MNPs by Fungi*

Fungi can be considered as an ancient group of living organisms with about 80,000–120,000 species described to date; although the total number of species is estimated at 1.5 million (Hawksworth 2001; Webster and Weber 2007). Fungi are eukaryotic and heterotrophic organisms which obtain their nutrients by extracellular digestion due to the activity of secreted hydrolytic and/or oxidative enzymes to the surrounding environment. Furthermore, fungi have the ability to uptake organic or inorganic solutes and ions from extremely dilute solutions from the environment, accumulating them 1000-fold or more against their concentration gradient (Griffin 1994; Webster and Weber 2007).

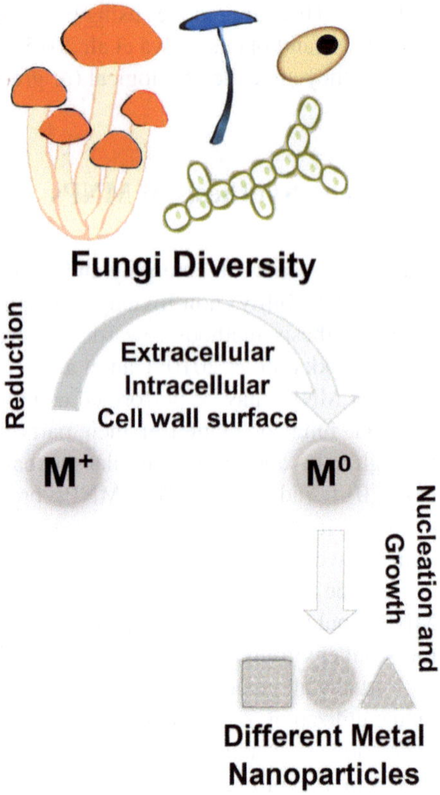

Fig. 4.1 Schematic illustration of the green synthesis of metal nanoparticles using fungi

The cell wall is an essential component to the growth, morphogenesis and survival of fungi. The components of the cell walls are glucans (e.g. β1-3, β1-3/β1-4, β1-6, α1-3 glucans, among others), chitin, chitosan, mannans and/or galactomannans and glycoproteins (Durán and Nombela 2004; Bowman and Free 2006; Free 2013). Thus, it is a dynamic structure that provides a protective barrier against the environmental changes such as heat, cold, desiccation, osmotic stress, microorganisms and metals (Durán and Nombela 2004; Bowman and Free 2006; Free 2013; Yadav et al. 2015).

Metals are directly or indirectly involved with growth, metabolism and differentiation of microorganisms, including fungi (Gadd 2010). However, the interaction of metals with microorganisms depends on the metal species, organism and environment. Structural components and metabolic activity also influence metal speciation and, therefore, solubility, mobility, bioavailability and toxicity (Gadd and Griffiths 1978; Gadd 1992, 2007, 2010). Indeed, important survival mechanisms of the microorganisms depend on changes in metal speciation leading to decreased or increased mobility such as the redox reactions, the production of metal-binding and chelating peptides and proteins (e.g. metallothioneins and phytochelatins), organic and inorganic precipitation, active transport, efflux and intracellular compartmentalization. Furthermore, cell walls and other structural components have metal-binding potential

(Mowll and Gadd 1984; White and Gadd 1998; Gadd 2007, 2010). In this way, fungi may have a high tolerance and even resistance to metals. In addition, several fungi groups (e.g yeasts, filamentous fungi and mushrooms) represent a great source to transform metal forms into MNPs (Table 4.1) (Durán et al. 2005; Fayaz et al. 2009; Li et al. 2012; Tarafdar and Raliya 2013; Devi and Joshi 2015; El-Baz et al. 2015). However, the mechanisms towards the green synthesis of MNPs by fungi are not yet completely elucidated.

MNPs synthesis by fungi can be intracellular (inside the fungal cell), extracellular (outside the fungal cell) or even on the surface of the cell; and can occur by enzymatic reduction as well as a cell wall bound process using different biomolecules (Figs. 4.2, 4.3, and 4.4) (Kashyap et al. 2013; Alghuthaymi et al. 2015; Moghaddam et al. 2015; Yadav et al. 2015). The extracellular synthesis of MNPs is faster and easier than intracellular synthesis because the former can secrete large amounts of enzymes and/or other compounds used for the synthesis process (Nayak et al. 2011; Kashyap et al. 2013) (Fig. 4.2). The extracellular synthesis of Ag-MNPs by several strains of *Fusarium oxysporum* were achieved by reduction of the silver ions which would probably occur by the presence of a nitrate-dependent reductase as well as a shuttle quinone extracellular process (Devi and Joshi 2015). Li et al. (2012) also observed a dependence on the Ag-MNPs synthesis on a NADH-dependent reductase from *Aspergillus terreus* (Li et al. 2012). Kumar et al. (2007) reported the enzymatic synthesis of Ag-MNPs using a nitrate reductase purified from *F. oxysporum*, $AgNO_3$, phytochelatin, 4-hydroxyquinoline and α-NADPH at 25 °C for 5 h under anaerobic conditions (Kumar et al. 2007). The Ag-MNPs synthesis requires the reduction of α-NADPH to α-$NADP^+$ and the hydroxyquinoline can act as an electron shuttle transferring the electron generated during the reduction of nitrate. Therefore, the action of hydroxyquinoline is probably similar to quinones in the electron transport taking place in the mitochondria or the chloroplast. Thus, the Ag-MNPs synthesis process by fungi suggests dependence of the reductase/electron shuttle relationships. Chan and Mashitah (2012) used different macrofungi (mushrooms) for the Ag-MNPs synthesis being suggested a possible action of a diketone compound as responsible for the process of silver ions reduction (Chan and Mashitah 2012). Other compounds can also be related with the MNPs synthesis such as the melanin secreted by *Yarrowia lipolytica* (Apte et al. 2013) and a glucan purified of *Pleurotus florida* (Sen et al. 2013).

Extracellular synthesis of nanoparticles using cell free extract (CFE) from *Saccharomyces boulardii* showed Ag-MNPs with a size range of 3–10 nm. The Fourier Transform Infrared spectroscopy (FTIR) analysis indicated the presence of proteins in the reduction and capping process besides the size of protein required for the synthesis is greater than 10 kDa. The CFE prepared from cell activities metabolically according to different phases of their growth (log phase, early or late stationary phase) where the early stationary phase of the cells is the most efficient for Ag-MNPs synthesis. Furthermore, temperatures greater than 40 °C fall the Ag-MNPs synthesis probably due to protein denaturation. Also, the Ag-MNPs synthesis increases when pH rises too but there is the formation of agglomerates. Thus, the authors confirmed the role of proteins in the Ag-MNPs synthesis and stabilization. Finally, the anticancer activity of Ag-MNPs on early-stage of breast cancer line

Table 4.1 Green synthesis of MNPs using fungi

Fungi species	Sample condition	Metal	Size (nm)	Shape	Applications	References
Alternaria alternaria	Cell free (filtration)	Ag	32.5	Spherical	Antimicrobial activity	Gajbhiye et al. (2009)
Aspergillus aculeatus	Inactive biomass (autoclaved)	Ni	5.89	–	Magnetic recording media (e.g. catalyst and medical fields)	Salvadori et al. (2014a, b)
Aspergillus niger	Cell free (filtration)	Ag	8.7±6	Spherical	–	Devi and Joshi (2015)
Aspergillus tamarii	Cell free (filtration)	Ag	3.5±3.3	Spherical	–	Devi and Joshi (2015)
Aspergillus terreus	Cell free (filtration)	Ag	1–20	Spherical	Antimicrobial activity	Li et al. (2012)
Aspergillus oryzae	Cell free (filtration)	Fe	10–24.6	Spherical	Agriculture, biomedical and engineering sector	Tarafdar and Raliya (2013)
Cryptococcus humicola	Active biomass with artificial magnetic fields	Fe	8–9	Regular	Drug delivery systems	Vainshtein et al. (2014)
Fusarium oxysporum	Active biomass	Ag	20–50	–	Antibacterial activity	Duran et al. (2011)
Ganoderma neo-japonicum	Inactive biomass (boiled and filtered)	Ag	5–8	Spherical	Anticancer	Gurunathan et al. (2013)
Hypocrea lixii	Inactive biomass (autoclaved)	Cu	24.5	Spherical	–	Salvadori et al. (2013)
Neurospora crassa	Active biomass	Au	3–200	Different shape	Biological sensing	Quester et al. (2013)
Penicllium ochrochloron	Cell free (filtration)	Ag	7.7±4.3	Spherical	–	Devi and Joshi (2015)
Penicillium purpurogenum	Cell free (filtration)	Ag	8–10	Spherical	Antimicrobial activity	Nayak et al. (2011)
Phanerochaete chrysosporium	Active biomass	Cd	2.56	Face-centered cubic	Bio-imaging, bio-labeling	Chen et al. (2014)

Rhodotorula mucilaginosa	Inactive biomass (autoclaved)	Cu	10.5	Spherical	Bioremediation process	Salvadori et al. (2014a, b)
Saccharomyces boulardii	Cell free (centrifugation)	Ag	3–10	–	Anticancer	Kaler et al. (2013)
Saccharomyces cerevisiae	Active biomass with artificial magnetic fields	Fe	–	Irregular	Drug delivery systems	Vainshtein et al. (2014)
Trichoderma viride	Active biomass (filtered)	Ag	5–40	Spherical	Antibacterial nanocoating for food products	Fayaz et al. (2009)
Trichosporon jirovecii	Cell suspension (UV use)	CdS	6–15	Spherical	Photonics and electronics	El-Baz et al. (2015)
Verticillium sp	Active biomass	Ag	25–12		–	Sastry et al. (2003)
Volvariella volvacea	Inactive biomass (boiled and filtered)	Au	20–150	Triangular spherical hexagonal	Therapeutic applications	Philip et al. (2011)
Volvariella volvacea	Inactive biomass (boiled and filtered)	Ag	15	Spherical	Therapeutic applications	Philip et al. (2011)

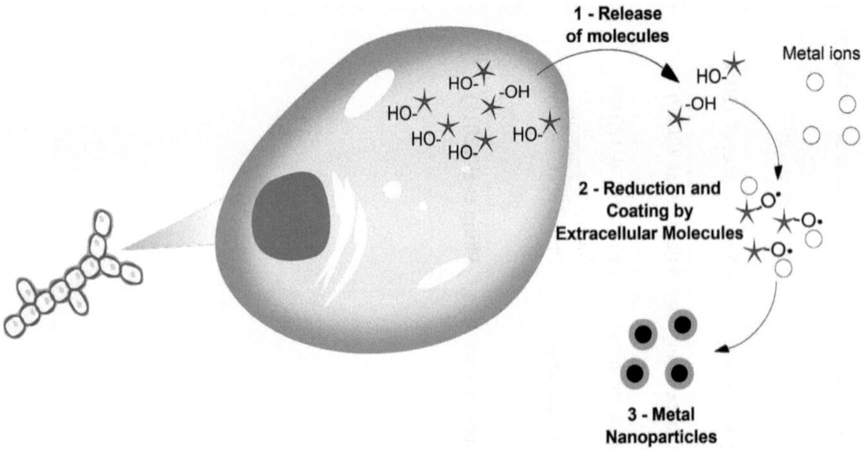

Fig. 4.2 Schematic illustration of the extracellular green synthesis of metal nanoparticles by fungi

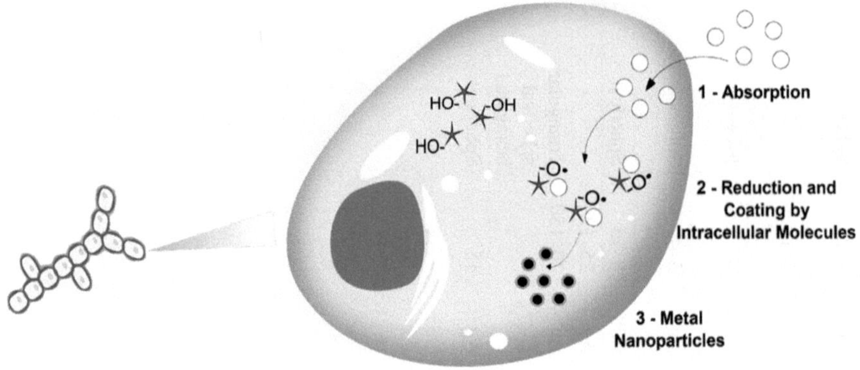

Fig. 4.3 Schematic illustration of the intracellular green synthesis of metal nanoparticles by fungi

MCF-7 cells was proposed (Kaler et al. 2013). In this case, Ag-MNPs synthesis from CFE of other different fungi like *Aspergillus niger*, *Aspergillus tamarii*, *Aspergillus terreus*, *Aspergillus oryzae*, *Penicillium ochrochloron*, *Penicillium purpurogenum* and *Saccharomyces boulardii* also used proteins for the green synthesis processes (Nayak et al. 2011; Li et al. 2012; Kaler et al. 2013; Tarafdar and Raliya 2013; Devi and Joshi 2015). In addition, the Ag-MNPs were synthesized using cell free supernatant from *Aspergillus terreus* and showed spherical shape and diameters ranging from 1 to 20 nm. In this case, the key factors related to the Ag-MNPs synthesis are (i) reaction was highly dependent on an active substance with a low molecular weight (<7000 Da) and (ii) NADH might be an important factor for the synthesis suggesting NADH-dependent reductase activity. Moreover, the Ag-MNPs

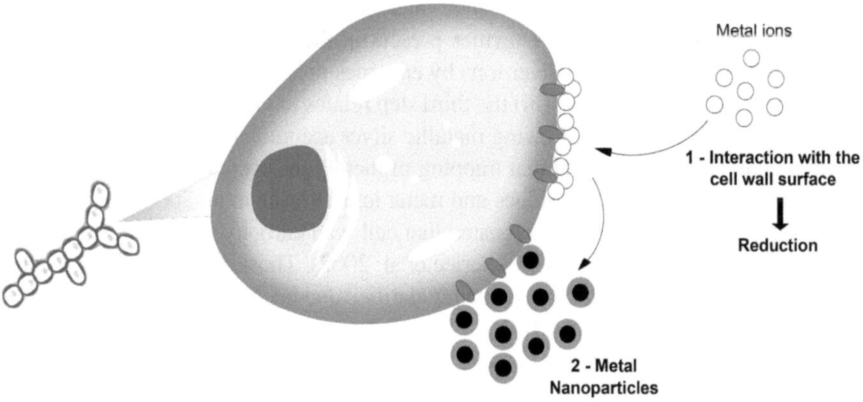

Fig. 4.4 Schematic illustration of the green synthesis of metal nanoparticles by interaction with fungi cell wall surface

showed a promising broad-spectrum of antimicrobial activity against *Staphylococcus aureus*, *Pseudomonas aeruginosa*, *Escherichia coli*, *Candida albicans*, *Candida krusei*, *Candida parapsilosis*, *Candida tropicalis*, *Aspergillus flavus* and *Aspergillus fumigatus* (Li et al. 2012).

Studies about the use of dead fungal biomass to MNPs synthesis have some advantages such as (i) limited toxicity, (ii) storage for a long period of time, (iii) it does not require growth media and nutrients for its maintenance (Salvadori et al. 2013, 2014a, b). The NiO-MNPs synthesis on film form using dead biomass from *Aspergillus aculeatus* as reducing agent represents an environmental-friendly nanotechnological innovation. So, live biomass was autoclaved to obtain the dead biomass and thus it was dried at 50 °C until becomes crispy. The dead biomass exhibited the highest capacity to produce NiO-MNPs. Furthermore, the proteins probably act as a capping agent for NiO-MNP and the fungi cell proteins are released during the autoclaving process. Finally, these NiO-MNPs formed a film on the biomass surface (Salvadori et al. 2014a, b).

Another example is the Ag-MNPs synthesis using inactivated biomass (boiling aqueous extracts) of the mushroom *Ganoderma neo-japonicu* with size ranging from 5 to 8 nm and spherical shape. The Ag-MNPs exhibited cytotoxic effects against breast cancer cells (MDA-MB-231) with apoptotic features and suggested that the generation of reactive oxygen species (ROS) have a significant role in apoptosis (Gurunathan et al. 2013). In addition, other promising examples for MNPs synthesis using inactive cells are those from *Rhodotorula mucilaginosa* and *Volvariella volvacea* suggesting that this strategy of sample preparation may be promising for the green synthesis of MNPs.

Despite intracellular synthesis of MNPs (Fig. 4.3) is not well clarified, some authors suggest the mechanisms involved in the process. Sastry et al. (2003) suggested an intracellular mechanism for Ag-MNP synthesis from fungi *Verticillium* sp. that consist of different steps: (i) the first step involves trapping of the Ag^{+} ions on the surface of the fungi cell walls possibly via electrostatic interaction between the Ag^+ and negatively charged carboxylate groups in enzymes present in the cell wall; (ii) the second step refers to the reduction of the silver ions by enzymes present in the cell wall leading to the formation of silver nuclei; (iii) the third step relates to growth of the nuclei by further reduction of Ag^+ ions following metallic silver accumulation (Sastry et al. 2003). Riddin et al. (2006) suggested that trapping of metal ions occurs through electrostatic interactions between lysine residues and metal ions (Riddin et al. 2006). On the other hand, Mukherjee et al. (2002) suggested that cell wall carbohydrates act as the major reducing agents of metal ions (Mukherjee et al. 2002). Thus, the intracellular mechanisms of MNPs synthesis need to be better clarified and need to extend these studies to the other metals (Yadav et al. 2015).

Some authors suggested a correlation between the size of synthesized MNPs with subcellular location, where the intracellular synthesis would form MNPs smaller than those obtained by extracellular synthesis (Thakkar et al. 2010; Narayanan and Sakthivel 2011; Yadav et al. 2015). In this case, the difference in size could be possibly limited by the dimensions of intracellular regions which lead to uniform particles and monodispersity (Nayak et al. 2011; Yadav et al. 2015). However, the examples of intracellular synthesis are limited and need more information to reinforce this correlation. The preference by extracellular synthesis occurs due to the fact that this synthesis is processed in an easier manner because fungi secrete large amounts of enzymes and/or other compounds and it is simpler to treat with the biomass (Mukherjee et al. 2002; Ahmad et al. 2003; Bhainsa and D'Souza 2006; Yadav et al. 2015). In contrast, since intracellular synthesis takes place inside the cell, its downstream process becomes difficult and hence the cost of MNPs synthesis increases (Gajbhiye et al. 2009; Yadav et al. 2015). Thus, the extracellular synthesis can be considered an important strategy for MNPs synthesis using fungi.

Finally, the mechanism of MNPs synthesis using fungi needs to be better understood, besides the MNPs synthesis can be influenced by several factors and can decrease the reproducibility of the synthesis and these factors will be described further.

3 Current Trends on Green Synthesis of MNPs Using Fungi

MNPs are mainly synthesized from noble metals such as gold or silver which share potential for medical applications. The Ag-MNPs can have antimicrobial activity against *Staphylococcus aureus*, *Pseudomonas aeruginosa*, *Escherichia coli*, *Candida albicans*, *Candida krusei*, *Candida parapsilosis*, *Candida tropicalis*, *Aspergillus flavus*, *Aspergillus fumigatus*, among others and combined synergistically with antibiotics or fungicides provide novel therapeutic platforms to the control of resistant microorganisms (Li et al. 2012). For example, Ag-MNPs synthesized by *Alternaria alternaria* associated with fluconazole (a fungicide) showed maximum inhibition against *Candida albicans* (Gajbhiye et al. 2009). Furthermore, Ag-MNPs can also be used for drug delivery to cancer treatment (Gurunathan et al. 2013; Moghaddam et al. 2015). Ag-MNPs and Au-MNPs kill cancer cells probably by induction of apoptosis but further studies are needed to understand their mechanisms of action (Kaler et al. 2013). Au-MNPs can also display antiviral activity as well as theranostic applications (Moghaddam et al. 2015).

Other important medical applications for nanoparticles synthesized by fungi are the bio-imaging and bio-labeling using metal nanoparticles and mainly quantum dots (QDs). In spite of the fact that QDs are not MNPs per se, they are semiconductor nanoparticles which possess unique electronic and optical properties and have been used as biological fluorescence markers, production of optoelectronic transistor components and solar batteries (Wegner and Hildebrandt 2015). Fluorogenic Cadmium sulfide (CdS) QDs synthesis by *Phanerochaete chrysosporium* showed nanoparticles with an average size of 2.56 nm and a pure blue emission with a full width at half-maximum narrower than 30 nm, indicating excellent optical properties. Thus, CdS QDs showed superior optical properties, low toxicity and applications in bio-imaging and bio-labeling (Chen et al. 2014).

Nanotechnology is an important field able to contribute in providing incremental solutions through green chemistry approaches for advancing food security (Kashyap et al. 2013). Thus, MNPs can be applied as antimicrobial agents to the management of plant pathogens serving as nano-pesticide (Roni et al. 2015), nanofungicide (Gajbhiye et al. 2009) and nanopackaging (Fayaz et al 2009). The use of protective nanostructured coating and suitable packaging is very important to increase shelf life of many edible products (Amini et al. 2014). The Ag-MNPs synthesis from *Trichoderma viride* and their incorporation into sodium alginate for vegetable (carrot—*Daucus carota*) and fruit (Pears—*Pyrus communis*) protection due to its antibacterial activity against both Gram-positive (*Staphylococcus aureus*) and Gram-negative (*Escherichia coli*) was investigated and suggested a great potential for active food packaging (Fayaz et al 2009). Thus, MNPs can help agriculture, food, feed and environment.

Nanomaterials can have important applications in different fields such as photoelectric materials, recording media, catalysts, sensors, ceramic materials and others (Patil et al. 2011; Salvadori et al. 2014a, b). Nickel oxide nanoparticles (NiO-MNPs) can exhibit particular catalytic, magnetic properties among others (Patil et al. 2011; Salvadori et al. 2014a, b). Salvadori et al. 2014a, 2014b reported for the first time the synthesis and uptake of NiO-MNPs organized in a nanostructured film form by dead biomass from *Aspergillus aculeatus* being a unique environmental-friendly process (Salvadori et al. 2014a, b). Another study using the dead biomass from *Rhodotorula mucilaginosa* may be considered an efficient bioprocess, being fast and low-cost to production of copper nanoparticles (Cu-MNPs) and a probable nano-adsorbent of this metal in wastewater for bioremediation (Salvadori et al. 2014a, b). CdS has many promising applications in fields such as photochemical catalysis, nonlinear optical materials, detectors for laser and infrared, gas sensor, solar cells and various luminescence devices (Hullavarad et al. 2008). Thus, the cadmium sulfide nanoparticles (CdS-NPs) synthesis has been considered a modern growing field of research due to its relevant physical, optical and chemical properties (Hullavarad et al. 2008). *Trichosporon jirovecii* was used as biological source for the synthesis of CdS-NPs being considered the first study related to eco-friendly biosynthesis of CdS-NPs using fungi (El-Baz et al. 2015).

4 Reproducibility Challenges for the Green Synthesis of MNPs Using Fungi

The MNPs synthesis using biological materials has a great challenge regarding the reproducibility. The factors that influence the MNPs synthesis by fungi which may impair their reproducibility are: (i) fungi species and/or strains may have genetic and/or biochemical variability which can result in changes in the cellular structure, metabolism and chemical composition (Griffin 1994; Treseder and Lennon 2015; Prasad et al. 2015). In addition, the location and collection conditions, the presence of stressor agents (biotic or abiotic) or even the cultivation conditions may be related with the alterations described previously. Some strategies can be used aiming biological material homogeneity and are described in Table 4.2. In this case, it can be observed a great difference between distinct fungi species towards MNPs synthesis like reported (Moghaddam et al. 2015; Yadav et al. 2015); (ii) cultivation in different substrates indicates that the fungi can be found in different ecosystems and its cultivation conditions can also be drastically different. However, a standardization of cultivation condition is essential for the establishment of ideal conditions for fungi growth and also to maintain the biological material homogeneity; (iii) sample preparation methods are

Table 4.2 Factors that influence MNPs synthesis using fungi and strategies to improve the reproducibility

Factors that influence the synthesis	Strategies that can improve the reproducibility
Species and/or strains	Selection of genetically stable species and/or strains
	Selection of species and/or strains that present a quick and easy cultivation as well as check the degree of pathogenicity and achieve selection according to application of MNPs
	Standardization of a preservation method more effective for the species and/or strains that seeks the significant reduction of cellular metabolic activity, allowing their preservation for years as the cryopreservation, lyophilization, among others
Cultivation condition of the fungi	Standardization of cultivation conditions such as the substrate type, pH, temperature, oxygenation, among others
	Determination of the fungi growth curve to choose the best phase for the MNPs' synthesis
Sample preparation condition	Determine the best condition of sample preparation in accordance with the goal of the study and its standardization
Metal species	Choice of metal species (regarding cost-benefit, potential reduction for synthesis as well as application)
	Standardization of metal species concentration
pH	Standardization of ideal pH for the MNPs synthesis
Temperature	Standardization of ideal temperature for the MNPs synthesis
Time	Standardization of ideal time of reaction for the MNPs synthesis

diversified including active filtered or centrifuged cells, dead cells by boiling or autoclaving, proteins or enzymes secreted to the external environment, among others. In this case, it can be observed great difference between MNPs' synthesis by fungi indicating that the form of preparation is also an important factor to the sample reproducibility (Salvadori et al. 2014a, b; Devi and Joshi 2015; Gurunathan et al. 2013; Quester et al. 2013; Vainshtein et al. 2014). The sample concentration used for the MNPs' synthesis can also be an important factor to be considered to the reproducibility of the final outcome; (iv) the metal species, the concentration of metal species, pH, incubation time and temperature are other essential factors for the MNPs synthesis using fungi (Kaler et al. 2013; Kashyap et al. 2013; Quester et al. 2013; Alghuthaymi et al. 2015; Kitching et al. 2015; Prasad et al. 2015). Thus, the uses of strategies fine-tuning the factors that influence the MNPs' synthesis are essential for different applications consequently improving the reproducibility of the process.

5 Scaling-Up Challenges for the Green Synthesis of MNPs Using Fungi

Scaling-up the green synthesis of MNPs using fungi or any other biological agent is a challenging process which requires technical and strategic understanding of a plethora of important and fundamental concepts of bioprocess engineering. These steps range from biological aspects of the fungal biochemistry like secondary metabolites and macromolecules content and concentration to engineering aspects such as the growth conditions and other variables. Several abiotic and biotic factors can affect fungal growth and consequently the production of macromolecules and secondary metabolites associated with the green synthesis of MNPs. Abiotic factors include humidity (water), pH (acid or basic), light, gases (CO_2, O_2), temperature and nutrients (macroelements and microelements); and biotic factors include the population size and cellular specializations.

6 Predictability Challenges for the Green Synthesis of MNPs Using Fungi

The most challenging topic in green synthesis of MNPs by fungi or any other biological source is the unpredictability of the reactions outcome on the basis of the initial properties (e.g. reagents concentration and ratio; temperature; pH; reaction time). In fact, the green synthesis of MNPs by fungi is an empirical and multifactorial process that is difficult to predict by computational methods, even with a small number of variables or reaction conditions. Indeed, the possibilities can grow exponentially when multiple factors are combined. Next years may soon bring good news but there is no current solution for this challenging issue that deserves to be cited here.

7 Risk Assessment and Risk Management Challenges for Applicability of MNPs Obtained by Green Synthesis Using Fungi

Evaluating the risks and benefits constitutes an essential step to be considered during the development of new nanomaterials, including MNPs (Cobb and Macoubrie 2004; Maynard et al. 2006; Besley et al. 2008; Kahan et al. 2009). Current approaches in risk assessment and management for using MNPs are generally based on the properties of their constituents and their hierarchical organization and architecture at nanoscale. Indeed, the use of greener methods commonly represents decreased risk to the environment and human health. However, this is a controversial issue and some scholars disagree with this opinion. But there is an apparent consensus that the choice of non pathogenic or harmless fungi (e.g. *Saccharomyces cerevisiae* and edible mushrooms) towards the green synthesis of MNPs can be an important point to be considered by researchers.

8 Regulatory Challenges for Applicability of MNPs Obtained by Green Synthesis Using Fungi

The regulatory challenges in nanotechnology field are commonly based on the risk management and safety (Besley et al. 2008). However, there is almost consensus that the regulation of green nanotechnology initiatives must be different from conventional nanotechnologies because they are based on green chemistry principles and practice. Clearly, the quest for greener MNPs must be conducted in a framework of the evaluation of impacts (positive or deleterious) to humans and environment and MNPs produced by fungi must be considered in this scenario during future discussions on regulatory streamlining.

9 Conclusions

Fungi are an excellent source of compounds useful for the green synthesis of MNPs to be applied in medical, agricultural and environmental areas. The MNPs synthesis by fungi can be easier and simpler than similar approaches with other biological sources mainly due to the fact that they produce and secrete high amounts of enzymes. The MNPs synthesis by fungi depend on many factors as the biological material (e.g. species and/or strains; cultivation and sample preparation) and synthesis conditions (e.g. metal species and concentration; pH; temperature; and time of incubation) being strategies necessary to improve the reproducibility of the MNPs synthesis. Anyway, the advantages and disadvantages (Table 4.3) in using fungi to synthesize MNPs must be clear and next years will prove (or not) fungi are good candidates for green synthesis approaches.

Table 4.3 Fungi as biological sources for the MNPs synthesis: advantages and disadvantages

Advantages	Disadvantages
High tolerance to metals	Contaminants
Cultivation in bioreactors	Standardization and maintenance growth
Low cost culture media including agroindustrial byproducts and waste	Maintenance of genetic stability
	Maintenance of growth
Standardization and control growth conditions	
Homogeneous samples	
Scalability to industrial process	
Extracellular synthesis:	
Enzymes end/or metabolites (e.g. secrete high amounts; easier purification)	
Active biomass	
Inactive biomass (e.g. storage for long period of time)	
MNPs' synthesis with different size and shape in control conditions	

References

Ahmad A, Senapati S, Khan MI, Kumar R, Ramani R, Srinivas V, Sastry M (2003) Intracellular synthesis of gold nanoparticles by a novel alkalotolerant actinomycete, *Rhodococcus* species. Nanotechnol 14:824–828

Albanese A, Tang PS, Chan WCW (2012) The effect of nanoparticle size, shape, and surface chemistry on biological systems. Annu Rev Biomed Eng 14:1–16

Albrecht MA, Evans CW, Raston CL (2006) Green chemistry and the health implications of nanoparticles. Green Chem 8:417–432

Alghuthaymi MA, Almoammar H, Rai M, Said-Galiev E, Abd-Elsalam K (2015) Myconanoparticles: synthesis and their role in phytopathogens management. Biotechnol Biotechnol Equip 29(2):221–236

Amini SM, Gilaki M, Karchani M (2014) Safety of nanotechnology in food industries. Electron Physician 6(4):962–968

Apte M, Girme G, Bankar A, Ravikumar A, Zinjarde S (2013) 3,4 Dihydroxy-L-phenylalanine-derived melanin from *Yarrowia lipolytica* mediates the synthesis of silver and gold nanostructures. J Nanobiotechnol 11:2

Aziz N, Faraz M, Pandey R, Sakir M, Fatma T, Varma A, Barman I, Prasad R (2015) Facile algae-derived route to biogenic silver nanoparticles: synthesis, antibacterial and photocatalytic properties. Langmuir 31:11605–11612

Besley JC, Kramer VL, Priest SH (2008) Expert opinion on nanotechnology: risks, benefits, and regulation. J Nanopart Res 10(4):549–558

Bhainsa KC, D'Souza SF (2006) Extracellular biosynthesis of silver nanoparticles using the fungus *Aspergillus fumigatus*. Colloids Surf B Biointerfaces 47(2):160–164

Bhuyan T, Mishra K, Khanuja M, Prasad R, Varma A (2015) Biosynthesis of zinc oxide nanoparticles from *Azadirachta indica* for antibacterial and photocatalytic applications. Mat Sci Semicon Proc 32:55–61

Bonatto CC, Silva LP (2014) Higher temperatures speed up the growth and control the size and optoelectrical properties of silver nanoparticles greenly synthesized by cashew nutshells. Ind Crop Prod 58:46–54

Bowman SM, Free SJ (2006) The structure and synthesis of the fungal cell wall. Bioessays 28(8):799–808

Brila SS, Tiwari VV, Gade AK, Ingle AP, Yadav AP, Rai MK (2009) Fabrication of silver nanoparticles by *Phoma glomerata* and its combined effect against *Escherichia coli*, *Pseudomonas aeruginosa* and *Staphylococcus aureus*. Appl Microbiol 48:173–179

Chan YS, Mashitah MD (2012) Instantaneous biosynthesis of silver nanoparticles by selected macro fungi. Aust J Basic Appl Sci 6(1):222–226

Chen G, Yi B, Zeng G, Niu Q, Yan M, Chen A, Du J, Huang J, Zhang Q (2014) Facile green extracellular biosynthesis of CdS quantum dots by white rot fungus *Phanerochaete chrysosporium*. Colloids Surf B Biointerfaces 117:199–205

Cobb MD, Macoubrie J (2004) Public perceptions about nanotechnology: risks, benefits and trust. J Nanopart Res 6(4):395–405

Devi LS, Joshi SR (2015) Ultrastructures of silver nanoparticles biosynthesized using endophytic fungi. J Microscopy Ultrastructure 3:29–37

Dubas ST, Pimpan V (2008) Green synthesis of silver nanoparticles for ammonia sensing. Talanta 76(1):29–33

Durán A, Nombela C (2004) Fungal cell wall biogenesis: building a dynamic interface with the environment. Microbiology 150:3099–30103

Durán N, Marcato PD, Alves De Souza OLGIH, Esposito E (2005) Mechanistic aspects of biosynthesis of silver nanoparticles by several *Fusarium oxysporum* strains. J Nanobiotechnol 3:8

Duran N, Marcato PD, Duran M, Yadav A, Gade A, Rai M (2011) Mechanistic aspects in the biogenic synthesis of extracellular metal nanoparticles by peptides, bacteria, fungi, and plants. Appl Microbiol Biotechnol 90(5):1609–1624

El-Baz AF, Sorour NM, Shetaia YM (2015) *Trichosporon jirovecii*-mediated synthesis of cadmium sulfide nanoparticles. J Basic Microbiol 55:1–11

Fayaz AM, Balaji K, Girilal M, Kalaichelvan PT, Venkatesan R (2009) Mycobased synthesis of silver nanoparticles and their incorporation into sodium alginate films for vegetable and fruit preservation. J Agric Food Chem 57(14):6246–6252

Fleischer T, Grunwald A (2008) Making nanotechnology developments sustainable. A role for technology assessment? Sustainable Nanotechnol Devel 16(8-9):889–898

Free SJ (2013) Fungal cell wall organization and biosynthesis. In: Friedmann T, Dunlap JC, Goodwin SF (eds) Advances in genetics. Elsevier, USA, pp 33–68

Fu G, Jiang X, Tao L, Chen Y, Lin J, Zhou Y (2013) High quality research in physical chemistry, chemical physics and biophysical chemistry. Chem Phys 15:3793–3802

Gadd GM (1992) Metals and microorganisms: a problem of definition. FEMS Microbiol Lett 100:197–204

Gadd GM (2007) Geomycology: biogeochemical transformations of rocks, minerals, metals and radionuclides by fungi, bioweathering and bioremediation. Mycol Res 111:3–49

Gadd GM (2010) Metals, minerals and microbes: geomicrobiology and bioremediation. Microbiology 156(3):609–643

Gadd GM, Griffiths AJ (1978) Microorganisms and heavy metal toxicity. Microb Ecol 4:303–317

Gajbhiye M, Kesharwani J, Ingle A, Gade A, Rai M (2009) Fungus-mediated synthesis of silver nanoparticles and their activity against pathogenic fungi in combination with fluconazole. Nanomedicine 5(4):382–386

Gan PP, Ng SH, Huang Y, Yau SS (2012) Green synthesis of gold nanoparticles using palm oil mill effluent (POME): a low-cost and eco-friendly viable approach. CESE 113:132–135

Griffin DH (1994) Fungal Physiology, 2nd edn. Wiley-Liss, New York

Gurunathan S, Raman J, Abd Malek SN, John PA, Vikineswary S (2013) Green synthesis of silver nanoparticles using *Ganoderma neo-japonicum* Imazeki: a potential cytotoxic agent against breast cancer cells. Int J Nanomedicine 8:4399–4413

Hawksworth DL (2001) The magnitude of fungal diversity: the 1.5 million species estimate revisited. Mycol Res 105:1422–1432

Hullavarad NV, Hullavarad SS, Karulkar PC (2008) Cadmium sulphide (CdS) nanotechnology: synthesis and applications. J Nanosci Nanotechnol 8(7):3272–3299

Hutchison JE (2008) Greener nanoscience: a proactive approach to advancing applications and reducing implications of nanotechnology. ACS Nano 2(3):395–402
Iravani S (2011) Green synthesis of metal nanoparticles using plants. Green Chem 13:2638–2650
Kahan DM, Braman D, Slovic P, Gastil J, Cohen G (2009) Cultural cognition of the risks and benefits of nanotechnology. Nature Nanotechnol 4:87–90
Kaler A, Jain S, Banerjee UC (2013) Green and rapid synthesis of anticancerous silver nanoparticles by *Saccharomyces boulardii* and insight into mechanism of nanoparticle synthesis. Biomed Res Int 1–8
Kashyap PL, Kumar S, Srivastava AK, Sharma AK (2013) Myconanotechnology in agriculture: a perspective. World J Microbiol Biotechnol 29(2):191–207
Katti K, Chanda N, Shukla R, Zambre A, Suibramanian T, Kulkami RR, Kannan R, Kattia KV (2009) Green nanotechnology from cumin phytochemicals: generation of biocompatible gold nanoparticles. Inter J Green Nanotechnol Biomed 1(1):39–52
Kitching M, Ramani M, Marsili E (2015) Fungal biosynthesis of gold nanoparticles: mechanism and scale up. Microb Biotechnol 8(6):904–917
Kumar AS, Abyaneh MK, Gosavi SW, Kulkarni SK, Pasricha R, Ahmad A, Khan MI (2007) Nitrate reductase-mediated synthesis of silver nanoparticles from $AgNO_3$. Biotechnol Lett 29(3):439–445
Li G, He D, Qian Y, Guan B, Gao S, Cui Y, Yokoyama K, Wang L (2012) Fungus-mediated green synthesis of silver nanoparticles using *Aspergillus terreus*. Int J Mol Sci 3(1):466–476
Liao H, Neh C, Hafner JH (2006) Biomedical applications of plasmon resonant metal nanoparticles. Nanomedicine 1(2):201–208
Liu J, Qin G, Raveendran P, Ikushima Y (2005) Facile green synthesis, characterization, and catalytic function of β-D-glucose-stabilized Au nanocrystals. Chem A Eur J 12(8):2131–2138
Matus KJM, Hutchison JE, Peoples R, Rung S, Tanguay R (2011) Green nanotechnology challenges and opportunities. ACS 1–32
Maynard AD, Aitken RJ, Butz T, Colvin V, Donaldson K, Oberdörster G, Philbert MA, Ryan J, Seaton A, Stone V, Tinkle SS, Tran L, Walker NJ, Warheit DB (2006) Safe handling of nanotechnology. Nature 444:267–269
Moghaddam BA, Namvar F, Moniri M, Tahir P, Azizi S, Mohamad R (2015) Nanoparticles biosynthesized by fungi and yeast: a review of their preparation, properties, and medical applications. Molecules 20(9):16540–16565
Mowll JL, Gadd GM (1984) Cadmium uptake by *Aureobasidium pullulans*. J Gen Microbiol 130:279–284
Mukherjee P, Senapati S, Mandal D, Ahmad A, Khan MI, Kumar R, Sastry M (2002) Extracellular synthesis of gold nanoparticles by the fungus *Fusarium oxysporum*. Chembiochem 3(5):461–463
Narayanan KB, Sakthivel N (2011) Green synthesis of biogenic metal nanoparticles by terrestrial and aquatic phototrophic and heterotrophic eukaryotes and biocompatible agents. Adv Colloid Interface Sci 169(2):59–79
Nath D, Benerjee P (2013) Green nanotechnology-A new hope for medical biology. Environ Toxicol Pharm 36(3):997–1014
Nayak RR, Pradhan N, Behera D, Pradhan KM, Mishra S, Sukla LB, Mishra BK (2011) Green synthesis of silver nanoparticle by *Penicillium purpurogenum* NPMF: the process and optimization. J Nanopart Res 13(8):3129–3137
Nune SK, Chanda N, Shukla R, Katti K, Kullkami RR, Thilakavathy S, Mekapothula S, Kannan R, Katti KV (2009) Green nanotechnology from tea: phytochemicals in tea as building blocks for production of biocompatible gold nanoparticles. J Mater Chem 19:2912–2920
Park Y, Hong YN, Weyers A, Kim YS (2011) Polysaccharides and phytochemicals: a natural reservoir for the green synthesis of gold and silver nanoparticles. Nanobiotechnology-IET 5(3):69–78
Patil VP, Pawar S, Chougule M, Godse P, Sakhare R, Sen S, Joshi P (2011) Effect of annealing on structural, morphological, electrical and optical studies of nickel oxide thin films. JSEMAT 1:35–41

Philip D (2009) Honey mediated green synthesis of gold nanoparticles. Spectrochim Acta Mol Biomol Spectroscopy 73(4):650–653

Philip D, Unni C, Aromal A, Vidnu VK (2011) *Murraya Koenigii* leaf-assisted rapid green synthesis of silver and gold nanoparticles. Spectrochim Acta Mol Biomol Spectroscopy 78(2):899–904

Prasad R (2014) Synthesis of silver nanoparticles in photosynthetic plants. J Nanoparticles, Article ID: 963961. doi:10.1155/2014/963961

Prasad R, Swamy VS (2013) Antibacterial activity of silver nanoparticles synthesized by bark extract of *Syzygium cumini*. J Nanoparticles. doi:10.1155/2013/431218

Prasad R, Kumar V, Prasad KS (2014) Nanotechnology in sustainable agriculture: present concerns and future aspects. Afr J Biotechnol 13(6):705–713

Prasad R, Pandey R, Barman I (2015) Engineering tailored nanoparticles with microbes: quo vadis. WIREs Nanomed Nanobiotechnol. doi:10.1002/wnan.1363

Quester K, Avalos-Borja M, Vilchis-Nestor AR, Camacho-López MA, Castro-Longoria E (2013) SERS properties of different sized and shaped gold nanoparticles biosynthesized under different environmental conditions by *Neurospora crassa* extract. PLoS One 9–8(10), e77486

Raveendran P, Fu J, Wallen SL (2003) Completely "Green" synthesis and stabilization of metal nanoparticles. J Am Chem Soc 125(46):13940–13941

Riddin TL, Gericke M, Whiteley CG (2006) Analysis of the inter- and extracellular formation of platinum nanoparticles by *Fusarium oxysporum* f. sp. lycopersici using response surface methodology. Nanotechnol 17(14):3482–3489

Roni M, Murugan K, Panneerselvam C, Subramaniam J, Nicoletti M, Madhiyazhagan P et al (2015) Characterization and biotoxicity of *Hypnea musciformis*-synthesized silver nanoparticles as potential eco-friendly control tool against *Aedes aegypti* and *Plutella xylostella*. Ecotoxicol Environ Saf 121:31–38

Salvadori MR, Lepre LF, Ando RA, Oller do Nascimento CA, Corrêa B (2013) Biosynthesis and uptake of copper nanoparticles by dead biomass of *Hypocrea lixii* isolated from the metal mine in the Brazilian Amazon Region. PLoS One 25:8–11

Salvadori MR, Ando RA, Oller do Nascimento CA, Corrêa B (2014) Intracellular biosynthesis and removal of copper nanoparticles by dead biomass of yeast isolated from the wastewater of a mine in the Brazilian Amazonia. PLoS One 29–9(1):e87968

Salvadori MR, Nascimento CA, Corrêa B (2014b) Nickel oxide nanoparticles film produced by dead biomass of filamentous fungus. Sci Rep 17(4):6404

Sastry M, Ahmad A, Islam MK, Kumar R (2003) Biosynthesis of metal nanoparticles using fungi and actinomycete. Curr Sci 85(2):162–170

Sathishkumar M, Sneha K, Won SW, Cho CW, Kim S, Yun S (2009) Cinnamon zeylanicum bark extract and powder mediated green synthesis of nano-crystalline silver particles and its bactericidal activity. Colloids Surf B Biointerfaces 73(2):332–338

Schmidt K (2007) Green nanotechnology: it's easier than you think. Technical Report. Project on Emerging Nanotechnologies

Sen IK, Mandal AK, Chakraborti S, Dey B, Chakraborty R, Islam SS (2013) Green synthesis of silver nanoparticles using glucan from mushroom and study of antibacterial activity. Int J Biol Macromol 62:439–449

Shah SAS, Park AR, Zhang K, Park JH, Yoo PJ (2012) Green synthesis of biphasic TiO2-reduced graphene oxide nanocomposites with highly enhanced photocatalytic activity. ACS Appl Mater Interfaces 4(8):3893–3901

Sharma VK, Yngard RA, Lin Y (2009) Silver nanoparticles: green synthesis and their antimicrobial activities. Adv Colloid Interface Sci 145(1–2):83–96

Silva LP, Reis IG, Bonatto CC (2015) Green synthesis of metal nanoparticles by plants: current trends and challenges. In: Basiuk VA, Basiuk EV (eds) Green processes for nanotechnology. Springer, Switzerland, pp 259–275

Sivaraman SK, Elango I, Kumar S, Santhanam V (2009) A green protocol for room temperature synthesis of silver nanoparticles in seconds. Curr sci 97(7):1055–1059

Suman PR, Jain VK, Varma A (2010) Role of nanomaterials in symbiotic fungus growth enhancement. Curr Sci 99:1189–1191

Tarafdar JC, Raliya R (2013) Rapid, low-cost, and ecofriendly approach for iron nanoparticle synthesis using *Aspergillus oryzae* TFR9. J Nanoparticles. doi:10.1155/2013/141274

Thakkar KN, Mhatre SS, Parikh RY (2010) Biological synthesis of metallic nanoparticles. Nanomedicine 6(2):257–262

Treseder KK, Lennon JT (2015) Fungal traits that drive ecosystem dynamics on land. Microbiol Mol Biol Rev 79(2):243–262

Vainshtein M, Belova N, Kulakovskaya T, Suzina N, Sorokin V (2014) Synthesis of magneto-sensitive iron-containing nanoparticles by yeasts. J Ind Microbiol Biotechnol 41(4):657–663

Venkatessham M, Ayodhya D, Madnusudhan A, Babu NV, Veerabhadra G (2014) A novel green one-step synthesis of silver nanoparticles using chitosan: catalytic activity and antimicrobial studies. Appl Nanoscale 4(1):113–119

Vigneshwaran N, Nachane RP, Balasubramanya RH, Varadarajan PV (2006) A novel one-pot 'green' synthesis of stable silver nanoparticles using soluble starch. Carbohydr Res 341(12):2012–2018

Webster J, Weber R (2007) Introduction to fungi, 3rd edn. Cambridge University Press, Cambridge

Wegner KD, Hildebrandt N (2015) Quantum dots: bright and versatile in vitro and in vivo fluorescence imaging biosensors. Chem Soc Rev 44(14):4792–4834

White C, Gadd GM (1998) Accumulation and effects of cadmium on sulphate-reducing bacterial biofilms. Microbiology 144:1407–1415

Yadav A, Kon K, Kratosova G, Duran N, Ingle AP, Rai M (2015) Fungi as an efficient mycosystem for the synthesis of metal nanoparticles: progress and key aspects of research. Biotechnol Lett 37(11):2099–2120

Yang X, Li Q, Wang H, Huang J, Lin L, Wang W, Sun D, Su Y, Opiyo JB, Hong L, Wang Y, He N, Less LJ (2010) Green synthesis of palladium nanoparticles using broth of *Cinnamomum camphora* leaf. J Nanopart Res 12(5):1589–1598

Chapter 5
Microbial Enzymes: Current Features and Potential Applications in Nanobiotechnology

Mohammadhassan Gholami-Shabani, Masoomeh Shams-Ghahfarokhi, Zeynab Gholami-Shabani, and Mehdi Razzaghi-Abyaneh

Abstract Nanobiotechnology is an immensely developing field of biotechnology due to its wide-ranging applications in different areas of science and technology. It is an integration of different fields of science which holds promise in the pharmaceutical industry, medicine, and agriculture. The synthesis of mono-dispersed nanoparticles with various sizes and shapes has been a big challenge in nanotechnology. Although different physical and chemical methods have been extensively used to produce mono-dispersed nanoparticles, these methods suffer from large limitations of toxicity and adverse reactions for the biological systems. In recent years, interest in employment of enzymatic systems like as fungal and bacterial enzymes as cell-free systems in production of nanoparticles with new biological activities has increased dramatically as efficient routes over traditional synthesis by whole organisms. Since various enzymes have different capacities for synthesis of nanoparticles in a diverse range of shapes and sizes, it is very important to find suitable enzymes for such purposes and improve the method for suitable conditions of nanoparticle synthesis. Enzymatically-synthesized nanoparticles have several advantages over those synthesized by microbial biomasses and culture supernatants. Besides meaningful decrease of the downstream steps needed for purification of produced nanoparticles, they have high potential for manufacturing applications as the enzymes can be immobilized for recycling in nanoparticle synthesis. Likewise, microbial enzymes have great importance in the progress of industrial bioprocesses with potential application in pulp and paper industries, detergents and textiles, pharmaceuticals, chemicals, food and beverages, biofuels, animal feed and personal care.

M. Gholami-Shabani • M. Razzaghi-Abyaneh (✉)
Department of Mycology, Pasteur Institute of Iran, Tehran 13164, Iran
e-mail: mrab442@pasteur.ac.ir; mrab442@yahoo.com

M. Shams-Ghahfarokhi
Department of Mycology, Faculty of Medical Sciences, Tarbiat Modares University, Tehran 14115-331, Iran

Z. Gholami-Shabani
Faculty of Aerospace, Science and Research Campus, Islamic Azad University, Tehran, Iran

R. Prasad (ed.), *Advances and Applications Through Fungal Nanobiotechnology*, Fungal Biology, DOI 10.1007/978-3-319-42990-8_5

Today, there is an urgent need for newly developed versatile enzymes in order to use in economically nanoparticle production processes. Microbial diversity and innovative molecular techniques, such as metagenomics and genomics, are being used to discover novel microbial enzymes whose major properties can be improved by different strategies based on rational, semi-rational and random directed evolution. Nearly all industrial enzymes are recombinant forms produced in bacteria and fungi. In this chapter, we highlight current status and future prospects of cell-free synthesis of biologically active nanomaterials using enzymes originated from fungi, bacteria and actinomycetes as an important part of biodiversity.

1 Introduction

Nanotechnology and nanoscience is the most innovative and advanced field of twenty-first century. Wide general research is going on for commercializing nanoproducts all over the world. Due to their unique properties, nanoparticles have gained significant importance compared to bulk counterparts (Guo 2012; Asmathunisha and Kathiresan 2013; Tran and Le 2013). Metal nanoparticles with fascinating chemical and physical properties are ideal building blocks for engineering and modifying nanoscale structures for specific technological applications (Mukherjee et al. 2012; Ghaseminezhad et al. 2012). Nanostructured metal colloids have been generated by both the so-called "top down" and "bottom up" approaches. However, in compare to "top-down" methods, "bottom-up" procedures allow an almost flexible and low-cost preparation, being therefore more intensively examined during the past two decades (Heath 1999; Xia et al. 2013). The word "nanotechnology" appears to propose something that belongs far in the future or in the area of our favorite sci-fi movies. But metal nanoparticles, one of the most important "building blocks" of nanotechnology are all around us right now, and have been all around us during human history. They were with us when human beings began production their first tools, and they are current in products we buy every day. They largely flew under the radar until electron microscopes become ordinary place several decades ago, but now, the more we turn our microscopes on everyday objects, the more nanoparticles we seem to find. Some metal ions might be adsorbed and more reduced to metal nanoparticles by microorganisms, biomass, plants, etc. Inspired via the natural environment model for the creation of metal nanoparticles, biosynthesis has appeared as novel and alternatively attractive synthetic procedures for metal nanoparticles. Metal nanoparticles are of great attention owing to their novel physicochemical, magnetic, and optoelectronic properties that are administered via their shape, size, and size distribution (Wei et al. 2012; Arjunan et al. 2012; Mittal et al. 2013). Several microorganisms have been successfully used for intracellular and extracellular biosynthesis of metal nanoparticles (Fig. 5.1). Kalishwaralal et al. (2009) reported biological synthesis of gold nanocubes from *Bacillus licheniformis*. Beveridge and Doyle (1989) reported for the first time that gold particles of

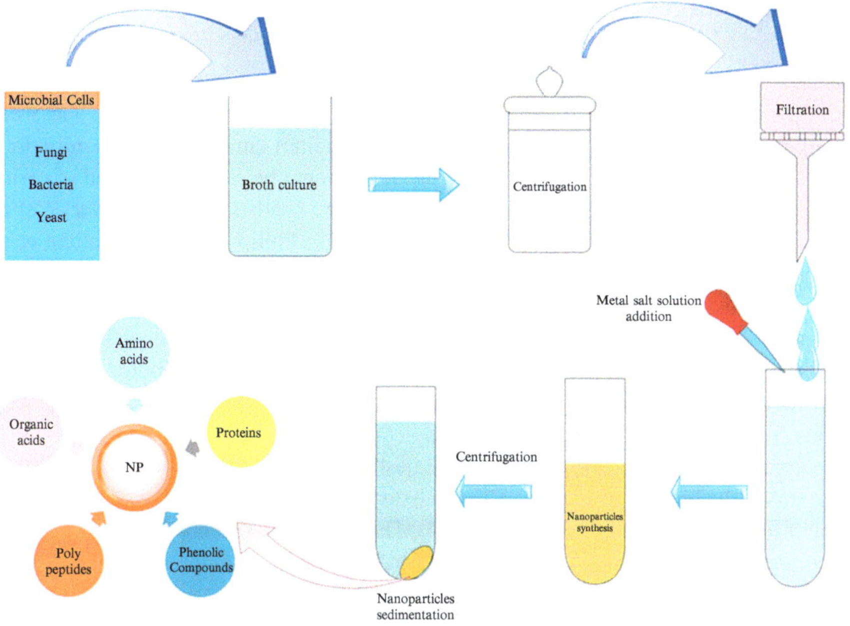

Fig. 5.1 Sequential steps of nanoparticle synthesis by various microorganisms

nanoscale dimensions could form inside the cell walls of *Bacillus subtilis*. Also, *Pseudomonas stutzeri* AG259 was the first bacterium that used for synthesis of silver nanoparticles (Klaus et al. 1999). The first biosynthetic approach including fungus-mediated methods for the metal nanoparticle synthesis was observed in the beginning of the twentieth century, and silver nanoparticles with diameter of 25 ± 12 nm were synthesized by the fungus *Verticillium* (Mukherjee et al. 2001). One of the first plants used as a source for the biosynthesis of metal nanoparticles was *alfalfa sprouts* (Gardea-Torresdey et al. 1999, 2003). The convergence between biology and nanotechnology has generated the innovative field of nano-biotechnology that includes the use of biological entities such as fungi, bacteria, actinomycetes algae, viruses, yeasts, plants, proteins and DNA in a number of biophysical and biochemical processes (Fig. 5.1). The biological synthesis via nanobiotechnology procedures have an important potential to improve nanoparticle production without the use of toxic, harsh and expensive chemicals commonly used in conventional chemical and physical procedures (Prasad 2014; Prasad et al. 2015). The aim of this chapter is to present a brief overview of the procedures used to characterize metal nanoparticles, microbial approaches for synthesis metal and metal oxide nanoparticles, use of plants for synthesis of metal nanoparticles, proteins/enzymes or the biomolecules extracted from microorganisms and plants, factors influencing the procedure of nanoparticle biosynthesis, possible mechanisms involved in nanoparticle synthesis and growth, and potential applications of metal nanoparticles synthesized by natural biological sources.

2 Advantages of Microbial Cell Factories

Different kinds of chemical and physical approaches are used for the synthesis of metal nanoparticles. The use of these production techniques needs both strong and weak chemical reducing agents and protecting agents (sodium citrate, sodium borohydride and alcohols) which are generally toxic, piceous, cannot be simply disposed of owing to environmental problems and also indicating a low synthesis rate (Mukherjee et al. 2001; Singaravelu et al. 2007; Saifuddin et al. 2009; Philip 2009; Gurunathan et al. 2015) Also, these methods are usually expensive and inefficient regard to the energy and materials used. Furthermore, in many cases, production is approved out at higher temperatures, which produce huge amounts of heat. Similar to thermal decomposition process, producing is carried out at very high temperature (Mohanpuria et al. 2008; Iravani et al. 2014). The biological systems for the green synthesis of metal nanoparticles engage use of natural agents like fungi, yeast, algae, bacteria, and actinomycetes (Reddy et al. 2010; Roopan et al. 2014; Rajeshkumar et al. 2014; Prasad et al. 2015). Therefore, the biological systems offer an extensive range of resources for the "green-synthesis" of metal nanoparticles (Fig. 5.2). The amount of reduction of metal ions via natural agents is creating to be much faster and also at room temperature and pressure conditions (Mishra et al. 2014). For example, in case of metal nanoparticle synthesis by *Aspergillus niger*, biosynthesis of silver nanoparticles was detected within 3–5 h of treatment of the fungal filtrate with silver nitrate ($AgNO_3$) solution (Gade et al. 2008). Therefore, the biological method needs smallest time for biosynthesis of metal nanoparticles. Size and shape–controlled metal nanoparticles could be produced via monitoring the pH or the temperature of the reaction mixture. In biosynthesis of metal nanoparticles, different shape morphologies (spheres, triangle, hexagons, cubic, and rods) can be obtained via controlling the pH of the reaction mixture (Sau and Murphy 2004; Gericke and Pinches 2006b; Grzelczak et al. 2008; Geng and Grove 2015). It has been shown that at high temperatures, lesser amounts of metal nanoparticles were produced, while at room temperature, higher amounts of metal nanoparticles were detected. The natural agents such as fungi and bacteria produce a large amount of proteins, which are able of hydrolyzing metals and so cause enzymatically reduction of metals ions (Geng and Grove 2015; Gholami-Shabani et al. 2015). In case of fungi and bacteria, the oxidoreductase enzymes are found to be responsible for the biosynthesis of metal nanoparticles (Gholami-Shabani et al. 2014, 2015). Besides oxidoreductases, some other enzymes like as α-amylase has been successfully used for synthesis of metal nanoparticles (Ahmad et al. 2015). The biomass used for the synthesis of metal nanoparticles is greener, simpler to use, gets easily prepared of in the environment and moreover has the easier downstream processes. Production can be carried out at room temperature and need no specialized pressure conditions as well as less amounts of chemical materials (Riddin et al. 2009; Gholami-Shabani et al. 2014). The biosynthesizing method is easier, low-cost, and nontoxic. Therefore, green synthesis of metal nanoparticles shows to be more efficient compared with the chemical and physical approaches owing to its environmentally-friendly phenomenon and the fact that each microbial cell can be used as a factory for synthesis metal nanoparticles (Fig. 5.3).

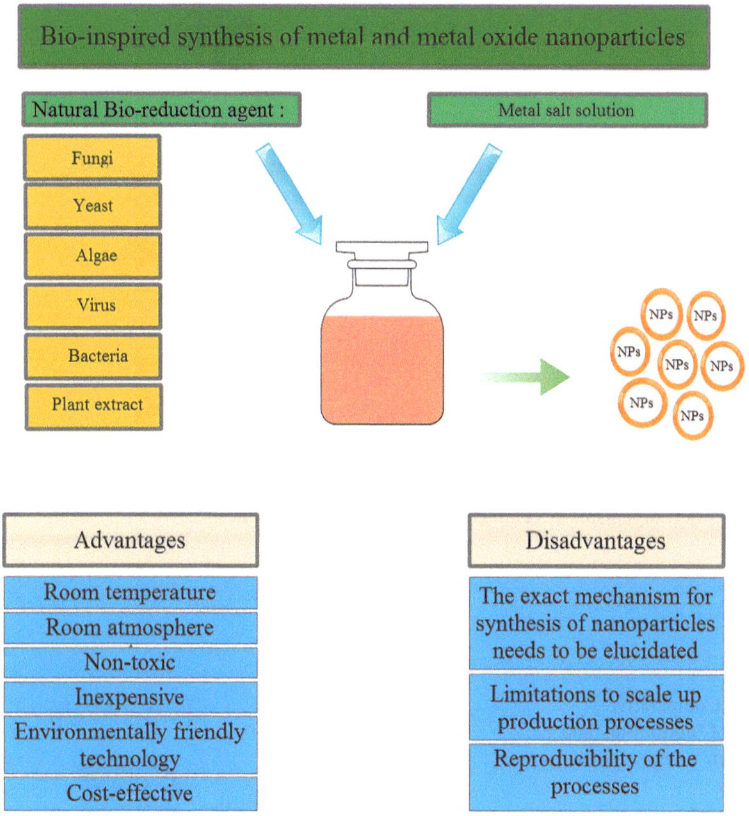

Fig. 5.2 Bio-inspired synthesis of nanoparticles: advantages and disadvantages

3 Characterization Methods

To date, various approaches are developed for synthesis metal and metal oxide nanoparticles. However, these methods are restricted to two general approaches and can be defined as either "top-down" or "bottom-up" methods (Wang and Xia 2004; Mijatovic et al. 2005). The "top-down" method begins with a material of interest, which then undergoes size decrease by chemical and physical procedures to synthesis nanoparticles (Fig. 5.4). Significantly, metal nanoparticles are highly depending on their shape, size, surface structure and processing tends to present surface limitations. These surface limitations can remarkably affect the overall metal nanoparticle surface physicochemical properties (Kelly et al. 2003; Noguez 2007). In the bottom-up approach (Fig. 5.4), nanoparticles are built-up atom by atom, molecules and smaller particles/monomers by self-organization or self-assembly (Weller 2003; Ananikov et al. 2007). In both methods, the resulting metal nanoparticles are characterized by different approaches to determine properties such as size distribution, particle size,

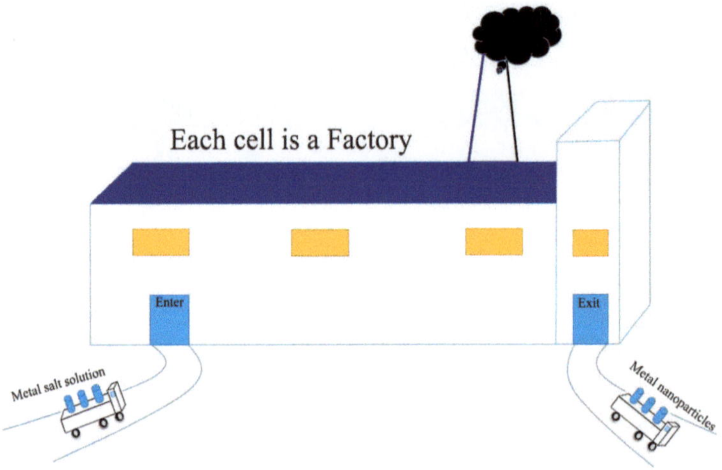

Fig. 5.3 Microorganisms as cell factories of nanoparticle synthesis

Fig 5.4 Comparative scheme of 'bottom-up' and 'top-down' nanoparticles synthesis

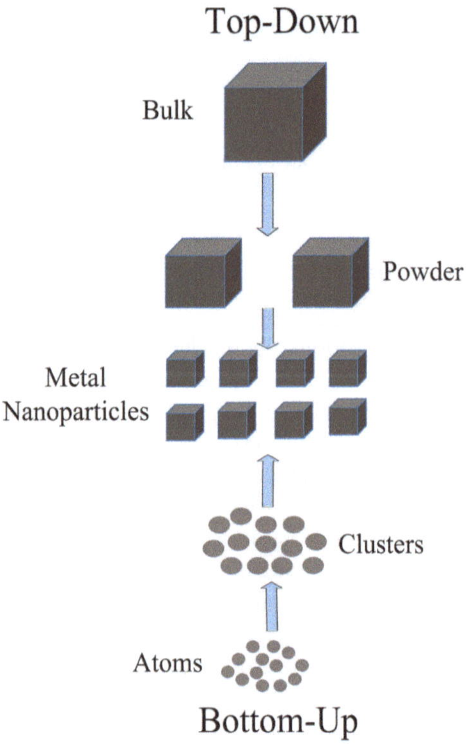

shape, and surface area. This is of specific significance if the properties of metal nanoparticles essential to be homogeneous for a specific application. In the example of biological and chemical synthesis of metal nanoparticles, the aqueous metal ion predecessors from metal salts are reduced and as a product, a color change happens in the reaction mixture. This is the first qualitative sign that metal nanoparticles are being built. One fascinating property of the colloidal particles in solution, owing to their shape and size is their capability to be seen when a laser beam passes through the colloidal solution. This impact is well-known as the "Tyndall effect" and is an easy and simple system that can be used to identify the existence of metal nanoparticles in solution (Poinern 2014). After the reaction, produced colloidal nanoparticles can be separated by high speed centrifugation and then can be observed via advanced "nano-characterization" systems. Metal nanoparticles are usually characterized via their size, shape, disparity, and surface area. The common devices used for characterization of nanoparticles are UV–visible spectroscopy, Fourier transform infrared (FT-IR) spectroscopy, dispersive X-ray spectroscopy (EDS), scanning electron microscopy (SEM), energy transmission electron microscopy (TEM), atomic force microscopy (AFM), X-ray diffraction (XRD), dynamic light scattering (DLS) and Raman spectroscopy. Microscopy-based methods such as TEM, SEM and AFM are considered as direct techniques of gaining data from images taken of the metal nanoparticles. In particular, both TEM and SEM have been extensively used to characterize morphology and size features of metal nanoparticles (Eppler et al. 2000; Feldheim and Foss 2002; Cao 2004; Poinern 2014). Spectroscopy-based methods such as UV–vis, FT-IR, DLS, EDS, XRD and Raman are considered indirect techniques of gaining data related to structure, composition, crystal phase and other properties of metal nanoparticles. UV–visible spectroscopy is a generally used system. The light wavelengths in the range of 200–700 nm are commonly used to characterize a range of metallic and metal oxide nanoparticles. Spectrophotometric absorption measurements in the wavelength ranges of 400–450, 500–550, and 300–400 nm are used to characterize silver, gold, and zinc oxide nanoparticles, respectively. FTIR spectroscopy is suitable for characterizing the surface chemistry of nanoparticles. Biological functional groups (e.g., carbonyls, hydroxyls) attached to the surface of metal nanoparticles and the other surface chemical residues are detected via FTIR spectroscopy. When the biosynthesis is completed, the shape, size and dispersion state of metal and metal oxide nanoparticles are typically measured using SEM, TEM, and AFM. SEM and TEM are used for morphological characterization on the nanometer to micrometer scale. TEM has a 1000-fold higher resolution than SEM. The elemental composition of nanoparticles is generally recognized using energy dispersive spectroscopy (EDS). AFM provides imaging in three dimensions. The resolution in the vertical, or *Z*-axis, is limited via the vibration environment of the tool, whereas the resolution in the horizontal, or *X*–*Y*-axis, is limited via the diameter of the tip used for scanning. AFM affords surface characterization on the atomic scale. XRD is applied for identification and properties of the crystal structure of metal nanoparticles. X-rays penetrate into the nanomaterials and the resulting diffraction pattern is compared with standards to gain structural information. DLS is useful to determine the surface charge and size distribution of nanoparticles suspended in a liquid.

4 Bio-Inspired Synthesis of Metal Nanoparticles

Microbial-nanotechnology is the border between 'Nanotechnology' and Microbiology (bacteriology and mycology) and has substantial potential, somewhat due to the varied species and variety of the microbes (Honary et al. 2012; Jain et al. 2013; Ramamurthy et al. 2013). Metal nanoparticles are observed as the important structure in nanotechnology (Nagajyothi et al. 2012; Kharissova et al. 2013). They are the beginning points for making a lot of nano-structured materials and machines. Presently, there is a developing essential to via biologically environmentally friendly metal nanoparticles that do not products toxic damages in their synthesis procedure. Recent studies have shown that the biological synthesis of nanoparticles using microorganisms and plants is much cheaper, greener, safer and environmentally friendly (Fig. 5.2) (Ayaseelan et al. 2012; Rajakumar et al. 2012; Asmathunisha and Kathiresan 2013; Prasad 2014; Prasad et al. 2015). Both microorganisms and plants are shown to have the capability to absorb and collect inorganic metallic ions from their close environment. These charming properties make many bio-production and bioremediation systems efficient biological factories able of considerable decreasing environmental toxic waste and recovering heavy metals from manufacturing waste. Essentially, the capability of a biological system to use its intrinsic biochemical procedures to change inorganic metallic ions into nanoparticles has shown to a reasonably new and mostly unknown field of research (Gholami-Shabani et al. 2012). Today, the capability of microorganisms to cooperate in extraction, and a mass metallic materials from their environments has been capitalized on in a number of biotechnological applications including "bioleaching" and "bioremediation" (Mann 2001; Dujardin and Mann 2002; Xu et al. 2007; Meldrum and Cölfen 2008). The capability of microorganisms to interact with their close environment, using structure of their lipid-built amphipathic membranes allows a variation of "oxidoreduction mechanisms" to happen and support of biochemical changes (Kumar et al. 2007; Gholami-Shabani et al. 2014, 2015, 2016). Previous studies have shown that prokaryotes and eukaryotes succeed both "intracellular" and "extracellular" synthesis of nanomaterials as shown in Table 5.1, and in the example of metal and metal oxide nanoparticle synthesis, culturable microorganisms in specific medium can also support them in helping to oxidoreduction occurrence (Gericke and Pinches 2006b; Narayanan and Sakthivel 2010; Mishra et al. 2014; Huang et al. 2015). The particular oxidoreduction mechanisms create the core and then, metallic nanoparticles are made. More information is needed for fully explain differences in nanoparticle morphology and size between various metals when produced using the similar microorganisms (Dhanasekar et al. 2015; Show et al. 2015). Metal nanoparticles can be produced in the periplasmic space between the cytoplasmic and outer membranes. A periplasm is also present in gram-positive bacteria between the cytoplasmic membrane and the peptidoglycan, on the cell-wall and external the cells. It is fictional that numerous microbial enzymes take a dynamic part in the bio-reduction procedure of transferring electrons from definite electron donors to metal electron acceptors for various microorganisms. The role of microbial enzymes in the production of nanoparticles has been widely studied in

Table 5.1 Biologically-synthesized metal nanoparticles produced by fungi and bacteria

Microorganism	Nanoparticles	Size (nm)	Morphology	Reference
Aspergillus clavatus	Silver	10–25	Spherical, hexagonal	Verma et al. (2010)
Aspergillus flavus	Silver	8.92±1.61	Irregular	Vigneshwaran et al. (2007)
Aspergillus fumigatus	Silver	5–25	Spherical	Bhainsa and D'Souza (2006)
Aspergillus niger	Silver	20	Spherical	Gade et al. (2008)
Cladosporium cladosporioides	Silver	10–100	Spherical	Balaji et al. (2009)
Alternaria alternata	Silver	20-60	Spherical	Gajbhiye et al. (2009)
Fusarium acuminatum	Silver	5–40	Spherical	Ingle et al. (2008)
Fusarium oxysporum	Silver	5–50	Spherical	Ahmad et al. (2003)
Fusarium semitectum	Silver	10–60	Spherical	Basavaraja et al. (2008)
Fusarium solani	Silver	5–35	Spherical	Gade et al. (2009)
Penicillium brevicompactum	Silver	23–105	Spherical	Shaligram et al. (2009)
Penicillium fellutanum	Silver	1–100	Spherical	Kathiresan et al. (2009)
Alternaria sp.	Gold	7–18	Quasi-spherical, Spherical	Dhanasekar et al. (2015)
Botrytis cinerea	Gold	1–100	Irregular	Castro et al. (2014)
Thermus scotoductus	Gold	ND	Triangular, Hexagonal	Erasmus et al. (2014)
Brevibacterium casei	Gold, silver	10–50	Spherical	Kalishwaralal et al. (2010)
Klebsiella pneumoniae	Gold	35–65	Spherical	Malarkodi et al. (2013)
Rhodopseudomonas capsulata	Gold	10–20	Irregular	He et al. (2007)

recent years. This is also true that use of biological proteins/enzymes can be important for the synthesis of metal nanoparticles (Gholami-Shabani et al. 2015; Huang et al. 2015; Sharma et al. 2015). Therefore, optimization of culturing parameters for example light, temperature, nutrients, and buffer strength can considerably raise microbial enzyme activity. Nanoparticles such as silver, gold, cadmium sulfite, copper oxide, platinum and zinc oxide have grown significant in recent years due to their basic and technological interest. Some fungi, yeasts and bacteria and currently microbial proteins play a significant role in emendation of toxic metals via reduction of the metal ions. Such as, biologically environmentally friendly microorganisms' enzymes could reduce the toxicity in the method of metal nanoparticle synthesis by reduction of the metal ions (Gholami-Shabani et al. 2014, 2015; Huang et al. 2015). While concentrating on the synthesis of nanoparticles using microbial enzymes, they were observed that metal nanoparticles of good sizes and well monodispersity could be produced (Kumar et al. 2007; Riddin et al. 2009; Gholami-Shabani et al. 2014, 2015). As microbes have the ability to spatter considerable amounts of proteins, they might affect the substantial mass productivity of metal nanoparticles. The microbial proteins are able of reducing metal salts. Furthermore, bacteria and fungi are easy to isolate and culture. Also, the downstream processes when using of microbial enzymes are less complex than the synthetic approaches (Gholami-Shabani et al. 2014). Recently, the employment of biological methods, mainly bacteria and fungi proteins, has used as a new system for the synthesis of metal nanoparticles (Gholami-Shabani et al. 2015, 2016). At the nanoscale, characterization of materials is considerably dissimilar from their macroscopic mass characterization (Wei et al. 2012; Arjunan et al. 2012; Mittal et al. 2013).

5 How Microorganisms Tolerate Toxic Metal Ions?

Biotechnology targets to produce value via transforming a cheap material into an expensive product. There are three parts for using heavy-metal resistance in biotechnology: first, adding metal resistance to a microorganism could facilitate a biotechnological procedure, which could or could not be linked to heavy metals. Second, heavy-metal-resistant fungi and bacteria could be used for any kind of biomining of expensive metals, directly on ores or by recovering metals from waste materials of industrial processes. Third, heavy metal-resistant fungi and bacteria could be used for bioremediation of metal-contaminated environments. How metal resistance can be gained by a microorganism of biotechnological use depends on the amount of control one has over the procedure, which itself depends on the increase of value the procedure produces. In a highly controlled fermentor reaction, adding of a heavy metal-resistance determinant into the chromosome of a particular microorganism is easily carried out by molecular biological approaches, if the toxic effect of a heavy metal has to be reduced. On the other hand, a sewage plant with restricted control over the cleaning process probably does not allow the use of a highly modified organism. However, in these cases, heavy-metal resistant fungi and bacteria may be recognized in the sewage plant, or plasmids with a broad host range of replication and metal-resistance expression could easily be introduced into the bacterial community. The presence of heavy metals will cause the plasmids to be stably maintained in the bacterial population. In all cases, determinants for efflux systems should be used, since detoxification by efflux is more economical for bacteria than binding, except in the case of mercury. Scientists have used the natural ability of microorganisms to convert metal-salt solutions into metal nanoparticles (Fig. 5.5). Klaus et al. (1999) reported *Pseudomonas stutzeri* AG259 used to produce silver nanoparticles. Microbial resistance against the toxicity of silver ions is necessary to produce silver nanoparticles and accumulate them outside the cytoplasmic membrane. Mukherjee et al. (2001) reported the use of *Verticillium* in the intracellular synthesis

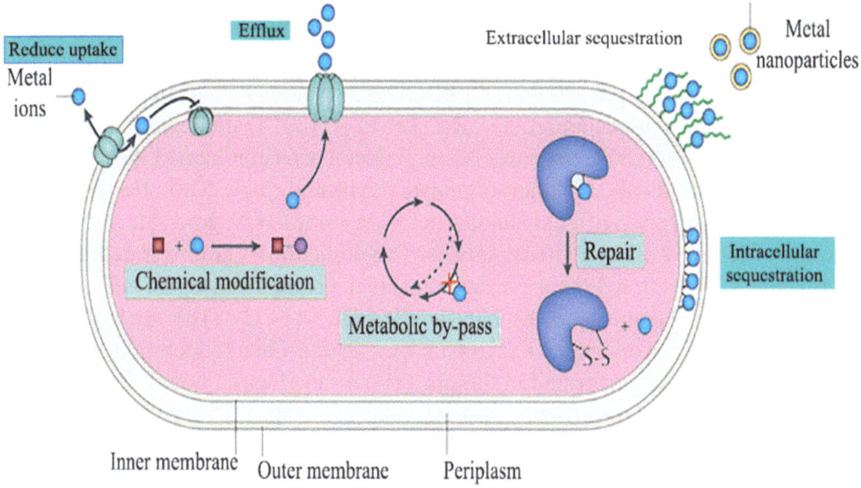

Fig. 5.5 Proposed mechanism of cellular conversion of metal-salt solutions to metal nanoparticles

of metal nanoparticles. They indicated that the toxic properties of silver ions could be decreased by reducing silver ions to silver nanoparticles inside the cells, which were alive and could more re-growth after reduction. Silver nanoparticles were also synthesized via silver nitrate reduction by *Lactobacillus* strains (Nair and Pradeep 2002). While the particular silver-resistant mechanism was not explained in this research, more studies related to the extracellular synthesis mechanism of silver nanoparticles using silver-resistant *Morganella* sp. established via three homologous genes *silP*, *silE* and *silS* which were closely related with silver-resistance (Parikh et al. 2008). Several proteins were found to be unknown external the cell during the growth and were proposed as being responsible for the reduction of silver ions and synthesis of silver nanoparticles in the extracellular micro-environments (Mukherjee et al. 2001; Das et al. 2012a, b). Ahmad et al. (2003) first reported the extracellular synthesis of nanoparticles by eukaryotic systems such as fungi. They presented that unknown enzymes are responsible in the reduction procedure. Extracellular production is valuable as the synthesized metal nanoparticles will not bind to the biomass and it is so potential to extend this method for the biosynthesis of metal nanoparticles over a range of chemical compositions, such as oxides, nitrides, and so far. Very recently, Lin et al. (2014) studied the biosynthesis of silver nanoparticles upon reduction of silver ions by the periplasmic "nitrate reductase" c-type cytochrome subunit NapC in a silver-resistant *Escherichia coli* in anaerobic conditions. They indicated that c-type cytochromes such as NapC found in the periplasm could reduce silver ions to silver nanoparticles. Produced silver and gold ions are less toxic to microbes. Das et al. 2012a, b established a complete research on the characterization of reduction method of gold ions in the fungus *Rhizopus oryzae* to form gold nanoparticles. Growing of the fungus in the existence of sub lethal gold ions concentrations (130–150 mM) induced stress response proteins which were associated with the synthesis of gold nanoparticles. Two cytoplasmic proteins of about 45 and 42 kDa were shown to be responsible for the bio-production of gold nanoparticles. Also, a protein with 80 kDa was reasonably suggested to work as a capping agent for gold nanoparticles. Ro¨sken et al. (2014) studied the time-related synthesis of in vivo gold nanoparticles in a cyanobacterium named *Anabaena* sp., using 0.8 mM gold salt. All microorganisms appeared to be dead after eight days owing to the absorption of gold nanoparticles. In some cases, gold salt could not be absolutely reduced using microorganisms. The reduction of gold salt necessary the involvement of electron donors as established in the rapid synthesis of gold nanoparticles in the periplasmic space by *Shewanella algae* ATCC 51181 via hydrogen as electron donor (Konishi et al. 2006, 2007a, b). The microbial reduction of palladium and platinum ions is performed in the presence of an electron donor. Reduction of palladium ions by *Desulfovibrio desulfuricans* NCIMB 8307 in the presence of hydrogen or formate as the electron donor is a good example in this regard (Yong et al. 2002). Palladium nanoparticles cannot be produced by bacteria without the addition of an electron donor. Likewise, platinum nanoparticles could be synthesized using *D. desulfuricans* NCIMB 8307 and hydrogen as electron donors. Using sodium lactate as an electron donor, platinum nanoparticles could also be produced from *S. algae* ATCC5118161 and *E. coli* MC410062 using hydrogen as the electron donor (Yong et al. 2007; Konishi et al. 2007a, b; Attard et al. 2012). Recently, Yates et al. (2013) showed that palladium nanoparticles synthesized outside the *Geobacter sulfurreducens* cells could decrease the toxicity of metal ions and allow the recovery of palladium nanoparticles without cell damage. However, the resistance of the associated microorganisms against palladium and platinum ions was little discussed; therefore, it is unclear whether silver or gold resistance is applicable to that of palladium and platinum ions. As much as the resistance against other metals is worried, a few reports are obtainable for the microbial reduction. A recent presentation pointed of the production of stable cooper nanoparticles by silver-resistant *Morganella morganii* RP42. It was claimed that cooper ions were up-taken by *M. morganii* using the similar proteins involved in silver ion uptake from previous studies (Ramanathan et al. 2011). The resulting cooper nanoparticles were then released into the media by the bacterial efflux structure. When compared to other kinds of microorganisms, their eco-friendliness and simplicity during study lead to increasing the use of fungi in "green synthesis". Another important reason for choosing the technique of synthesis is the reaction speed. First reported of rapid production using fungi was using *Aspergillus fumigatus* that allowed gaining mono-dispersed silver nanoparticles within 10 min. Also, one of the most common molds *Aspergillus fumigatus* was used to production silver nanoparticles in a matter of minutes, when silver ions entered into interaction with the cell filtrate. These research were clear examples describing appropriateness and the potential of using fungi for mass synthesis of nanoparticles. More recently, nanoparticles synthesized using *Aspergillus flavus* were combined with antibiotics to improve the biocidal efficiency against multidrug-resistant bacteria (Roy and Das 2015).

In addition, the metal-resistance abridged above, proteins and enzymes show vital types in the microbial reduction of metal salts. While some innovative studies were presented the identification of amino acids and enzymes involved and accepting of their properties in microbial reduction remains an important challenge. Table 5.2 summarized some cases of enzymes involved in the microbial reduction of metal salts. Intracellular production of nanoparticles is usually achieved by bacteria. Synthesis of gold nanoparticles via reduction of gold salt (gold chloride hydrate) by *Lactobacillus* strains is one of the first reports on microbial synthesis of nanomaterials (Nair and Pradeep 2002). Authors believed that gold salts were reduced by enzymes and sugars on bacterial cell walls. However, no experimental indication was provided. Ahmad et al. (2003) established the use of individual-spore bacteria (*Thermomonospora* sp.) for synthesis gold nanoparticles at 50 °C.

Table 5.2 Enzymatically-synthesized metal nanoparticles produced by fungi and bacteria

Microorganisms	NP type	Size (nm)	Morphology	Enzyme	Reference
Fusarium oxysporum	Silver	20–40	Spherical	NADH-dependent reductase	Mukherjee et al (2002)
Fusarium oxysporum	Silver	50	Spherical	Nitrate reductase	Gholami-Shabani et al. (2014)
Fusarium oxysporum	Gold	7–20	Spherical	NADPH-dependent sulfite reductase	Kumar et al. (2007)
Fusarium oxysporum	Platinum	70–180	Spherical	Hydrogenase	Govender et al. (2009)
Rizopus oryzae	Gold	15	Icosahedral	Cytoplasmic proteins	Das et al. (2012a, b)
Lentinula edodes	Gold	5–50	Spherical	Laccase Tyrosinase Peroxidase	Vetchinkina et al. (2014)
Thermomono spora sp.	Gold	2–6	Spherical	Sulfite reductase	Khan and Ahmad (2014)
Phanerochaete chrysosporium	Gold	10-100	Spherical	Extracellular enzymes	Sanghi et al. (2011)
Desulfovibrio desulfuriacans	Palladium	ND	Spherical	Hydrogenase Cytochrome C3	Mabbett et al. (2004)
Morganella psychrotolerans	Silver	100–150	Triangular, Hexagonal	Ag reductase	Ramanathan et al. (2011)
Stenotrophomonas maltophilia	Silver	93	Cubic	Chromium reductase	Oves et al. (2013)
Shewanella oneidensis	Silver	24.4±0.8	Spherical	c-Type cytochromes	Ng et al. (2013)
Escherichia coli	Silver	5–70	Spherical	Nitrate reductase	Lin et al. (2014)
Escherichia coli	Gold	10	Spherical	Sulfite reductase	Gholami-Shabani et al. (2015)

ND not determined

The significant proteins playing a significant character in the reduction of gold salts also were suitable for the survival of the thermophilic bacteria. The reduction of palladium ions to synthesis palladium nanoparticles with *D. desulfuriacans* ATCC 29577 was shown in the presence of sodium pyruvate, formate or hydrogen as electron donors (Mabbett et al. 2004). Hydrogenase and cytochrome C_3 were claimed to be possibly involved in synthesis of palladium nanoparticles (Mabbett et al. 2004). Intracellular synthesis of metal nanoparticles using fungi was first established by Mukherjee et al. (2001). Intracellular bio-mineralization mechanism of gold via a zygomycete *Rhizopus oryzae* has been reported by Das et al. (2012a, b). The authors indicated that the main part of gold was transported into the fungal cytoplasm where reduction to gold nanoparticles facilitated by cytoplasmic proteins (oxidoreductase). It has been recently reported that spherical gold nanoparticles can be produced inside the mycelia cells of *L. edodes* (Vetchinkina et al. 2014). The results of enzyme assays indicated that the intracellular phenol-oxidizing enzymes (tyrosinases, laccases, and peroxidases) were involved in gold salt reduction to give electrostatically stabilized colloidal solutions. Extracellular production of metal nanoparticles using fungi was also first reported by Mukherjee et al. (2002) for the specific case of gold nanoparticles which could be extracellular synthesized by *Fusarium oxysporum*. Results also indicated that *F. oxysporum* was capable to release a large number of coenzyme (NADH)-depended proteins to reduce gold ions. Oxidoreductase were a class of characteristic enzymes of fungus *F. oxysporum*, while intra- or extra-cellular reduction of gold ions could not be achieved by other strains such as *F. moniliforme*. Protein-bound gold nanoparticles through the amino groups from cysteine acid and lysine residues provided gold nanoparticles with long-term stability. These studies showed that the use of fungi has the extra advantages of handling and simplicity in processing of biomass for extracellular production (Mukherjee et al. 2002). Founded on the same reduction mechanism, extracellular synthesis of gold, silver, gold–silver and other nanoparticles with *F. oxysporum* was more investigated at the National Chemical Laboratory of India. Studies of metal nanoparticles synthesis and size control obviously established that cell walls of microorganisms generally provide binding sites for metal ions and favored nucleation sites for the synthesis of metal nanoparticles in the presence of enzymes. In the intracellular synthesis of metal nanoparticles, electrostatic interactions between gold or silver ions and the charged groups (such as cysteine or lysine residues) of enzymes within the cell wall of *Verticillium* sp. can lead to entrapping of metal ions on the cell surface, where enzymes helped the reduction of metal ions to metal nanoparticles (Ro¨sken et al. (2014)) Similarly, gold ions initially bound by the cell surface of the fungus *R. oryzae* could be reduced to intermediate gold–protein complexes and eventually to gold nanoparticles (Das et al. 2012a, b). Bacterial cells were capable of providing enzymes as reducing agents in the reduction of palladium ions *by D. desulfuricans* NCIMB 8307 using formate-hydrogen complex as electron donors (Yong et al. 2002). Recently, Ro¨sken et al. (2014) reported a time-dependent study on gold nanoparticles growth via X-ray powder diffraction (XRD) and transmission electron microscopy (TEM) pointed out that the foundation of gold nanoparticles was started at the heterocyst polysaccharide layer (HEP) of heterocysts (HCs).

At longer times, the vegetative cells were the most important area of synthesis. Even after one day, the number of gold nanoparticles inside the HCs also inside HEP was inferior as compared to that in VCs. Microorganisms used for the extracellular production of metal nanoparticles need to be extensively screened (Huang et al. 2009). Metal nanoparticles were produced intracellular and extracellular in most cases, with the associated problems in controlling their particle size. Furthermore, microorganisms have a hierarchical cell structure, which is essentially detrimental for the production of metal nanoparticles with a narrow particle size distribution. Adjusting the artificial conditions can potentially facilitate particle size control in metal nanoparticles synthesis. As an example, the formation rate of intracellular gold nanoparticles using two kinds of fungi (*V. luteoalbum*) could be influenced by modifications in pH value, reaction temperature, gold salt solution concentration and reduction reaction time, affecting the size of gold nanoparticles (Gericke and Pinches 2006a, b). A genetic approach was recently proposed by Ng et al. (2013) comparing the particle size of the extracellular metal nanoparticles from wild type *Shewanella oneidensis* and its mutant. Results indicated that the mutant missing outer membrane c-type cytochromes synthesized significantly smaller metal nanoparticles with respect to the wild type.

6 Mechanisms of Microbial Nanoparticle Synthesis

The clear and accurate mechanism for the biosynthesis of metal nanoparticles using natural agents has not been shown yet as different biological methods respond differently with metal ions and moreover, there are various biomolecules accountable for the biosynthesis of metal nanoparticles. Also, the mechanism for intracellular and extracellular production of metal nanoparticles is dissimilar in numerous biological agents. The cell wall of the microorganisms plays an important role in the intracellular production of metal nanoparticles. The enzymes existing within the cell wall reduce the metal ions to metal nanoparticles, and lastly the lesser sized metal nanoparticles get dispersed of through the cell wall. The fungal cells surface when comes in interaction with metal ions act together and traps the ions and then, the enzymes existing in the cell wall reduce the metal ions (Fig. 5.5). At the end, aggregation of metal particles and biosynthesis of metal nanoparticles take places (Mukherjee et al. 2001; Nair and Pradeep 2002). According to the literature, every time pH increases, more competition follows between protons and metal ions for negatively charged binding sites. The mechanism of extracellular production of metal nanoparticles by fungi and bacteria is mostly found to be involving the action of oxidoreductases mainly via two enzymes named nitrate reductase and sulfite reductase. The oxidoreductases secreted by the fungi and bacteria helps in the bioreduction of metal ions and biosynthesis of metal nanoparticles. A number of researchers supported oxidoreductase for cell-free synthesis of metal nanoparticles (Kumar et al. 2007; Gholami-Shabani et al. 2014, 2015; Prasad et al. 2015). It has been shown that the oxidoreductase is responsible for the reduction of metal ions and the following synthesis of metal nanoparticles. When nitrate reductase is used; the color of the mixture turned reddish from white when tested with fungal filtrate demonstrating the existence of nitrate reductase. Therefore, it can be suggested that an oxidoreductase is responsible for the reduction of metal ions to nanoparticles in fungi. Also, a similar mechanism was reported in the case of cell-free synthesis of gold nanoparticles using *E. coli* (Gholami-Shabani et al. 2015). This bacterium is known to secrete cofactor NADH and NADH-dependent enzymes. The bio-reduction of gold ions was found to be initiated via the electron transfer from the NADPH or NADH via NADPH- or NADH-dependent oxidoreductase as electron carrier. In next steps, the gold ions get electrons and are converting to gold nanoparticles. It has been shown also that biosynthesis of cadmium sulfite nanoparticles using yeast has been considered to be dependent on a stress protein response (Park et al. 2015).

All the organisms not have ability to synthesis all metal nanoparticles. As before said, those organisms which include the "Silver-resistance system" can synthesize silver nanoparticles if that the concentration of the silver ions does not pass the "threshold amount". The resistance machinery varies with organisms. Sources from organisms may performance both as reducing and capping agents in nanoparticles synthesis. The reduction of metal ions by mixtures of biomolecules found in these sources such as proteins/enzymes, amino acids, vitamins and polysaccharides is biologically safe and environmentally, yet chemically complex. But, the mechanism which is commonly believed for the synthesis of silver nanoparticles is the existence of enzyme called "Nitrate reductase" (Kumar et al. 2007; Kalimuthu et al. 2008; Gholami-Shabani et al. 2014). This enzyme is an enzyme in the nitrogen cycle responsible for the transformation of "nitrate to nitrite" (Duran et al. 2005). The reduction mediated via the presence of the enzyme in the organisms has been found to be responsible for the production. The use of a particular enzyme NADPH-dependent nitrate reductase in the in vitro production of nanoparticles is significant because this would do away with the downstream processing necessary for the use of these metal nanoparticles in similar catalysis and other applications such as medicine and nonlinear optics. During the synthesis, nitrate is changed to nitrite, and an electron will be transported to the incoming silver ions. This has been excellently explained in the microorganism *B. licheniformis*. This microorganism is identified to produce NADH and NADH-dependent enzymes, particularly nitrate reductase that might be responsible for the bio-reduction of silver ions to silver nanoparticles. Although all these are theory, direct indication was provided by Kumar et al. (2007), Gholami-Shabani et al. (2014) who directly used the purified nitrate reductase from the *Fusarium oxysporum* for the synthesis of silver nanoparticle in laboratory. Their reaction combination included only the enzyme nitrate reductase, silver nitrate, gelatin and NADPH. Bit by bit, the reaction combination turned yellow-brown with all the characteristics of silver nanoparticles. These are the first direct indication for the involvement of "nitrate reductase" in the production of silver nanoparticles. Although silver nanoparticles production is considered as an "ability" of the microorganism, it is primarily considered as a resistance-mechanism by the microorganisms to the incoming very reactive silver ions. Fascinating facts about silver nanoparticle production can be understood when the true-mechanism involved in the antimicrobial activity of silver ions is identified (Silver et al. 2006). Silver ions are very sensitive and are identified to bind with numerous vital mechanisms of the cells inducing cell death. Fascinatingly, "apoptosis" is a tool which is related to both of multicellular and unicellular microorganisms (Engelberg-Kulka et al. 2006). The resulting are the properties by which silver ions show their antimicrobial functions. This resistance mechanism is appropriate to various metals, where the difference happens only in the respective enzyme. In *B. licheniformis*, the nitrate reductase is establishing at the cell-membrane as respiratory nitrate reductase. Hence, it can be observed as that in most of the microorganisms recognized to production silver nanoparticles, an oxidoreductase called nitrate reductase will be a part and piece of the microorganism. Also, when the condition of the silver nanoparticle synthesis is basic, the synthesis will be quicker than in acidic conditions. Synthesis boosts as the pH rises towards basic region and reaches the maximum at pH 10 after which the speed of the metal nanoparticle synthesis decreases. This indicates that the production of silver nanoparticles will be ideal by the basic environment. At basic conditions there is no requirement of shaking the mixture for the synthesis of silver nanoparticles and all the silver ions complete will be changed to silver nanoparticles even within 30 min. The proteins/enzymes involved in the synthesis may bind with silver at thiol-regions (–SH) making a –S–Ag bond, a pure sign of which supports the change of silver ions to silver nanoparticles (Prasad and Swamy 2013). Furthermore, the basic ion (–OH) is very much essential for the reduction of metal ions. It takes 3–4 days for the making of silver ions in usual conditions whereas it is very much less than an hour when the pH is made basic. Besides, under basic conditions the capability of the enzyme responsible (not only nitrate reductase) for the production of silver nanoparticles increases (Sanghi and Verma 2009). Gholami-Shabani et al. (2015) used a cell-free viable technique for synthesis of gold nanoparticles by α-NADPH-dependent sulfite reductase purified from *Escherichia coli*. The enzyme was purified by ion exchange chromatography on DEAE Sephadex A-50. Molecular weight of the enzyme was determined by gel filtration on Sephacryl S-300 equal to 116 kDa composed of two subunits of 75 and 41 kDa which was successfully used for cell-free synthesis of spherical gold nanoparticles with an average size of 10 nm and a zeta potential of -30 ± 0.2. Gold nanoparticles showed strong antifungal activity against a wide range of human pathogenic fungi (Asghari et al. 2016). Evaluation of the in vitro cytotoxicity of gold nanoparticles showed no toxicity for two cell lines i.e. Vero and Hep 2 at the concentrations ranged from 0.31 to 10%. In case of fungi, the enzyme oxidoreductase is found to be responsible for the production of metal nanoparticles (Kumar et al. 2007; Gholami-Shabani et al. 2014, 2015). Also Govender et al. (2009) reported a mechanism for the bio-reduction of H2PtCl6 and PtCl2 into platinum nanoparticles by a hydrogenase enzyme from *Fusarium oxysporum*. Octahedral H_2PtCl_6 is too large to fit into the active region of the enzyme and, under conditions optimum for nanoparticle formation (pH 9, 65 °C), undergoes a two-electron reduction to $PtCl_2$ on the molecular surface of the enzyme. This smaller molecule is transported through hydrophobic channels within the enzyme to the active region where, under conditions optimal for hydrogenase activity (pH 7.5, 38 °C) and undergoes a second two-electron reduction to Pt(0). H_2PtCl_6 was unreactive at pH 7.5, 38 °C; PtCl2 was unreactive at pH 9, 65 °C.

Biomass once used for the biosynthesis of metal nanoparticles cannot be used again also the proteins/enzymes in free form cannot be used again. But when the proteins/enzymes are immobilized by a support the proteins/enzymes can be used for a large number of reactions. Fundamentally, the enzymes should be immobilized upon the surface of particle of immobilization to lay out the diffusion limitation (Clark 1994; Gholami-Shabani et al. 2014, 2015). The present request of sustainable green approaches has improved the use of enzymatic technology in industrial methods. Employment of enzyme as biocatalysts suggestions the benefits of mild reaction conditions, biodegradability and catalytic efficiency. The harsh conditions of industrial procedures, however, increase tendency of enzyme destabilization, shortening their industrial lifespan. As a result, the technology of enzyme immobilization provides a successful means to circumvent these concerns via enhancing enzyme catalytic properties and also unravel downstream processing and improve operational stability. There are several methods used to immobilize the enzymes, which range from reversible physical adsorption and ionic connections, to the irreversible stable covalent bonds. These approaches produce immobilized enzymes of variable stability due to modifications in the surface microenvironment and degree of multisite fitment. Therefore, it is mandatory to gain information about the configuration of the enzyme protein following interaction with the support surface plus interactions of the enzymes with other proteins. Characterization technologies at the nanoscale level to research enzymes immobilized on surfaces are essential to obtain valuable qualitative and quantitative information, including morphological imagining of the immobilized enzymes. These tools and technologies are relevant to assess efficacy of an immobilization system and progress of future enzyme immobilization approaches (Mohamad et al. 2015).

7 A Practical Approach to Synthesis of Metal Nanoparticles

7.1 Simple Downstream Steps for the Purification of Metal Nanoparticles

Think of a small judgment. Will it be simple to purify the complex with large number of impurities or with very less impurities? Clearly the second is the better. Use of purified protein/enzymes will create the downstream steps much easier as the enzyme will mostly be the impurity, but when immobilized protein/enzymes are used this can also be ruled out where the downstream steps can be very much reduced.

7.2 Simple Way for Large-Scale Synthesis

Use of biomass for the large-scale production needs numerous power-energy approaches such as worry and collecting the biomass. Furthermore, the purification of the metal nanoparticles is a boring method. Application of immobilized enzymes such as oxidoreductases will also decrease most of the power energy processes and downstream processing steps.

7.3 *Separation of Metal Nanoparticles by Electrophoresis Conforming to Shape and Size*

Hanauer et al. (2007) and Surugau and Urban (2009) reported a simple technique for separation of the metal nanoparticles using electrophoresis. To use size-dependent and shape-dependent material properties, for example quantum detention or plasmon resonances, it is important to have metal nanoparticles with the lowest size and shape scattering. An alternative to the high yield biosynthesis of metal nanoparticles with ultra-narrow size scattering is the post-synthetic separation of particles similar to cleaning techniques in biological synthesis. Electrophoresis is usually used to isolate biomolecules, but this method is also used in the isolation of metal nanoparticles according to their size. The isolation is mediated by the number of polymer chains involved and these are characterized via the strong colors made by plasmon resonance. The strong effect of size and shape on the occurrence or wavelength of the plasmon resonance would create it favorable to get mono-disperse metal nanoparticles for various applications. Compared to the other separation systems, electrophoresis has the advantage of allowing several runs in parallel on the same gel, which is a considerable advantage at the step of optimizing conditions and understanding mechanisms. To stabilize the metal nanoparticles, they cover with a layer of capping agent such as polyethylene glycol (PEG) or gelatin, which is covalently linked at one end to the metal surface via a thiol group. The other end of the polymer chain may transfer different functional groups, which we activity for controlling the overall particle control and mobility. The shape and size of the metal nanoparticles gained are used to synthesize particles of various shapes and these have been analyzed using electrophoresis.

7.4 *Scaling-up of Metal Nanoparticle Biosynthesis*

One of the significant challenges in metal nanoparticle synthesis is scaling-up laboratory procedures to the industrial scale. Mathematical modeling is an essential factor of our research approach, both for process scale-up and strategy and for method optimization and control. Scaled-up biosynthesis of mono-disperse colloidal of metal nanoparticles has become a significant research topic in recent years. In this context, consecutive flow reactors are commonly preferred over batch reactors. Analysis of metallic precursors in biological solvents is one of the most generally used methods to metal nanoparticles because the small size distribution can be achieved in such reaction methods. Rapid injection of precursors into a heated mixture of solvent, organizing ligands, and other precursors is often necessary. A batch method is appropriate for those reactions, but often limited to small scale production due to the low synthesis yield of metal nanoparticles and the time-consuming nature of the procedure. On the other hand, a flow reactor can production products on a consecutive basis once the reaction achieves steady state and is more suitable for a large scale synthesis than the batch reactor. Also, targeted reaction temperatures can be reached in second or even millisecond time scales in a micro-reactor. The advantage of tube-shaped reactor is that it can be simply scaled up via increasing the length of the reactor (synthesis of silver nanoparticles in a consecutive flow Tube-shaped Micro-reactor). Based on the above approaches, two types of method design can be created for the synthesis of metal nanoparticles by the biomass and the enzyme. Using the previous design, the biomass biosynthesized can be collected and used for the biosynthesis of metal nanoparticles (Fig. 5.6a). But it may need some more extra downstream steps for example purification of the metal nanoparticle from the other biomolecules. But when the immobilized protein/enzymes are used, the downstream steps can be very less and will be advantageous as the nitrate reductase can be reused for the biosynthesis of metal nanoparticles (Fig. 5.6b). In general, large-scale biosynthesis of metal nanoparticles has constantly been a big challenge. The biosynthesis of metal nanoparticles by biomass is a boring procedure. Thus, the second choice of metal nanoparticle biosynthesis by enzymes will be a better option. Scaling-up of silver nanoparticle biosynthesis should start from the first time of purification of an enzyme like as nitrate reductase. Vaidyanathan et al. (2010) reported optimized conditions for the production of nitrate reductase by the microorganism *B. licheniformis*. This optimization led to the improved biosynthesis of silver nanoparticles. The synthesis was found to be dependent on the enzyme activity. Since all the culture supernatants were provided with same

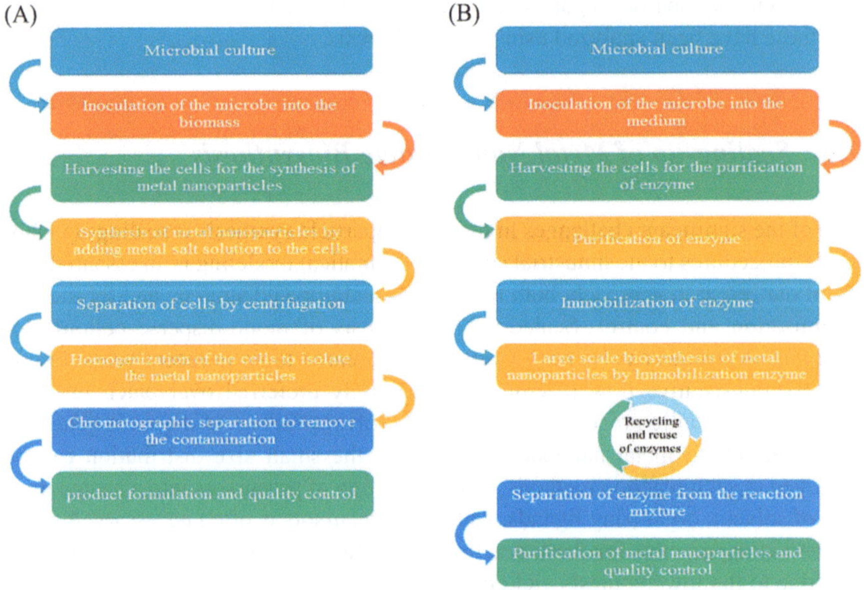

Fig. 5.6 Sequential steps of nanoparticle synthesis using microbial biomass (**a**) and microbial enzymes (**b**)

concentration of silver ions, there is no possibility of extra silver ions coming into image. This has shown that increase in the volume of enzyme increases the speed of reaction. Although there are no reports on very large amount of biosynthesis of metal nanoparticles, some examples have been designed in consecutive flow reactors similar to chemical approaches. The *Cinnamomum camphora* leaf extract is an example of a plant material which has been used to synthesize silver nanoparticles in a consecutive flow tubular micro-reactor (Huang et al. 2008).

7.5 Biocontrol of Nanoparticles Size via pH and Temperature

Although chemical production supports the size control over the production of nanoparticles, size control can also be successful biological approaches. A study by Gurunathan et al. (2009a, b) indicated that by controlling the environment of metal nanoparticle synthesis, silver nanoparticles of different sizes and shapes could be synthesized. Silver nanoparticles were synthesized with a size of 50 nm at room temperature, whereas at 60 °C they were produced with a size of 15 nm. Also at acidic pH solution, the size of the metal nanoparticles was 45 nm but at pH 10 the size was 15 nm. Even when the size was 2–20 nm, silver nanoparticles could be synthesized by microorganisms such as the fungus *Verticillium* sp. intracellularly (Mukherjee et al. 2001). The size-bio-controlled production of silver nanoparticles via controlling the environment is owing to the creation of many seed crystals. At lower temperatures and acidic pH there will be less nucleation for silver crystal creation on which new entering silver atoms deposited to structure larger sized metal nanoparticles. But as the temperature and pH increase, the dynamics of the metal ions growth and more nucleation parts are made appropriate to the accessibility of –OH ions and improved temperature.

8 Applications of Metal Nanoparticles and Biologically Inspired Models

8.1 Antimicrobial Activity of Metal Nanoparticles

The ever growing resistance of diseases towards antibiotics has been caused serious health problems in the recent decade. It has been indicated that by linking recent technologies such as material science and nanotechnology with natural antimicrobial activity of the metals, innovative applications for these materials could be recognized. Metal and metal oxide nanoparticles comprise a group of materials which are important in respect to their antimicrobial effects. Silver nanoparticles are the most common metal nanoparticles used as antifungal and antibacterial agents (Zinjarde 2012; Aziz et al. 2015; Asghari et al. 2016). This property of silver is widely advanced it to use in the numerous inoculation molded plastic produces, cloths, textiles and covering-based usages (Egger et al. 2009). Silver nanoparticles also have a variety of biomedical applications (Malarkodi et al. 2014; Prasad 2014; Prasad et al. 2014). It has been shown that silver nanoparticles indication a high and good antimicrobial activity comparable with its metal ionic form (Jo et al. 2009). It has also been established that silver nanoparticles are possible antimicrobial agents versus drug-resistant fungi or bacteria (Allahverdiyev et al. 2011). According to the recent studies, antibacterial action of silver nanoparticles results from damage of the bacteria external membrane (Lok et al. 2006). Some researchers believe that silver nanoparticles can induce gaps and pits in the bacterial membrane and then disjoint the cell (Yun et al. 2013; Iavicoli et al. 2013). Silver ions interact with sulfhydryl or disulfide groups of protein/enzymes that lead to cut of metabolic procedures which in turn reason the cell death (Egger et al. 2009). Jo et al. (2009) studied the effect of size-reduction on the antimicrobial result of silver nanoparticles. They used silver nanoparticles to control *Magnaporthe grisea* and *Bipolarissor okiniana*. Likewise, they also assessed the ability of silver nanoparticles on different kinds of human and plant pathogens for example soil-borne fungi which rarely products spores. They proposed that silver nanoparticles had a high potential for use in controlling spore-generating fungal human and plant pathogens. These metal nanoparticles might be safe and less toxic than artificial fungicides. In the other research, Mie et al. (2014) established the antimicrobial activity of their synthesized silver nanoparticles with size 19 nm against eight microorganisms by the disk diffusion method. Silver nanoparticles showed antimicrobial activity against gram-negative bacteria. Hernández-Sierra et al. (2008) reported reasonable bactericidal activity of silver nanoparticles, silver, and zinc oxide on *Streptococcus mutans*. Their results showed that silver nanoparticles demonstrated the most activity for controlling *S. mutans*. These researchers proposed that silver nanoparticles could be used in dental caries since which usually caused by *S. mutans*. Similarly, Besinis et al. (2014) studied the antibacterial effect of silver nanoparticles on *S. mutans*. Their results indicated that the antibacterial effect of silver nanoparticles against *S. mutans* was greater than that of chlorhexidine. Zarei et al. (2014) assessed antibacterial effect of silver nanoparticles against four food-borne pathogens namely *Listeria monocytogenes*, *Escherichia coli*, *Vibrio parahaemolyticus* and *Salmonella typhimurium*. Silver nanoparticles had excessive antibacterial effect on these bacterial pathogens. They determined that silver nanoparticles could be a good alternative for cleaning and sterilization of tools and surfaces in the food-related environments. Pal et al. (2007) reported the shape-dependent antibacterial activity of silver nanoparticles. According to their results, truncated triangular-shape nanoparticles were more sensitive due to their high-atom-density surfaces, and so they showed higher antimicrobial activity. Bera et al. (2014) stated the shape and size-dependent antimicrobial activity of fluorescent silver nanoparticles with size 1–5 nm against gram-positive bacteria (*Bacillus megaterium, Staphylococcus epidermidis*, *Bacillus megaterium*) and gram-negative bacteria (*Pseudomonas aeruginosa*). They concluded that size and shape of the metal nanoparticles control their activity. Conforming to these surveys, the smaller particles were easily entered the cell-wall and indicated the better-quality antimicrobial activity. These researchers concluded that these silver nanoparticles could be used for various determinations for example bio-adhesives, clinical wound dressing, biofilm control and the covering of biomedical materials. Bahrami et al. (2014) prepared silver–gold alloy nanoparticles to evaluate their antimicrobial effect against *Staphylococcus aureus*. The antibacterial activity of silver-gold alloy nanoparticles was intensified when they joined with penicillin G and piperacillin.

Gold nanoparticles are considered to be so noteworthy in the progress of antibacterial agents due to their non-toxicity, high capability to polyvalent properties, standardization, and ease of finding and photo-thermal activity (Tiwari et al. 2011; Zhou et al. 2012; Lima et al. 2013; Lolina and Narayanan 2013). Although production of "reactive oxygen species" is mainly involved in cellular death for the most antibiotics and antibacterial nanoparticles; however, antimicrobial activity of gold nanoparticles does not persuade any reactive oxygen species-depended mechanism (Cui et al. 2012). Cui et Al. (2012) showed that antimicrobial activity of the gold nanoparticles was attributable to add-on of these metal nanoparticles to the bacteria membrane and ATP level reduction and deterrence of tRNA compulsory to the ribosome. Tiwari et al. (2011) examined the antifungal and antibacterial activities of the gold nanoparticles functionalized by 5-fluorouracil against, *Pseudomonas aeruginosa*, *S. aureus*, *E. coli*, *Micrococcus luteus*, *Aspergillus niger* and *Aspergillus fumigatus*. These researchers reported that synthesized nanoparticles had less activity on gram positive bacteria than gram negative counterparts due to their easier enterance into the gram negative bacteria. These nanoparticles showed antifungal activity against *A. niger* and *A. fumigatus* as well. Zhou et al. (2012) studied antibacterial activity of silver and gold nanoparticles against *Bacillus Calmette-Guérin* and *E. coli*. They asserted that silver and gold nanoparticles showed meaningful antibacterial activity against both the Gram positive (*Bacillus*) and Gram negative (*E. coli*) bacteria. They also functionalized gold nanoparticles with a strongly bound capping agent (polyallylamine hydrochloride) and a weakly bound capping agent (citrate). Poly-allylamine hydrochloride could straight interaction with the bacteria cell membrane due to its positively charged nature (Goodman et al. 2004). Furthermore, gold nanoparticles functionalized with a strongly bound capping agent could self-assembly into 4–5 μm long chains (Kundu and Liang 2008). Zhou et al. (2012) described that these two mentioned methods facilitate the transfer of a large number of gold nanoparticles on the bacteria cell wall. Lima et al. (2013) described antimicrobial effect of gold nanoparticles with 5 nm size against *Salmonella typhi* and *E. coli*. They showed that synthesized nanoparticles reduced 90–95 % of *S. typhi* and *E. coli* colonies. These researchers highlighted that the main factors that influenced the biocidal effects were the roughness and the scattering of the gold nanoparticles on the culture medium. It appears that gold nanoparticles are safer and non-toxic to the mammalian cells than the other nanoparticles due to the reactive oxygen species-independent mechanism of their antimicrobial activity. Moreover, high capability of gold nanoparticles for standardization makes them ideal nanoparticles to be useful as targeted antimicrobial agents.

8.2 Other Applications of Metal Nanoparticles

The frequently developing field of nanotechnology is attended to need an important amount of optimized and functional nanoparticles. A wide range of formal physicochemical methods has been used to produce a wide range of metal nanoparticles. These metal nanoparticles have been used in a various range of applications (Fig. 5.7) such as targeted drug delivery, cancer treatments, biosensors, diagnostics and therapeutics, pesticides, and antimicrobials (Salata 2004; Neuberger et al. 2005; Guo et al. 2007; Tang et al. 2008; Boisselier and Astruc 2009; Das et al. 2009; Gholami-Shabani et al. 2012). However, nanoparticles synthesized by environment-friendly natural and biologically entities have only been used in relatively few applied applications. Silver nanoparticles have involved significant research interest owing to their characteristic antimicrobial activities in an extensive range of commercially presented medical and consumer products (Riddin et al. 2006; Perelshtein et al. 2008; Ilic et al. 2009; Rai et al. 2009). Another developing application of metal nanoparticles in specific is in crop safety and the controlling of agricultural plant diseases (Mukherjee

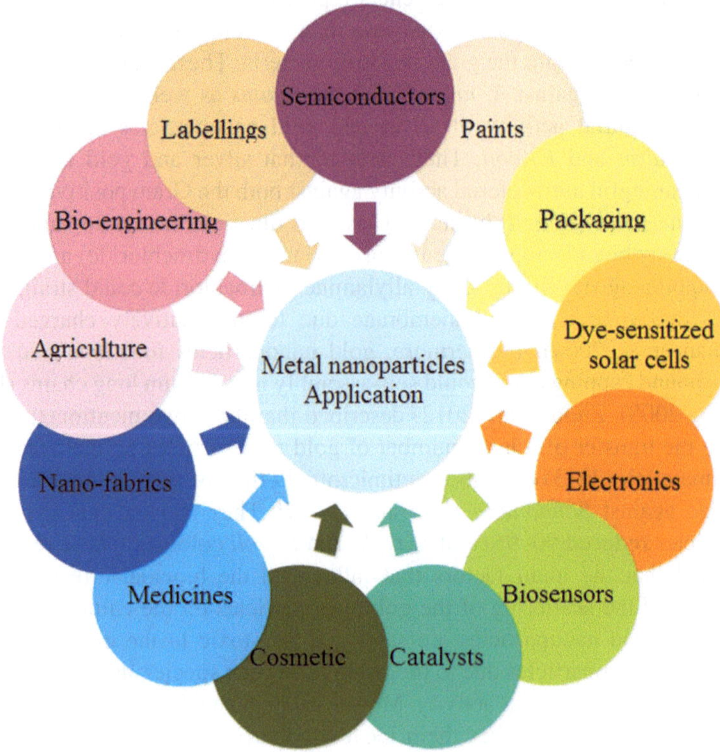

Fig. 5.7 Principal applications of nanoparticles in biology, medicine and industry

et al. 2001; Sugunan et al. 2007; Han et al. 2007). Recent researches have indicated the antimicrobial effects of silver nanoparticles (Vivek et al. 2011; Gholami-Shabani et al. 2014; Muthukrishnan et al. 2015). Furthermore, silver nanoparticles can be used to control some plant pathogens in a nontoxic way compared to standard fungicides (Park et al. 2006) and also silver nanoparticles can be active against plasmodial pathogens and cancer cells (Sukirtha et al. 2011; Seil and Webster 2012; Ponarulselvam et al. 2012; Subramanian 2012). Traditionally, gold has been used in some medical applications. Gold nanoparticles have attracted important interest over the last period as a medicinal material in treatment of tumors. For example, gold nanoparticles have the capability to inactively accumulate in tumors due to their size. Due to their unique optical and chemical properties, they can be used in thermal treatment ways (Hirsch et al. 2003; Zheng and Sache 2009). Moreover, biocompatible gold nanoparticles were successfully used as transporter platforms for the targeted delivery of anti-cancer drugs thus progress delivery and reducing treatment periods and side effects (Paciotti et al. 2006; Cai et al. 2008; Cheng et al 2010). Also, recent studies indicated that gold nanoparticles are active antimicrobial agents against a number of fungal and bacterial strains (Pissuwan et al. 2009; Poinern et al. 2013; Gholami-Shabani et al. 2015). In the other studied, copper and copper oxide nanoparticles have been used as strong antimicrobial agents and their fumigate properties against a number of infectious organisms means they can be used as effective fungicides and bactericides to cover hospital tools (Stoimenov et al. 2002; Akhavan and Ghaderi 2012; Hassan et al. 2012; Subhankari and Nayak 2013). Platinum nanoparticles were used in water electrolysis applications (Soundarrajan et al. 2012). Titanium oxide nanoparticles, for the reason of their antimicrobial activity, have been used in antimicrobial coverings and wastewater sterilization processes (Zhang and Chen 2009; Allahverdiyev et al. 2011; Miller et al. 2012). Zinc oxide nanoparticles show good antimicrobial activity and have been used in wastewater treatments and food packing (Perez-Espitia et al. 2012; Duran and Seabra 2012). Bacteria, viruses and bacteriophages are interesting assemblers for engineering one-dimensional structure into regular arrays. For example, the tobacco mosaic virus used successfully to synthesis gold, silver and platinum nanoparticles (Dujardin et al. 2003). Likewise, filamentous bacteriophages are used to production of nanotubes and silica fibres (Wang et al. 2008; Li et al. 2011, 2012a, b, c). Entities for example silk serein are used to produce nanofibrous complexes that direct the creation of hydroxyapatite particles (Yang et al. 2015) and boost osteogenic properties of human bone marrow cells (Yang et al. 2014). Moreover, films including gold nanoparticles have been used from genetically engineered filamentous viruses and bacteria to synthesize cadmium sulfide quantum dots (Mi et al. 2011) and colorimetric sensors (Liu et al. 2009; Oh et al. 2014). Recent studies have shown that viral nanofibers covered with magnetic iron oxide nanoparticles can be used for the finding of human serum antibody biomarkers (Wang et al. 2015). Nanoparticles and nanoparticle made assemblies have the potential to be used in an extensive variation of applications as conversed above, especially if they can be synthesized using biological agents that can ensure clean, eco-friendly and nontoxic approaches of production. The synthesis of metal nanoparticles via an extensive variation of biological agents, as conversed above, has been actively followed in recent years as a substitute bottom up method to self-assemble atoms by atom and then grow into nanometer scale particles. However, several reasons exist that can importantly impact the capability of this eco-friendly method for synthesizing metal nanoparticles. The most important known reasons being shape, particle size control and size distribution. These causes are all straight influenced via reactant moieties, reaction medium pH, reaction time, reactant concentrations, and temperature. As described above, even small differences in these factors can importantly affect particle shape, size, and size distribution. For example, in the sample of plant extracts, there can be obvious differences in the chemical composition of extracts taken at various times and locations around the world for the same species. This compositional difference can often lead to various laboratories making different results from the similar plant extract and metal salt solution. This can be a serious problem in using plant extracts to synthesis metal nanoparticles with stable physical and chemical properties. Clearly, even with the present limitations, biosynthesis suggestions various advantages and has the potential to deliver metal nanoparticles with prearranged properties. For example, Shankar et al. (2004), using efficient quality control and closely adaptable the reactant concentrations, reaction medium pH, reaction time, and temperature through production were able to reduce large quantities of triangular-shape gold nanoprisms using *Cymbopogon flexuosus* extract. More than 45 % of the total gold nanoparticles reduced from the aqueous chloroaurate ions ($AuCl_4^-$) and extract solution were composed of gold triangles nanoparticles. The triangles showed reduced vertices like to those seen for triangular silver and gold nanoprisms synthesized by chemical and photochemical approaches. In addition, successive centrifugation (3000–4000 × g), washing, and re-scattering of the reaction medium importantly improved the output of triangle nanoparticles numbers (up to 80–90 %). Interestingly, even with recent progresses in conventional chemical and physical approaches, many physical approaches still need relatively expensive tools, materials and have effective requirements for example pressurized gases, vacuum and high temperatures. While most chemical approaches tend to use toxic resources such as reducing agents, organic solvents and stabilizers. These cost-effective and toxicity related emphasizes the significance and necessity for further study into eco-friendly biosynthesis approaches factors further over the more traditional metal nanoparticle synthesis processes.

9 Concluding Remarks

Large-scale synthesis of biogenic nanoparticles from organisms has been a great challenge for several years. Synthesis of metal nanoparticles using the biological approaches especially purified enzymes is environmentally friendly compared with the routine physicochemical approaches, but there are sure key areas of study which need to be pointed out. Limited information has been reported on green synthesis of metal nanoparticles by enzymes. Currently, the majority of nanoparticles are synthesis by using fungal, bacterial and plant cells but not purified enzymes. This may induce several problems particularly when the method needs to be scaled-up for industrial applications. The main problem in cell-based methods is binding of the synthesized metal nanoparticles to the microbial biomass. This means that extra steps of separation of the microbial cells and subsequent separation of metal nanoparticles are needed for purification of metal nanoparticles. Also, metal ions have limitations for applying in the biological methods because they are toxic for the biomass at concentrations over the threshold (Kalimuthu et al. 2008). To overcome these problems, microbial culture supernatants have been successfully used for metal nanoparticle synthesis by several scientists. While culture supernatants are superior to microbial biomass in metal nanoparticle synthesis owing to the lower-costs and simplicity for application and maintenance, but they suffer from large limitations counting the limited numbers of microorganisms possessing necessary secretory proteins for metal nanoparticle synthesis as well as necessity for downstream methods for purification and cleansing the ending product. In compare to purified enzymes, both microbial biomass and culture supernatant cannot be recycled which makes the methods of metal nanoparticle synthesis non-economic specifically in industrial application (Clark 1994; Gholami-Shabani et al. 2014, 2015). Enzymatically produced metal nanoparticles have several advantages over those usually produced by culture supernatants and microbial biomasses. Besides significant reduction of the downstream steps necessary for purification of synthesized metal nanoparticles as final products, they have high potential for industrial uses as they can be immobilized for recycling in metal nanoparticle synthesis. Enzymatically synthesis of metal nanoparticles not only reduces downstream steps for purification of the metal nanoparticles from the other proteins/enzymes, but also provides a constant source of safe biologically-active nanoparticles for large scale production with potential application in medicine and agriculture. From this point of view, there are growing reports of green synthesis of biologically active metal nanoparticles by a wide array of enzymes such as nitrate reductase, sulfite reductase, hydrogenase, cytoplasmic proteins, laccases, tyrosinases, peroxidases, chromium reductase and α-amylase. It is hopeful that these enzymatically-based nanomaterials become commercialized to treat different types of human diseases in the near future.

References

Ahmad A, Senapati S, Khan MI, Kumar R, Sastry M (2003) Extracellular biosynthesis of monodisperse gold nanoparticles by a novel extremophilic actinomycete, *Thermomonospora* sp. Langmuir 19:3550–3553. doi:10.1021/la0267721

Ahmad R, Mohsin M, Ahmad T, Sardar M (2015) Alpha amylase assisted synthesis of TiO_2 nanoparticles: Structural characterization and application as antibacterial agents. J Hazard Mater 283:171–177. doi:10.1016/j.jhazmat.2014.08.073

Akhavan O, Ghaderi E (2012) Cu and CuO nanoparticles immobilized by silica thin films as antibacterial materials and photocatalysts. Surf Coat Technol 205:219–223. doi:10.1016/j.surfcoat.2010.06.036

Allahverdiyev AM, Abamor ES, Bagirova M, Rafailovich M (2011) Antimicrobial effects of TiO_2 and Ag_2O nanoparticles against drug-resistant bacteria and leishmania parasites. Future Microbiol 8:933–940. doi:10.2217/fmb.11.78

Ananikov VP, Orlov NV, Beletskaya IP, Khrustalev VN, Antipin MY, Timofeeva TV (2007) New approach for size-and shape-controlled preparation of Pd nanoparticles with organic ligands. Synthesis and application in catalysis. J Am Chem Soc 129:7252–7253. doi:10.1021/ja071727r

Arjunan NK, Murugan K, Rejeeth C, Madhiyazhagan P, Barnard DR (2012) Green synthesis of silver nanoparticles for the control of mosquito vectors of malaria, filariasis, and dengue. Vect Bor Zoonotic Dis 12:262–268. doi:10.1089/vbz.2011.0661

Asghari F, Jahanshiri Z, Imani M, Shams-Ghahfarokhi M, Razzaghi-Abyaneh, M (2016) Antifungal nanomaterials: Synthesis, properties and applications. In: Alexandru Mihai Grumezescu (eds) Nanobiomaterials in Antimicrobial Therapy - Applications of Bionanomaterials, Vol 6, 1st Edition, Chapter 10, Elsevier, pp 343–383.

Asmathunisha N, Kathiresan K (2013) A review on biosynthesis of nanoparticles by marine organisms. Coll Surf B Biointer 103:283–287. doi:10.1016/j.colsurfb.2012.10.030

Attard G, Casadesu´s M, Macaskie LE, Deplanche K (2012) Biosynthesis of platinum nanoparticles by *Escherichia coli* MC4100: can such nanoparticles exhibit intrinsic surface enantioselectivity? Langmuir 28:5267–5274. doi:10.1021/la204495z

Ayaseelan C, Rahuman AA, Kirthi AV, Marimuthu S, Santhoshkumar T, Bagavan A, Rao KB (2012) Novel microbial route to synthesize ZnO nanoparticles using *Aeromonas hydrophila* and their activity against pathogenic bacteria and fungi. Spectrochim Acta A Mol Biomol Spectrosc 90:78–84. doi:10.1016/j.saa.2012.01.006

Aziz N, Faraz M, Pandey R, Sakir M, Fatma T, Varma A, Barman I, Prasad R (2015) Facile algae-derived route to biogenic silver nanoparticles: synthesis, antibacterial and photocatalytic properties. Langmuir 31:11605–11612. doi:10.1021/acs.langmuir.5b03081

Bahrami K, Nazari P, Nabavi M, Golkar M, Almasirad A, Shahverdi AR (2014) Hydroxyl capped silver-gold alloy nanoparticles: characterization and their combination effect with different antibiotics against *Staphylococcus aureus*. Nanomed J 1:155–161. doi:10.7508/nmj.2014.03.005

Balaji DS, Basavaraja S, Deshpande R, Mahesh DB, Prabhakar BK, Venkataraman A (2009) Extracellular biosynthesis of functionalized silver nanoparticles by strains of *Cladosporium cladosporioides*. Coll Surf B Biointer 68:88–92. doi:10.1016/j.colsurfb.2008.09.022

Basavaraja S, Balaji SD, Lagashetty A, Rajasab AH, Venketaraman A (2008) Extracellular biosynthesis of silver nanoparticles using the fungus *Fusarium semitectum*. Mater Res Bull 43:1164–1170. doi:10.1016/j.materresbull.2007.06.020

Bera RK, Mandal SM, Retna Raj C (2014) Antimicrobial activity of fluorescent Ag nanoparticles. Lett Appl Microbiol 58:520–526. doi:10.1111/lam.12222

Besinis A, De Peralta T, Handy RD (2014) The antibacterial effects of silver, titanium dioxide and silica dioxide nanoparticles compared to the dental disinfectant chlorhexidine on *Streptococcus* mutants using a suite of bioassays. Nanotoxicology 8:1–16. doi:10.3109/17435390.2012.742935

Beveridge TJ, Doyle RJ (1989) Metal Ions and Bacteria. Wiley, New York

Bhainsa KC, D'Souza SF (2006) Extracellular biosynthesis of silver nanoparticles using the fungus *Aspergillus fumigatus*. Coll Surf B Biointer 47:160–164. doi:10.1016/j.colsurfb.2005.11.026

Boisselier E, Astruc D (2009) Gold nanoparticles in nanomedicine: preparations, imaging, diagnostics, therapies and toxicity. Chem Soc Rev 38:1759–1782. doi:10.1039/B806051G

Cai W, Gao T, Hong H, Sun J (2008) Applications of gold nanoparticles in cancer nanotechnology. Nanotechnol Sci Appl 1:17–32

Cao G (2004) Nanostructures and nanomaterials: synthesis, properties and applications. Imperial College Press, London

Castro ME, Cottet L, Castillo A (2014) Biosynthesis of gold nanoparticles by extracellular molecules produced by the phytopathogenic fungus *Botrytis cinerea*. Mater Lett 115:42–44. doi:10.1016/j.matlet.2013.10.020

Cheng Y, Samia AC, Li J, Kenney ME, Resnick A, Burda C (2010) Delivery and efficacy of a cancer drugs a function of the bond to the gold nanoparticle surface. Langmuir 26:2248–2255. doi:10.1021/la902390d

Clark DS (1994) Can immobilization be exploited to modify enzyme activity? Trends Biotechnol 12:439–443. doi:10.1016/0167-7799(94)90018-3
Cui Y, Zhao Y, Tian Y, Zhang W, Lü X, Jiang X (2012) The molecular mechanism of action of bactericidal gold nanoparticles on *Escherichia coli*. Biomaterials 33:2327–2333. doi:10.1016/j.biomaterials.2011.11.057
Das SK, Das AR, Guha AK (2009) Gold nanoparticles: microbial synthesis and applications in water hygiene management. Langmuir 25(14):8192–8199. doi:10.1021/la900585p
Das SK, Liang J, Schmidt M, Laffir F, Marsili E (2012a) Biomineralization mechanism of gold by zygomycete fungi *Rhizopous oryzae*. ACS Nano 6:6165–6173. doi:10.1021/nn301502s
Das SK, Dickinson C, Lafir F, Brougham DF, Marsili E (2012b) Synthesis, characterization and catalytic activity of gold nanoparticles biosynthesized with *Rhizopus oryzae* protein extract. Green Chem 14:1322–1334. doi:10.1039/C2GC16676C
Dhanasekar NN, Rahul G, Narayanan KB, Raman G, Sakthivel N (2015) Green chemistry approach for the synthesis of gold nanoparticles using the fungus *Alternaria* sp. J Microbiol Biotechnol 25:1129–1135. doi:10.4014/jmb.1410.10036
Dujardin E, Mann S (2002) Bio-inspired materials chemistry. Adv Mater 14:1–14
Dujardin E, Peet C, Stubbs G, Culver JN, Mann S (2003) Organization of metallic nanoparticles usingtobacco mosaic virus templates. Nano Lett 3:413–417. doi:10.1021/nl034004o
Duran N, Seabra AB (2012) Metallic oxide nanoparticles: state of the art in biogenic syntheses and their mechanisms. Appl Microbiol Biotechnol 95:275–288. doi:10.1007/s00253-012-4118-9
Duran N, Marcato PD, Alves OL, DeSouza G, Esposito E (2005) Mechanistic aspects of biosynthesis of silver nanoparticles by several *Fusarium oxysporum* strains. J Nanobiotechnol 3:1–8. doi:10.1186/1477-3155-3-8
Egger S, Lehmann RP, Height MJ, Loessner MJ, Schuppler M (2009) Antimicrobial properties of a novel silver-silica nanocomposite material. Appl Environ Microbiol 75:2973–2976. doi:10.1128/AEM.01658-08
Engelberg-Kulka H, Amitai S, Kolodkin-Gal I, Hazan R (2006) Bacterial programmed cell death and multicellular behavior in bacteria. PLoS Genet 10:135. doi:10.1371/journal.pgen.0020135
Eppler AS, Rupprechter G, Anderson EA, Somorjai GA (2000) Thermal and chemical stability and adhesion strength of Pt nanoparticle arrays supported on silica studied by transmission electron microscopy and atomic force microscopy. J Phys Chem B 104:7286–7292. doi:10.1021/jp0006429
Erasmus M, Cason ED, van Marwijk J, Botes E, Gericke M, van Heerden E (2014) Gold nanoparticle synthesis using the thermophilic bacterium *Thermus scotoductus* SA-01 and the purification and characterization of its unusual gold reducing protein. Gold Bull 47:245–253. doi:10.1007/s13404-014-0147-8
Feldheim DL, Foss CA (2002) Metal nanoparticles: synthesis, characterization, and applications. CRC/Taylor & Francis, Boca Raton
Gade AK, Bonde P, Ingle AP, Marcato PD, Duran N, Rai MK (2008) Exploitation of *Aspergillus niger* for synthesis of silver nanoparticles. J Biobased Mat Bioenerg 2:243–247. doi:10.1166/jbmb.2008.401
Gade A, Ingle A, Bawaskar M, Rai M (2009) *Fusarium solani*: a novel biological agent for the extracellular synthesis of silver nanoparticles. J Nanopart Res 11:2079–2085. doi:10.1007/s11051-008-9573-y
Gajbhiye M, Kesharwani J, Ingle A, Gade A, Rai M (2009) Fungus-mediated synthesis of silver nanoparticles and their activity against pathogenic fungi in combination with fluconazole. Nanomed Nanotechnol Biol Med 5:382–386. doi:10.1016/j.nano.2009.06.005
Gardea-Torresdey JL, Tiemann KJ, Gamez G, Dokken K, Tehuacanero S, Jose-Yacaman M (1999) Gold nanoparticles obtained by bio-precipitation from gold (III) solutions. J Nanopart Res 1:397–404. doi:10.1023/A:1010008915465
Gardea-Torresdey JL, Gomez E, Peralta-Videa JR, JG Parsons JG, Troiani H, Jose-Yacaman M (2003) *Alfalfa sprouts*: a natural source for the synthesis of silver nanoparticles. Langmuir 19:1357–1361. doi:10.1021/la020835i

Geng X, Grove TZ (2015) Repeat protein mediated synthesis of gold nanoparticles: effect of protein shape on the morphological and optical properties. RSC Adv 5:2062–2069. doi:10.1039/C4RA12014K

Gericke M, Pinches A (2006a) Microbial production of gold nanoparticles. Gold Bull 39:22–28. doi: 10.1007/BF03215529

Gericke M, Pinches A (2006b) Biological synthesis of metal nanoparticles. Hydrometallurgy 83:132–140. doi:10.1016/j.hydromet.2006.03.019

Ghaseminezhad SM, Hamedi S, Shojaosadati SA (2012) Green synthesis of silver nanoparticles by a novel method: comparative study of their properties. Carbo Polym 89:467–472. doi:10.1016/j.carbpol.2012.03.030

Gholami-Shabani MH, Akbarzadeh A, Mortazavi M, Emadzadeh MK (2012) Evaluation of the antibacterial properties of silver nanoparticles synthesized with *Fusarium oxysporum* and *Escherichia coli*. New Cell Mol Biotechnol J 2:27–33

Gholami-Shabani M, Akbarzadeh A, Norouzian D, Amini A, Gholami-Shabani Z, Imani A, Chiani M, Riazi G, Shams-Ghahfarokhi M, Razzaghi-Abyaneh M (2014) Antimicrobial activity and physical characterization of silver nanoparticles green synthesized using nitrate reductase from *Fusarium oxysporum*. Appl Biochem Biotechnol 172:4084–4098. doi:10.1007/s12010-014-0809-2

Gholami-Shabani M, Imani A, Shams-Ghahfarokhi M, Gholami-Shabani Z, Pazooki A, Akbarzadeh A, Riazi G, Razzaghi-Abyaneh M (2016) Bioinspired synthesis, characterization and antifungal activity of enzyme-mediated gold nanoparticles using a fungal oxidoreductase. J Iran Chem Soc doi:10.1007/s13738-016-0923-x

Gholami-Shabani M, Shams-Ghahfarokhi M, Gholami-Shabani Z, Akbarzadeh A, Riazi G, Ajdari S, Amani A, Razzaghi-Abyaneh M (2015) Enzymatic synthesis of gold nanoparticles using sulfite reductase purified from *Escherichia coli*: a green eco-friendly approach. Proc Biochem 50:1076–1085. doi:10.1016/j.procbio.2015.04.004

Goodman C, McCusker C, Yilmaz T, Rotello V (2004) Toxicity of gold nanoparticles functionalized with cationic and anionic side chains. Bioconjug Chem 15:897–900. doi:10.1021/bc049951i

Govender Y, Riddin T, Gericke M, Whiteley CG (2009) Bioreduction of platinum salts into nanoparticles: a mechanistic perspective. Biotechnol Lett 31:95–100. doi:10.1007/s10529-008-9825-z

Grzelczak M, Pérez-Juste J, Mulvaney P, Liz-Marzán LM (2008) Shape control in gold nanoparticle synthesis. Chem Soc Rev 37:1783–1791. doi:10.1039/B711490G

Guo KW (2012) Green nanotechnology of trends in future energy: a review. Int J Energy Res 36:1–17. doi:10.1002/er.1928

Guo C, Boullanger P, Jiang L, Liu T (2007) Highly sensitive gold nanoparticles biosensor chips modified with a self-assembled bilayer for detection of Con A. Biosens Bioelectron 22:1830–1834. doi:10.1016/j.bios.2006.09.006

Gurunathan S, Lee KJ, Kalishwaralal K, Sheikpranbabu S, Vaidyanathan R, Eom SH (2009a) Antiangiogenic properties of silver nanoparticles. Biomaterials 30:6341–6350. doi:10.1016/j.biomaterials.2009.08.008

Gurunathan S, Kalishwaralal K, Vaidyanathan R, Venkataraman D, Pandian SRK, Muniyandi J, Hariharan N, Eom SH (2009b) Biosynthesis, purification and characterization of silver nanoparticles using *Escherichia coli*. Coll Surf B Biointer 74:328–335. doi:10.1016/j.colsurfb.2009.07.048

Gurunathan S, Jeong JK, Han JW, Zhang XF, Park JH, Kim JH (2015) Multidimensional effects of biologically synthesized silver nanoparticles in *Helicobacter pylori*, *Helicobacter felis*, and human lung (L132) and lung carcinoma A549 cells. Nanoscale Res Lett 10:1–17. doi:10.1186/s11671-015-0747-0

Han G, Ghosh P, De M, Rotello MV (2007) Drug and gene delivery using gold nanoparticles. Nanobiotechnology 3:40–45. doi:10.1007/s12030-007-0005-3

Hanauer M, Lotz A, Pierrat S, Sonnichsen C, Zins I (2007) Separation of nanoparticles by gel electrophoresis according to size and shape. Nano Lett 7:2881–2885. doi:10.1021/nl071615y

Hassan MS, Amna T, Yang OB, El-Newehy MH, Al-Deyab SS, Khil MS (2012) Smart copper oxide nanocrystals: synthesis, characterization, electrochemical and potent antibacterial activity. Coll Surf B Biointer 97:201–206. doi:10.1016/j.colsurfb.2012.04.032

He S, Guo Z, Zhang Y, Zhang S, Wang J, Gu N (2007) Biosynthesis of gold nanoparticles using the bacteria *Rhodopseudomonas capsulata*. Mater Lett 61:3984–3987. doi:10.1016/j.matlet.2007.01.018

Heath JR (1999) Nanoscale materials. Acc Chem Res 32:388. doi:10.1021/ar990059e

Hernández-Sierra JF, Ruiz F, Cruz Pena DC, Martínez-Gutiérrez F, Martínez AE, de Jesús Pozos Guillén A, Tapia-Pérez H, Martínez Castañón G (2008) The antimicrobial sensitivity of *Streptococcus mutans* to nanoparticles of silver, zinc oxide, and gold. Nanomed: Nanotech Biol Med 4:237–240. doi:10.1016/j.nano.2008.04.005

Hirsch LR, Stafford RJ, Bankson JA, Sershen SR, Rivera B, Price RE, Hazle JD, Halas NJ, West JL (2003) Nanoshell-mediated near-infrared thermal therapy of tumours under magnetic resonance guidance. Proc Natl Acad Sci U S A 100:13549–13554. doi:10.1073/pnas.2232479100

Honary S, Barabadi H, Gharaei-Fathabad E, Naghibi F (2012) Green synthesis of copper oxide nanoparticles using *Penicillium aurantiogriseum*, *Penicillium citrinum* and *Penicillium waksmanii*. Dig J Nanomater Bios 7:999–1005

Huang J, Lin L, Li Q, Sun D, Wang Y, Lu Y, He N, Yang K, Yang X, Wang H, Wang W, Lin W (2008) Continuous-flow biosynthesis of silver nanoparticles by lixivium of Sundried *Cinnamomum camphora* leaf in tubular microreactors. Ind Eng Chem Res 47:6081–6090. doi:10.1021/ie701698e

Huang JL, Wang WT, Lin LQ, Li QB, Lin WS, Li M, Mann S (2009) A general strategy for the biosynthesis of gold nanoparticles by traditional Chinese medicines and their potential application as catalysts. Chem Asian J 4:1050–1054. doi:10.1002/asia.200900064

Huang J, Lin L, Sun D, Chen H, Yang D, Li Q (2015) Bio-inspired synthesis of metal nanoparticles and application. Chem Soc Rev 44:6330–6374

Iavicoli I, Fontana L, Leso V (2013) Bergamaschi A (2013) The Effects of Nanomaterials as Endocrine Disruptors. Int J Mol Sci 14:16732–16801. doi:10.3390/ijms140816732

Ilic V, Saponjı´c Z, Vodnik V, Potkonjak B, Jovancic P, Nedeljkovı´c J, Radetic M (2009) The influence of silver content on antimicrobial activity and color of cotton fabrics functionalized with Ag nanoparticles. Carbohydr Polym 78:564–569. doi:10.1016/j.carbpol.2009.05.015

Ingle A, Gade A, Pierrat S, Seonnichsen C, Rai M (2008) Mycosynthesis of silver nanoparticles using the fungus *Fusarium acuminatum* and its activity against some human pathogenic bacteria. Curr Nanosci 4:141–144. doi:10.2174/157341308784340804

Iravani HK, Mirmohammadi SV, B Zolfaghari S (2014) Synthesis of silver nanoparticles: chemical, physical and biological methods. Res Pharmaceutical Sci 9:385–406

Jain N, Bhargava A, Tarafdar JC, Singh SK, Panwar J (2013) A biomimetic approach towards synthesis of zinc oxide nanoparticles. Appl Microbiol Biotechnol 97:859–869. doi:10.1007/s00253-012-3934-2

Jo YK, Kim BH, Jung G (2009) Antifungal activity of silver ions and nanoparticles on phytopathogenic fungi. Plant Dis 93:1037–1043. doi:10.1094/PDIS 93 10 1037

Kalimuthu K, Babu RS, Venkataraman D, Bilal M, Gurunathan S (2008) Biosynthesis of silver nanocrystals by *Bacillus licheniformis*. Coll Surf B Biointer 65:150–153. doi:10.1016/j.colsurfb.2008.02.018

Kalishwaralal K, Deepak V, Pandian SRK, Gurunathan S (2009) Biological synthesis of gold nanocubes from *Bacillus licheniformis*. Bioresource Technol 100:5356–5358. doi:10.1016/j.biortech.2009.05.051

Kalishwaralal K, Deepak V, Pandian SRK, Kottaisamy M, BarathManiKanth S, Kartikeyan B, Gurunathan S (2010) Biosynthesis of silver and gold nanoparticles using *Brevibacterium casei*. Coll Surf B Biointer 77:257–262. doi:10.1016/j.colsurfb.2010.02.007

Kathiresan K, Manivannan S, Nabeel MA, Dhivya B (2009) Studies on silver nanoparticles synthesized by a marine fungus, *Penicillium fellutanum* isolated from coastal mangrove sediment. Coll Surf B Biointer 71:133–137. doi:10.1016/j.colsurfb.2009.01.016

Kelly KL, Coronado E, Zhao LL, Schatz GC (2003) The optical properties of metal nanoparticles: the influence of size, shape, and dielectric environment. J Phys Chem B 107:668–677. doi:10.1021/jp026731

Khan SA, Ahmad A (2014) Enzyme mediated synthesis of water-dispersible, naturally protein capped, monodispersed gold nanoparticles; their characterization and mechanistic aspects. RSC Adv 4:7729–7734. doi:10.1039/C3RA43888K

Kharissova OV, Dias HR, Kharisov BI, Pérez BO, Pérez VMJ (2013) The greener synthesis of nanoparticles. Trends Biotechnol 31:240–248. doi:10.1016/j.tibtech.2013.01.003

Klaus T, Joerger R, Olsson E, Granqvist CG (1999) Silver-based crystalline nanoparticles, microbially fabricated. Proc Natl Acad Sci U S A 96:13611–13614. doi:10.1073/pnas.96.24.13611

Konishi Y, Tsukiyama T, Ohno K, Saitoh N, Nomura T, Nagamine S (2006) Intracellular recovery of gold by microbial reduction of $AuCl_4^-$ ions using the anaerobic bacterium *Shewanella algae*. Hydrometallurgy 81:24–29. doi:10.1016/j.hydromet.2005.09.006

Konishi YK, Ohno N, Saitoh T, Nomura S, Nagamine H, Hishida Y, Takahashi T, Konishi Y, Ohno K, Saitoh N, Nomura T, Nagamine S, Hishida H, Uruga, T (2007) Bioreductive deposition of platinum nanoparticles on the bacterium *Shewanella* algae. Uruga. J Biotechnol 128:648–653. doi:10.1016/j.jbiotec.2006.11.014

Konishi Y, Tsukiyama T, Tachimi T, Saitoh N, Nomura T, Nagamine S (2007b) Microbial deposition of gold nanoparticles by the metal-reducing bacterium *Shewanella algae*. Electrochim Acta 53:186–192. doi:10.1016/j.electacta.2007.02.073

Kumar SA, Abyaneh MK, Gosavi SW, Kulkarni SK, Ahmad A, Khan MI (2007) Sulfite reductase-mediated synthesis of gold nanoparticles capped with phytochelatin. Biotechnol Appl Biochem 47:191–195. doi:10.1042/BA20060205

Kundu S, Liang H (2008) Polyelectrolyte-mediated non-micellar synthesis of monodispersed 'aggregates' of gold nanoparticles using a microwave approach. Coll Surf A Physicochem Eng Asp 330:143–150. doi:10.1016/j.colsurfa.2008.07.043

Li WR, Xie XB, Shi QS, Duan SS, Ouyang YS, Chen YB (2011) Antibacterial effect of silver nanoparticles on *Staphylococcus aureus*. Biometals 24:135–141. doi:10.1007/s10534-010-9381-6

Li D, Mathew B, Mao C (2012a) Biotemplated synthesis of hollow double-layered core/shell titania/silica nanotubes under ambient conditions. Small 8:3691–3697. doi:10.1002/smll.201200421

Li D, Newton SMC, Klebba PE, Mao C (2012b) Flagellar display of bone- protein-derived peptides for studying peptide-mediated biomineralization. Langmuir 28:16338–16346. doi:10.1021/la303237u

Li D, Qu X, Newton SMC, Klebba PE, Mao C (2012c) Morphology-controlled synthesis of silica nanotubes through pH and sequence-responsive morphological change bacterial flagellar biotemplates. J Mater Chem 22:15702–15709. doi:10.1039/C2JM31034A

Lima E, Guerra R, Lara V, Guzmán A (2013) Gold nanoparticles as efficient antimicrobial agents for *Escherichia coli* and *Salmonella typhi*. Chem Cent J 7:1–7

Lin WS, Lok CN, Che CM (2014) Biosynthesis of silver nanoparticles from silver (i) reduction by the periplasmic nitrate reductase c-type cytochrome subunit NapC in a silver-resistant *E. coli*. Chem Sci 5:3144–3150. doi:10.1039/C4SC00138A

Liu A, Abbineni G, Mao C (2009) Nanocomposite films assembled from genetically engineered filamentous viruses and gold nanoparticles: Nanoarchitecture- and humidity-tunable surface plasmon resonance spectra. Adv Mater 21:1001–1005. doi:10.1002/adma.200800777

Lok CN, Ho CM, Chen R, He QY, Yu WY, Sun H, Tam PKH, Chiu JF, Che CM (2006) Proteomic analysis of the mode of antibacterial action of silver nanoparticles. J Proteome Res 5:916–924. doi:10.1021/pr0504079

Lolina S, Narayanan V (2013) Antimicrobial and anticancer activity of gold nanoparticles synthesized from grapes fruit extract. Chem Sci Trans 2:S105–S110. doi:10.7598/cst2013.22

Mabbett AN, Yong P, Farr JPG, Macaskie LE (2004) Reduction of Cr (VI) by "palladized" biomass of *Desulfovibrio desulfuricans* ATCC 29577. Biotechnol Bioeng 87:104–109. doi:10.1002/bit.20105

Malarkodi C, Rajeshkumar S, Vanaja M, Paulkumar K, Gnanajobitha G, Annadurai G (2013) Eco-friendly synthesis and characterization of gold nanoparticles using *Klebsiella pneumoniae*. J Nanostruct Chem 3:1–7. doi:10.1186/2193-8865-3-30

Malarkodi C, Rajeshkumar S, Paulkumar K, Vanaja M, Gnanajobitha G, Annadurai G (2014) Biosynthesis and antimicrobial activity of semiconductor nanoparticles against oral pathogens. Bioinorg Chem Appl 2014:1–10. doi:10.1155/2014/347167

Mann S (2001) Biomineralization. Oxford University Press, Oxford

Meldrum FC, Cölfen H (2008) Controlling mineral morphologies and structures in biological and synthetic systems. Chem Rev 108:4332–4432. doi:10.1021/cr8002856

Mi C, Wang Y, Zhang J, Huang H, Xu L, Wang S, Fang X, Fang J, Mao C, Xu S (2011) Biosynthesis and characterization of CdS quantum dots in genetically engineered *Escherichia coli*. J Biotechnol 153:125–132. doi:10.1016/j.jbiotec.2011.03.014

Mie R, Samsudin MW, Din LB, Ahmad A, Ibrahim N, Adnan SNA (2014) Synthesis of silver nanoparticles with antibacterial activity using the lichen *Parmotrema praesorediosum*. Int J Nanomed 9:121–127. doi:10.2147/IJN.S52306

Mijatovic D, Eijkel JCT, Van Den Berg A (2005) Technologies for nanofluidic systems: top-down vs. bottom-up—a review. Lab Chip 5:492–500. doi:10.1039/B416951D

Miller RJ, Bennett S, Keller AA, Pease S, Lenihan HS (2012) TiO_2 nanoparticles are phototoxic to marine phytoplankton. PLoS Biol 7:1–7. doi:10.1371/journal.pone.0030321

Mishra A, Kumari M, Pandey S, Chaudhry V, Gupta KC, Nautiyal CS (2014) Biocatalytic and antimicrobial activities of gold nanoparticles synthesized by *Trichoderma* sp. Bioresource Technol 166:235–242. doi:10.1016/j.biortech.2014.04.085

Mittal AK, Chisti Y, Banerjee UC (2013) Synthesis of metallic nanoparticles using plant extracts. Biotechnol Adv 31:346–356. doi:10.1016/j.biotechadv.2013.01.003

Mohamad NR, Marzuki NHC, Buang NA, Huyop F, Wahab RA (2015) An overview of technologies for immobilization of enzymes and surface analysis techniques for immobilized enzymes. Biotechnol Biotechnol Equip 29:205–220. doi:10.1080/13102818.2015.1008192

Mohanpuria P, Rana NK, Yadav SK (2008) Biosynthesis of nanoparticles: technological concepts and future applications. J Nanopart Res 10(3):507–517. doi:10.1007/s11051-007-9275-x

Mukherjee P, Ahmad A, Mandal D, Senapati S, Sainkar SR, Khan MI, Parishcha R, Ajaykumar PV, Alam M, Kumar R, Sastry M (2001) Fungus-mediated synthesis of silver nanoparticles and their immobilization in the mycelial matrix: a novel biological approach to nanoparticle synthesis. Nano Lett 1:515–519. doi:10.1021/nl0155274

Mukherjee P, Senapati S, Mandal D, Ahmad A, Khan MI, Kumar R, Sastry M (2002) Extracellular synthesis of gold nanoparticles by the fungus *Fusarium oxysporum*. Chem Bio Chem 3:461–463. doi:10.1002/1439-7633(20020503)3:5<461::AID-CBIC461>3.0.CO;2-X

Mukherjee S, Sushma V, Patra S, Barui AK, Bhadra MP, Sreedhar B, Patra CR (2012) Green chemistry approach for the synthesis and stabilization of biocompatible gold nanoparticles and their potential applications in cancer therapy. Nanotechnology 23:455103

Muthukrishnan S, Bhakya S, Kumar TS, Rao M (2015) Biosynthesis, characterization and antibacterial effect of plant-mediated silver nanoparticles using *Ceropegia thwaitesii*–an endemic species. Ind Crops Prod 63:119–124. doi:10.1016/j.indcrop.2014.10.022

Nagajyothi PC, Lee SE, An M, Lee KD (2012) Green synthesis of silver and gold nanoparticles using *Lonicera japonica* flower extract. Bull Korean Chem Soc 33:2609–2612. doi:10.5012/bkcs.2012.33.8.2609

Nair B, Pradeep T (2002) Coalescence of nanoclusters and formation of submicron crystallites assisted by *Lactobacillus* strains. Cryst Growth Des 2:293–298. doi:10.1021/cg0255164

Narayanan KB, Sakthivel N (2010) Biological synthesis of metal nanoparticles by microbes. Adv Colloid Interface Sci 156:1–13. doi:10.1016/j.cis.2010.02.001

Neuberger T, Schopf B, Hofmann H, Hofmann M, von Rechenberg B (2005) Super paramagnetic nanoparticles for biomedical applications: possibilities and limitations of a new drug delivery system. J Magn Mater 293:483–496. doi:10.1016/j.jmmm.2005.01.064

Ng CK, Sivakumar K, Liu X, Madhaiyan M, Ji L, Yang L, Tang C, Song H, Kjelleberg S, Cao B (2013) Influence of outer membrane c-type cytochromes on particle size and activity of extracellular nanoparticles produced by *Shewanella oneidensis*. Biotechnol Bioeng 110:1831–1837. doi:10.1002/bit.24856

Noguez C (2007) Surface plasmons on metal nanoparticles: the influence of shape and physical environment. J Phys Chem C 111:3806–3819. doi:10.1021/jp066539m

Oh JW, Chung WJ, Heo K, Jin HE, Lee BY, Wang E, Zueger C, Wong W, Meyer J, Kim C, Lee SY, Kim WG, Zemla M, Auer M, Hexemer A, Lee SW (2014) Biomimetic virus-based colourimetric sensors. Nat Commun 5:3043. doi:10.1038/ncomms4043

Oves M, Khan MS, Zaidi A, Ahmed AS, Ahmed F, Ahmad E, Sherwani A, Owais M, Azam A (2013) Antibacterial and cytotoxic efficacy of extracellular silver nanoparticles biofabricated from chromium reducing novel OS4 strain of *Stenotrophomonas maltophilia*. PLoS One 8, e59140. doi:10.1371/journal.pone.0059140

Paciotti GF, Mayer L, Weinreich D, Goia D, Pavel N, McLaughlin RE, Tamarkin L (2006) Colloidal gold: a novel nanoparticle vector for tumour directed drug delivery. Drug Deliv 11:169–183. doi:10.1080/10717540490433895

Pal S, Tak YK, Song JM (2007) Does the antibacterial activity of silver nanoparticles depend on the shape of the nanoparticle? A study of the gram-negative bacterium *Escherichia coli*. Appl Environ Microbiol 73:1712–1720. doi:10.1128/AEM.02218-06

Parikh RY, Singh S, Prasad BLV, Patole MS, Sastry M, Shouche YS (2008) Extracellular synthesis of crystalline silver nanoparticles and molecular evidence of silver resistance from *Morganella* sp.: towards understanding biochemical synthesis mechanism. Chem Bio Chem 9:1415–1422. doi:10.1002/cbic.200700592

Park HJ, Kim SH, Kim HJ, Choi SH (2006) A new composition of nanosized silica-silver for control of various plant diseases. Plant Pathol J 22:295–302

Park TJ, Lee KG, Lee SY (2015) Advances in microbial biosynthesis of metal nanoparticles. Appl Microbiol Biotechnol 1–14. doi:10.1007/s00253-015-6904-7

Perelshtein I, Applerot G, Perkas N, Guibert G, Mikhailov S, Gedanken A (2008) Sonochemical coating of silver nanoparticles on textile fabrics (nylon, polyester and cotton) and their antibacterial activity. Nanotechnology 19:1–6

Perez-Espitia PJ, Ferreira-Soares NF, dos Reis Coimbra JS, de Andrade NJ, Cruz RS, Medeiros EAA (2012) Zinc oxide nanoparticles: synthesis, antimicrobial activity and food packaging applications. Food Bioproc Technol 5:1447–1464. doi:10.1007/s11947-012-0797-6

Philip D (2009) Biosynthesis of Au, Ag and Au–Ag nanoparticles using edible mushroom extract. Spectrochim Acta A Mol Biomol Spectrosc 73:374–381. doi:10.1016/j.saa.2009.02.037

Pissuwan D, Cortie CH, Valenzuela SM, Cortie MB (2009) Functionalised gold nanoparticles for controlling pathogenic bacteria. Trends Biotechnol 28:207–213. doi:10.1016/j.tibtech.2009.12.004

Poinern GEJ (2014) A laboratory course in nanoscience and nanotechnology, 1st edn. CRC/Taylor & Francis, Boca Raton

Poinern GEJ, Le X, Chapman P, Fawcett D (2013) Green biosynthesis of gold nanometre scale plate using the leaf extracts from an indigenous Australian plant *Eucalyptus macrocarpa*. Gold Bull 46:165–173. doi:10.1007/s13404-013-0096-7

Ponarulselvam S, Panneerselvam C, Murugan K, Aarthi N, Kalimuthu K, Thangamani S (2012) Synthesisof silver nanoparticles using leaves of *Catharanthus roseus* Linn. G. Don and their antiplasmodial activities. Asian Pac J Trop Biomed 2:574–580. doi:10.1016/S2221-1691(12)60100-2

Prasad R (2014) Synthesis of silver nanoparticles in photosynthetic plants. J Nanoparticles, Article ID: 963961. doi:10.1155/2014/963961

Prasad R, Swamy VS (2013) Antibacterial activity of silver nanoparticles synthesized by bark extract of *Syzygium cumini*. J Nanoparticles. doi:10.1155/2013/431218

Prasad R, Kumar V, Prasad KS (2014) Nanotechnology in sustainable agriculture: present concerns and future aspects. Afr J Biotechnol 13(6):705–713

Prasad R, Pandey R, Barman I (2015) Engineering tailored nanoparticles with microbes: quo vadis. WIREs Nanomed Nanobiotechnol. doi:10.1002/wnan.1363

Rai M, Yadav A, Gade A (2009) Silver nanoparticles as a new generation of antimicrobials. Biotechnol Adv 27:76–83. doi:10.1016/j.biotechadv.2008.09.002

Rajakumar G, Rahuman AA, Roopan SM, Khanna VG, Elango G, Kamaraj C, Velayutham K (2012) Fungus-mediated biosynthesis and characterization of TiO_2 nanoparticles and their activity against pathogenic bacteria. Spectrochim Acta A Mol Biomolecul Spectrosc 91:23–29. doi:10.1016/j.saa.2012.01.011

Rajeshkumar S, Malarkodi C, Paulkumar K, Vanaja M, Gnanajobitha G, Annadurai G (2014) Algae mediated green fabrication of silver nanoparticles and examination of its antifungal activity against clinical pathogens. Int J Met 1–8. doi:10.1155/2014/692643

Ramamurthy CH, Padma M, Mareeswaran R, Suyavaran A, Kumar MS, Premkumar K, Thirunavukkarasu C (2013) The extra cellular synthesis of gold and silver nanoparticles and their free radical scavenging and antibacterial properties. Coll Surf B Biointer 102:808–815. doi:10.1016/j.colsurfb.2012.09.025

Ramanathan R, O'Mullane AP, Parikh RY, Smooker PM, Bhargava SK, Bansal V (2011) Bacterial kinetics-controlled shape-directed biosynthesis of silver nanoplates using *Morganella psychrotolerans*. Langmuir 27:714–719. doi:10.1021/la1036162

Reddy AS, Chen CY, Chen CC, Jean JS, Chen HR, Tseng MJ, Wang JC (2010) Biological synthesis of gold and silver nanoparticles mediated by the bacteria *Bacillus subtilis*. J Nanosci Nanotechnol 10:6567–6574. doi:10.1166/jnn.2010.2519

Riddin TL, Gericke M, Whiteley CG (2006) Analysis of the inter- and extracellular formation of platinum nanoparticles by *Fusarium oxysporum* f sp *lycopersici* using response surface methodology. Nanotechnology 17:3482–3489

Riddin TL, Govender Y, Gericke M, Whiteley CG (2009) Two different hydrogenase enzymes from sulphate-reducing bacteria are responsible for the bioreductive mechanism of platinum into nanoparticles. Enzyme Microb Technol 45:267–273. doi:10.1016/j.enzmictec.2009.06.006

Ro¨sken LM, Ko¨rsten S, Fischer CB, Scho¨nleber A, van Smaalen S, Geimer S, Wehner S (2014) Time-dependent growth of crystalline Au0-nanoparticles in cyanobacteria as self-reproducing bioreactors: 1. *Anabaena* sp. J Nanopart Res 16:2370. doi:10.1007/s11051-014-2370-x

Roopan SM, Surendra TV, Elango G, Kumar SHS (2014) Biosynthetic trends and future aspects of bimetallic nanoparticles and its medicinal applications. Appl Microbiol Biotechnol 98:5289–5300. doi:10.1007/s00253-014-5736-1

Roy S, Das TK (2015) A review on biosynthesis of silver nanoparticles using *Aspergillus* species. Adv Sci Engin Medi 7:729–738. doi:10.1166/asem.2015.1760

Saifuddin N, Wong CW, Nuryasumira AA (2009) Rapid biosynthesis of silver nanoparticles using culture supernatant of bacteria with microwave irradiation. Eur J Chem 6:61–70

Salata OV (2004) Application of nanoparticles in biology and medicine. J Nanobiotechnol 2:3–9

Sanghi R, Verma P (2009) Biomimetic synthesis and characterisation of protein capped silver nanoparticles. Bioresour Technol 100:501–504. doi:10.1186/1477-3155-2-3

Sanghi R, Verma P, Puri S (2011) Enzymatic formation of gold nanoparticles using *Phanerochaete chrysosporium*. Adv Chem Engin Sci 1:154. doi:10.4236/aces.2011.13023

Sau TK, Murphy CJ (2004) Room temperature, high-yield synthesis of multiple shapes of gold nanoparticles in aqueous solution. J Am Chem Soc 126:8648–8649. doi:10.1021/ja047846d

Seil JT, Webster TJ (2012) Antimicrobial applications of nanotechnology: methods and literature. Int J Nanomed 7:2767–2781. doi:10.2147/IJN.S24805

Shaligram NS, Bule M, Bhambure R, Singhal RS, Singh SK, Szakacs G, Pandey A (2009) Biosynthesis of silver nanoparticles using aqueous extract from the compactin producing fungal strain. Process Biochem 44:939–943. doi:10.1016/j.procbio.2009.04.009

Shankar SS, Rai A, Ankamwar B, Singh A, Ahmad A, Sastry M (2004) Biological synthesis of triangular gold nanoprisms. Nat Mater 3:482–488. doi:10.1038/nmat1152

Sharma VK, Filip J, Zboril R, Varma RS (2015) Natural inorganic nanoparticles–formation, fate, and toxicity in the environment. Chem Soc Rev 44:8410–8423. doi:10.1039/C5CS00236B

Show S, Tamang A, Chowdhury T, Mandal D, Chattopadhyay B (2015) Bacterial (BKH1) assisted silica nanoparticles from silica rich substrates: a facile and green approach for biotechnological applications. Coll Surf B Biointer 126:245–250. doi:10.1016/j.colsurfb.2014.12.039

Silver S, Phung LT, Silver G (2006) Silver as biocides in burn and wound dressings and bacterial resistance to silver compounds. J Ind Microbiol Biotechnol 33:627–634. doi:10.1007/s10295-006-0139-7

Singaravelu G, Arockiamary JS, Ganesh Kumar V, Govindaraju K (2007) A novel extracellular synthesis of mondisperse gold nanoparticles using marine alga, *Sargassum wightii* Greville. Coll Surf B Biointer 57:97–101. doi:10.1016/j.colsurfb.2007.01.010

Soundarrajan C, Sankari A, Dhandapani P, Maruthamuthu S, Ravichandran S, Sozhan G, Palaniswamy N (2012) Rapid biological synthesis of platinum nanoparticles using *Ocimum sanctum* for water electrolysis applications. Bioproc Biosys Eng 35:827–833. doi:10.1007/s00449-011-0666-0

Stoimenov PK, Klinger RL, Marchin RL, Klabunde KJ (2002) Metal oxide nanoparticles as bactericidal agents. Langmuir 18:6679–6686. doi:10.1021/la0202374

Subhankari I, Nayak PL (2013) Synthesis of copper nanoparticles using *Syzygium aromaticum* (Cloves) aqueous extract by using green chemistry. World J Nano Sci Technol 2:14–17. doi:10.5829/idosi.wjnst.2013.2.1.21134

Subramanian V (2012) Green synthesis of silver nanoparticles using *Coleus amboinicus* lour, antioxitant activity and in vitro cytotoxicity against Ehrlich's Ascite carcinoma. J Pharm Res 5:1268–1272

Sugunan A, Melin P, Schn€urer J, Hilborn JG, Dutta J (2007) Nutrition-driven assembly of colloidal nanoparticles: growing fungi assemble gold nanoparticles as microwires. Adv Mater 19:77–81. doi:10.1002/adma.200600911

Sukirtha R, Priyanka KM, Antony JJ, Kamalakkannan S, Thangam R, Gunasekaran P, Krishnan M, Achiraman S (2011) Cytotoxic effect of green synthesized silver nanoparticles using *Melia azedarach* against in vitro HeLa cell lines and lymphoma mice model. Proc Biochem 47:273–279. doi:10.1016/j.procbio.2011.11.003

Surugau N, Urban PL (2009) Electrophoretic methods for separation of nanoparticles. J Sep Sci 32:1889–1906. doi:10.1002/jssc.200900071

Tang T, Krysmann JM, Hamley WI (2008) In situ formation of gold nanoparticles with a thermoresponsive block copolymer corona. Colloids Surf A Physicochem Eng Asp 317:764–767. doi:10.1016/j.colsurfa.2007.11.032

Tiwari PM, Vig K, Dennis VA, Singh SR (2011) Functionalized gold nanoparticles and their biomedical applications. Nanomaterials 1:31–63. doi:10.3390/nano1010031

Tran QH, Le AT (2013) Silver nanoparticles: synthesis, properties, toxicology, applications and perspectives. Adv Nat Sci Nanosci Nanotechnol 4:033001

Vaidyanathan R, Gopalram S, Kalishwaralal K, Deepak V, Pandian SRK, Gurunathan S (2010) Enhanced silver nanoparticle synthesis by optimization of nitrate reductase activity. Coll Surf B Biointer 75:335–341. doi:10.1016/j.colsurfb.2009.09.006

Verma VC, Kharwar RN, Gange AC (2010) Biosynthesis of antimicrobial silver nanoparticles by the endophytic fungus *Aspergillus clavatus*. Nanomedicine 5:33–40. doi:10.2217/nnm.09.77

Vetchinkina EP, Loshchinina EA, Burov AM, Dykman LA, Nikitina VE (2014) Enzymatic formation of gold nanoparticles by submerged culture of the *basidiomycete Lentinus edodes*. J Biotechnol 82:37–45. doi:10.1016/j.jbiotec.2014.04.018

Vigneshwaran N, Ashtaputre NM, Varadarajan PV, Nachane RP, Paralikar KM, Balasubramanya RH (2007) Biological synthesis of silver nanoparticles using the fungus *Aspergillus flavus*. Mater Lett 66:1413–1418. doi:10.1016/j.matlet.2006.07.042

Vivek M, Kumar PS, Steffi S, Sudha S (2011) Biogenic silver nanoparticles by *Gelidiella acerosa* extract and their antifungal effects. Avicenna J Med Biotechnol 3:143–148

Wang Y, Xia Y (2004) Bottom-up and top-down approaches to the synthesis of monodispersed spherical colloids of low melting-point metals. Nano Lett 4:2047–2050. doi:10.1021/nl048689j

Wang F, Li D, Mao C (2008) Genetically modified flagella as templates for silica fibers: From hybrid nanotubes to 1D periodic nanohole arrays. Adv Funct Mater 18:4007–4013. doi:10.1002/adfm.200800889

Wang Y, Ju Z, Cao B, Gao X, Zhu Y, Qiu P, Xu H, Pan P, Bao H, Wang L, Mao C (2015) Ultrasensitive rapid detection of human serum antibody biomarkers by biomarker-capturing viral nanofibers. ACS Nano 9:4475–4483. doi:10.1021/acsnano.5b01074

Wei X, Luo M, Li W, Yang L, Liang X, Xu L, Liu H (2012) Synthesis of silver nanoparticles by solar irradiation of cell-free *Bacillus amyloliquefaciens* extracts and $AgNO_3$. Biores Technol 103:273–278. doi:10.1016/j.biortech.2011.09.118

Weller H (2003) Synthesis and self–assembly of colloidal nanoparticles. Philos Trans R Soc Lond A Math Phys Eng Sci 361:22–240. doi:10.1098/rsta.2002.1136

Xia Y, Xia X, Wang Y, Xie S (2013) Shape-controlled synthesis of metal nanocrystals. Bull MRS 38:335–344. doi:10.1557/mrs.2013.84

Xu AW, Ma Y, Cölfen H (2007) Biomimetic mineralization. J Mater Chem 17:415–449. doi:10.1039/B611918M

Yang M, Shuai Y, Zhang C, Chen Y, Zhu L, Mao C, OuYang H (2014) Biomimetic nucleation of hydroxyapatite crystals mediated by *Antheraea pernyi* silk sericin promotes osteogenic differentiation of human bone marrow derived mesenchymal stem cells. Biomacromolecules 15:1185–1193. doi:10.1021/bm401740x

Yang M, Zhou G, Shuai Y, Wang J, Zhu L, Mao C (2015) Ca^{2+} induced self-assembly of *Bombyx mori* silk sericin into a Nanofibrous network-like protein matrix for directing controlled nucleation of hydroxyapatite nano-needles. J Mater Chem B 3:2455–2462. doi:10.1039/C4TB01944J

Yates MD, Cusick RD, Logan BE (2013) Extracellular palladium nanoparticle production using *Geobacter sulfurreducens*. ACS Sustainable Chem Eng 1:1165–1171. doi:10.1021/sc4000785

Yong P, Rowson A, Farr JPG, Harris IR, Macaskie LE (2002) Bioreduction and biocrystallization of palladium by *Desulfovibrio desulfuricans* NCIMB 8307. Biotechnol Bioeng 80:369–379. doi:10.1002/bit.10369

Yong P, Paterson-Beedle M, Mikheenko IP, Macaskie LE (2007) From bio-mineralisation to fuel cells: biomanufacture of Pt and Pd nanocrystals for fuel cell electrode catalyst. Biotechnol Lett 29:539–544. doi:10.1007/s10529-006-9283-4

Yun H, Kim JD, Choi HC, Lee CW (2013) Antibacterial activity of CNT-Ag and GO-Ag nanocomposites against gram-negative and gram-positive bacteria. Bull Korean Chem Soc 34:3261. doi:10.5012/bkcs.2013.34.11.3261

Zarei M, Jamnejad A, Khajehali E (2014) Antibacterial effect of silver nanoparticles against four foodborne pathogens. Jundishapur J Microbiol 7(1):e8720. doi:10.5812/jjm.8720

Zhang H, Chen G (2009) Potent antibacterial activities of Ag/TiO2 nanocomposite powders synthesized by aone-Pot sol-gel method. Environ Sci Technol 43:2905–2910. doi:10.1021/es803450

Zheng Y, Sache L (2009) Gold nanoparticles enhance DNA damage induced by anti-cancer drugs and radiation. Radiat Res 172:114–119

Zhou Y, Kong Y, Kundu S, Cirillo JD, Liang H (2012) Antibacterial activities of gold and silver nanoparticles against *Escherichia coli* and bacillus Calmette-Guérin. J Nanobiotechnol 10:19

Zinjarde SS (2012) Bio-inspired nanomaterials and their applications as antimicrobial agents. Chron Young Sci 3:1–74. doi:10.4103/2229-5186.94314

Chapter 6
The Effect of Mycobiota on the Biointerface of Polyaniline Surface

Elena Binkauskienė, Vitalija Jasulaitienė, Ausra Selskienė, and Albinas Lugauskas

Abstract An exceptional combination of mycobiota/polyaniline surface interaction process seems to have great promise for change of the chemical, morphological and redox properties of polyaniline, which can detect unique biocomplexity and assessment of surface. The detecting of mycobiota metabolic activity on polyaniline surface is presented. Experiments are performed under modeling condition. The polyaniline samples are exposed to Petri dishes with a pure mycobiota culture, after that the electrochemical measurements are performed. The electrochemical analysis of the biomodified polymers completed by X-ray photoelectrons spectroscopy (XPS) and scanning electron microscopy (SEM). The SEM micrographs showed difference of topography of polyaniline treated by different mycobiota. The attachment to the polyaniline surface via the metabolic product and a conspicuous difference oxalate impurities at the polyaniline surface can be indicated. This work shows significant progress in chemical research for metabolic activity definition.

1 Characterization of Polyaniline

Conjugated polymers is a new group of substances, whose electronic properties are close to those of metals and semiconductors. Polyaniline is one of the extensively studied conducting polymers and has been suggested for many applications (transistor, sensors, energy storage, electrochromic displays or smart windows, electrocatalysis, organic electrochemistry, bioelectrochemistry, etc.) (Inzelt 2012; Stejskal et al. 2015). The long-time active fungi growth within the polyaniline surface is a reliable source of information about bioactive new combinations

---

E. Binkauskienė (✉) • V. Jasulaitienė • A. Selskienė • A. Lugauskas
State Research Institute Center for Physical Sciences and Technology, LT-10257, Saulėtekio Ave. 3, Vilnius, Lithuania
e-mail: elenbink@ktl.mii.lt

R. Prasad (ed.), *Advances and Applications Through Fungal Nanobiotechnology*, Fungal Biology, DOI 10.1007/978-3-319-42990-8_6

and chemically novel compounds (Binkauskienė et al. 2009, 2013). For these investigations the morphological characteristics of the studied samples of polyaniline are important.

1.1 General Information on Structure and Properties of Polyaniline

Polyaniline can be interconverted between three idealized oxidation state: leucoemeraldine (LE, fully reduced form), emeraldine (EM, partially oxidized form), pernigraniline (PNA, fully oxidized form) (Scheme 1). The nitrogen atoms are amines (–NH–) and imines (=N–).

Polyaniline has conjugated double bonds which promote delocalization of electrons. Polyaniline exhibits both the electronic and ionic conductivity. The polymer can achieve its conductive state either through protonation of the imine nitrogens in its EM oxidation state, or through the oxidation of the amine nitrogens in its reduced LM state. The mechanism of conduction is based on the presence of delocalized polarons and protons. Only the EM salt is conducting.

Polyaniline is known as redox agents for some redox couple behavior. It is established that polyaniline can participate in bacterial electron transfer. The influence of the slowly developed mycobiota on polyaniline electrochemical behavior in comparison to fast-acting bacteria is less investigated (Langer et al. 2007).

Polyaniline is prepared by the oxidative chemical or electrochemical oxidations of aniline monomer in acidic aqueous media. For investigations of the impact of biomodification on the electrochemical and redox properties of the polymer surface the morphological characteristics of the studied samples are essential. Cui et al. (1993) showed, that electrochemical polyaniline coatings are more suitable for surface morphological investigations than chemical. The electrochemical cyclic voltammetric polymer films have morphological and adhesion properties which make them more attractive for practical purposes than potentiostatic films.

1 – $-(\text{–B–NH–B–NH–B–NH–})_n$, LE – base;

2 – $-(\text{–B–N=Q=N–B–NH–B–NH–})_n$, EM – base;

3 – $-(\text{–B–N=Q=N–B–NH–B–NH–})_n^{+}$, EM^{+}, – salt;

4 – $-(\text{–B–N=Q=N–B–N=Q=N–})_n$, PNA – base

Scheme 1 Three idealized oxidation states of polyaniline, B and Q denote the C_6H_4 rings in the benzenoid form and quinonoid form, respectively

1.2 Polyaniline Electrosynthesis

The fabrication of a modified polyaniline film electrode is well controlled by cyclic voltammetry (CV). Figure 6.1 shows a voltammetric profile for ten successive cycles of electro synthesis of polyaniline. In the first cycles a sharp anodic current (I) peak attributed to the oxidation of aniline monomer in the region of potential (E) 0.9 V can be seen. After 5 cycles this peak diminished gradually. Three oxidation I peaks (in the 0.2–0.8 V range) and a larger reduction peak (in the 0.3–0.6 V range) are seen on the tenth scan. The first oxidation I peak is attributed to the transformation from the reduced LE state to a partially oxidized EM state, the second one—to the redox reaction of degradation products and the third one—to the transition of the polyaniline from EM to the PNA state (Arsov et al. 1998).

2 PANI Surface Biomodification

2.1 Growth of Mycobiota on PANI Surface

Fungal growth experiments are performed under modeling condition. The PANI samples are exposed to Petri dishes with a pure fungi culture on the malt extract agar and incubated at 26 °C (Fig. 6.2). The reference samples are exposed to medium of malt extract without mycobial treatment.

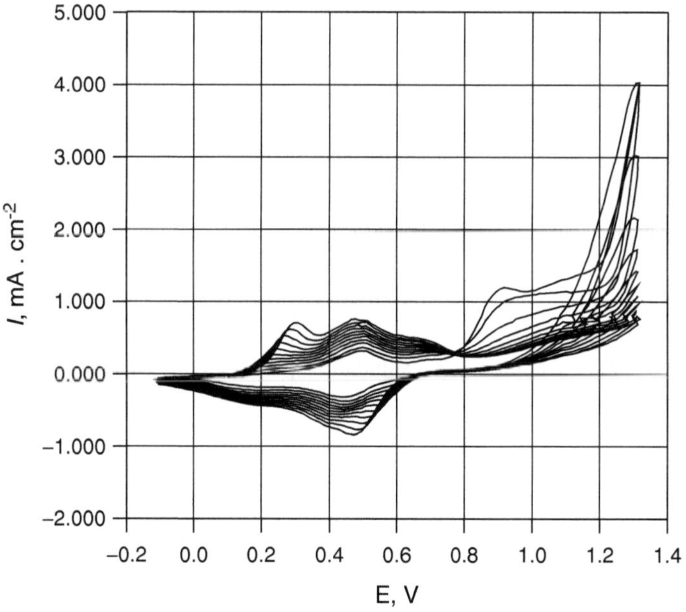

Fig. 6.1 Representative voltammograms for electrodeposition of polyaniline on nickel from 0.3 M $H_2C_2O_4$, containing 0.1 M aniline (potential cycling between E=−0.1 and 1.3 V, 0.05 Vs^{-1}, t = 10 min), vs. Ag/AgCl (From Binkauskiene et al. 2008)

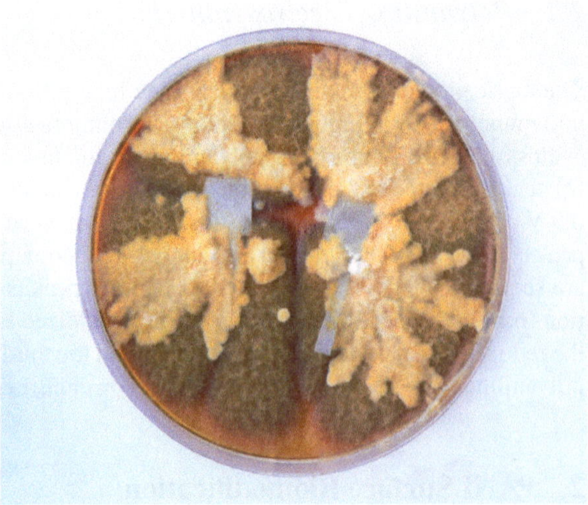

Fig. 6.2 The polyaniline electrodes exposed to Petri dishes with a pure micromycete

Table 6.1 Intensity of fungi growth (according to the five-point grading scale) on polyanilne surface, using 4-point schema and CV electrochemical characteristics

Fungal species	Biofilm surface coverage, on the five-point scale				CV electrochemical characteristics after 56 days exposition	
	7 days	14 days	32 days	56 days	ΔE, V	I_{pa}/I_{pb}
Trichoderma spp.	0	0	0	1.5	0.11	0.77
Botrytis cinerea	0	0.5	2.0	3.0	0.14	0.66
Chrysosporium pannorum	1.0	1.5	3.0	3.5	0.11	1.31
Alternaria alternata	2.0	3.0	3.0	3.5	0.16	0.86
A. rodicina	2.0	4.0	4.0	4.0	0.13	0.77
Ch. merdarium	3.5	4.5	4.5	4.5	0.13	1.53
Rhizomucor pusillus	4.0	5.0	4.5	4.5	0.25	0.72

The electrochemical characteristics are evaluated after incubation and wash

The extents of fungal growth are assessed by the naked eye and a light microscope. Biofilm modified surface coverage evaluated using a 5 point corrosion test (EN ISO 10289:1999) and checked using a standard 4-point scheme (Lugauskas et al. 2010). Comparison of biofilms formed by eight different fungi on the polyaniline surface has shown that the rate of biofilm formation markedly differs (Table 6.1).

2.2 *Morphological Characterization of Biomodified Polyaniline*

The effect of biomodification during interaction of fungi with the polyaniline surface significantly causes its morphological characteristics.

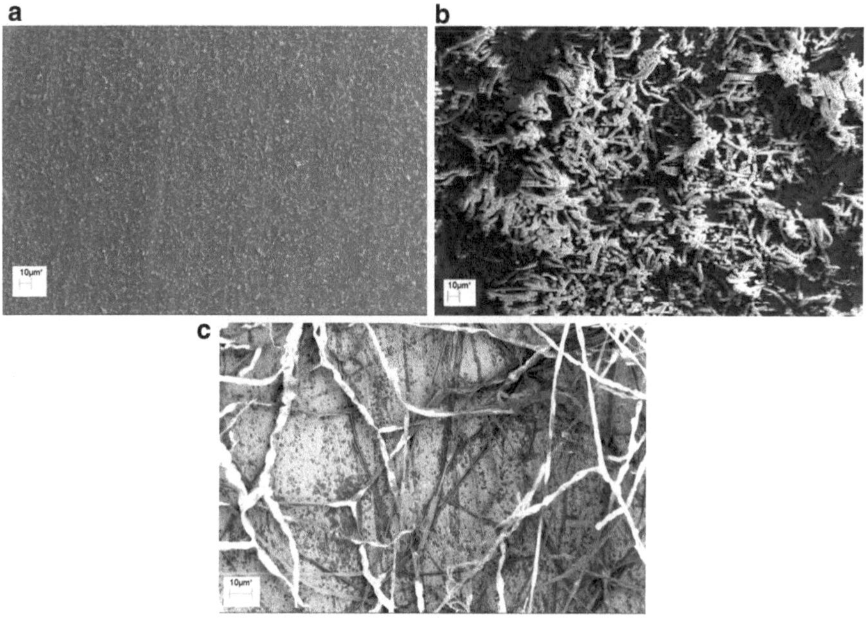

Fig. 6.3 SEM micrographs of polyaniline film: (**a**) reference, (**b**) after 14 days of exposure under the influence of *Aspergillus niger*, (**c**) under the influence of *Ch. merdarium* (From Lugauskas et al. 2008)

The mycobial reactions study shows the formation of fungal mats on the polymer surface (Fig. 6.3). Such intensive surface growth can testify good adhesion of the treated microorganisms.

Figure 6.4 illustrates the image of treated polyaniline sample by SEM after wash. The metabolizing *A. niger* L-10 biofilm has not oriented structure, is rough, crossed by the trenches. Not the growth of a new generation of *A. niger* L-10 are observed on the surface.

3 Electrochemical and Redox Behavior of Biomodified Polyanilne Surface

The reactivity of samples are studied by CV in the monomer-free acid electrolyte. The changes in the *I* peak and *E* values of an appropriate redox couple for biologically treated and untreated polyaniline surfaces are measured and the electrochemical effect of fungi metabolism is evaluated. The same film thickness must be

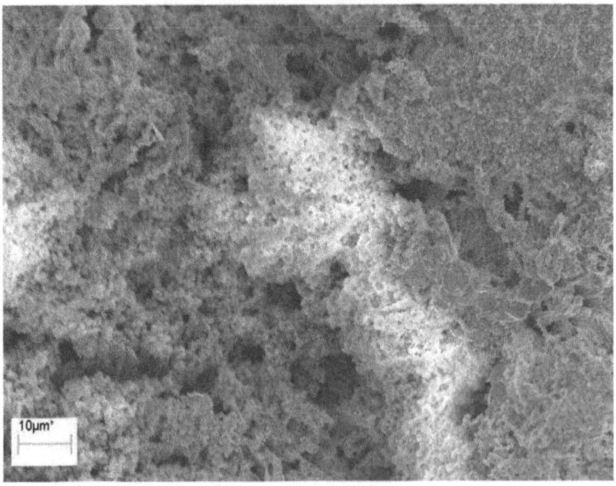

Fig. 6.4 SEM micrographs of polyaniline film after 36 days of exposure under the influence of *A. niger* L-10 after wash (From Binkauskienė et al. 2009)

performed during deposition of polymer in order to obtain similar characteristic of all the polyaniline modified electrodes. The reference polyaniline electrodes are examined after the same time of exposure.

3.1 Electrochemical Reactivity of Biomodified Polyanilne Surface

The CV of a polyaniline film in a 0.05 M H_2SO_4 solution is shown in Fig. 6.5 by a solid line. Two oxidation peaks and a sharp reduction peak ($E = 0.5$ V) are observed. This can be attributed to the proton elimination-addition and anion doping-undoping reaction on the polyaniline surface. Electrochemical incorporation of anions into the polymer matrix of a conjugated polymer increased the conductivity, activity and special catalytic properties (Inzelt et al. 2000; Binkauskiene and Binkauskas 2009). Unmodified nickel electrode (broken line) does not show electrochemical activity in relation to polyaniline modified nickel electrode (solid line).

Because of the treated surface shows reversible redox behavior and electronic conductivity, the polymers are also involved in the oxidation of metabolites. Additional to these functions, the polymer layers serves as a mixed product with varying electrode activity (Fig. 6.6).

The gap of the first anodic peak E between the treated and untreated electrodes (ΔE) on the CVs shows the level of oxidation of the surface. The comparison of the values of the first anodic *I* peak before (I_{pb}) and after (I_{pa}) the contact with fungi

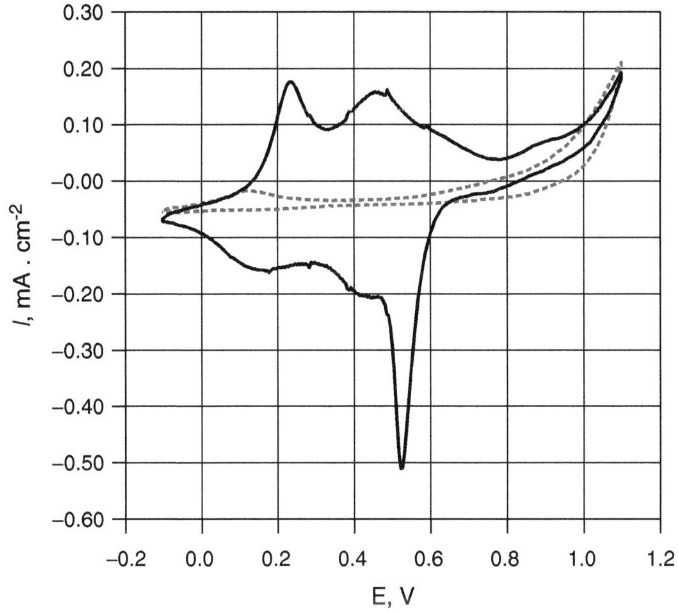

Fig. 6.5 Representative voltammograms for: (*solid line*) polyaniline film and (*broken line*) bare nickel recorded in 0.05 M H_2SO_4 solution (potential cycling between $E = -0.1$ and 1.1 V, 0.02 Vs^{-1}), vs. Ag/AgCl (From Binkauskiene et al. 2008)

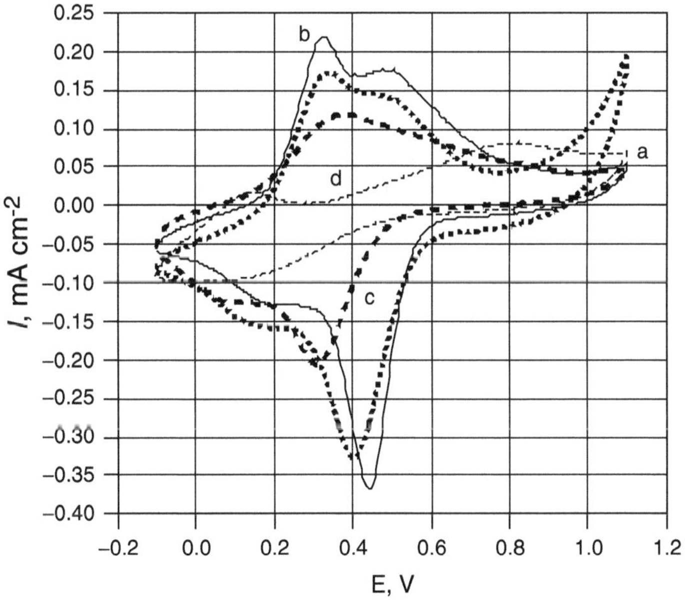

Fig. 6.6 Representative voltammograms for polyaniline recorded in 0.05 M H_2SO_4: (**a**) after 28 days of exposure to the influence of *A. niger* L-10 and (**b**) reference polyaniline film, (**c**) after 36 days of exposure to the influence of *A. niger* L-10 and (**d**) reference polyaniline film (potential cycling between $E = -0.1$ and 1.1 V, 0.02 Vs^{-1}), vs. Ag/AgCl (From Binkauskiene et al. 2008)

made it possible to evaluate the polymer electroactivity (Ruotolo and Gubulin 2003). The CVs electrochemical characteristics of polyaniline treated with the some fungi can be seen in the Table 6.1.

3.2 Redox Activity of Biomodified Polyanilne Surface

Polyaniline shows an electrocatalytic activity towards the oxidation of one of the most thoroughly investigated organic species hydroquinone to quinone (Rajendra Prasad and Munichandraiah 2002; Förch et al. 2009). Often oxalic acid constitutes the final product in the oxidative process of organics. Whereas many of the by-products of microbial metabolism, including oxalic acid are damaging, the studies of the redox behavior of the hydroquinone/quinone couple are extremely important (Polcaro et al. 2010).

Ours investigation (Binkauskiene et al. 2013) confirm a conspicuous difference in the oxidation levels and an effect towards hydroquinone electrooxidation at the polymer surface layer treated by investigated fungi (Fig. 6.7). For example, *Ch. merdarium* shows the bioelectrocatalytic effect on polyaniline electrochemical response (Fig. 6.7, curve 4). *Rh. pusillus* shows a suppression effect towards hydroquinone electrooxidation (Fig. 6.7, curve 10).

4 Structural XPS Analysis

The XPS results confirmed that the colonization of fungi on the polyaniline surface induced changes in the elemental composition in the outermost layer of polyaniline surface (Binkauskienė et al. 2008, 2009, 2013). The most significant feature resulting from mycobial colonization of the surface was the depletion of carbon and nitrogen and the enrichment of oxygen content as compared to untreated polyaniline films. The attachment of polyaniline surface via the metabolic product shows a noticeable difference in the O/C ratios between untreated and treated polyaniline. The carbon and oxygen spectrum for treated surface display three peaks (Fig. 6.8). C3 peak is attributed to the carbon atoms C_{ox} of the oxalate. The O2 peak shows an O_{ox} of the oxalate groups (Table 6.2). C3/O2 ratio corresponds to the theoretical C_{ox}/O_{ox} ratio of the oxalate ion.

The results show, that *Rh. pusillus* is a high level oxalic acid accumulator. During growth of *Ch. merdarium* on polyaniline no accumulation of oxalate has been observed. It can be explained by a general physiological trait of brown-rot and white-rot basidiomycetes that do not accumulate oxalic acid but metabolize and/or decompose it (Gray and Goddard 2012).

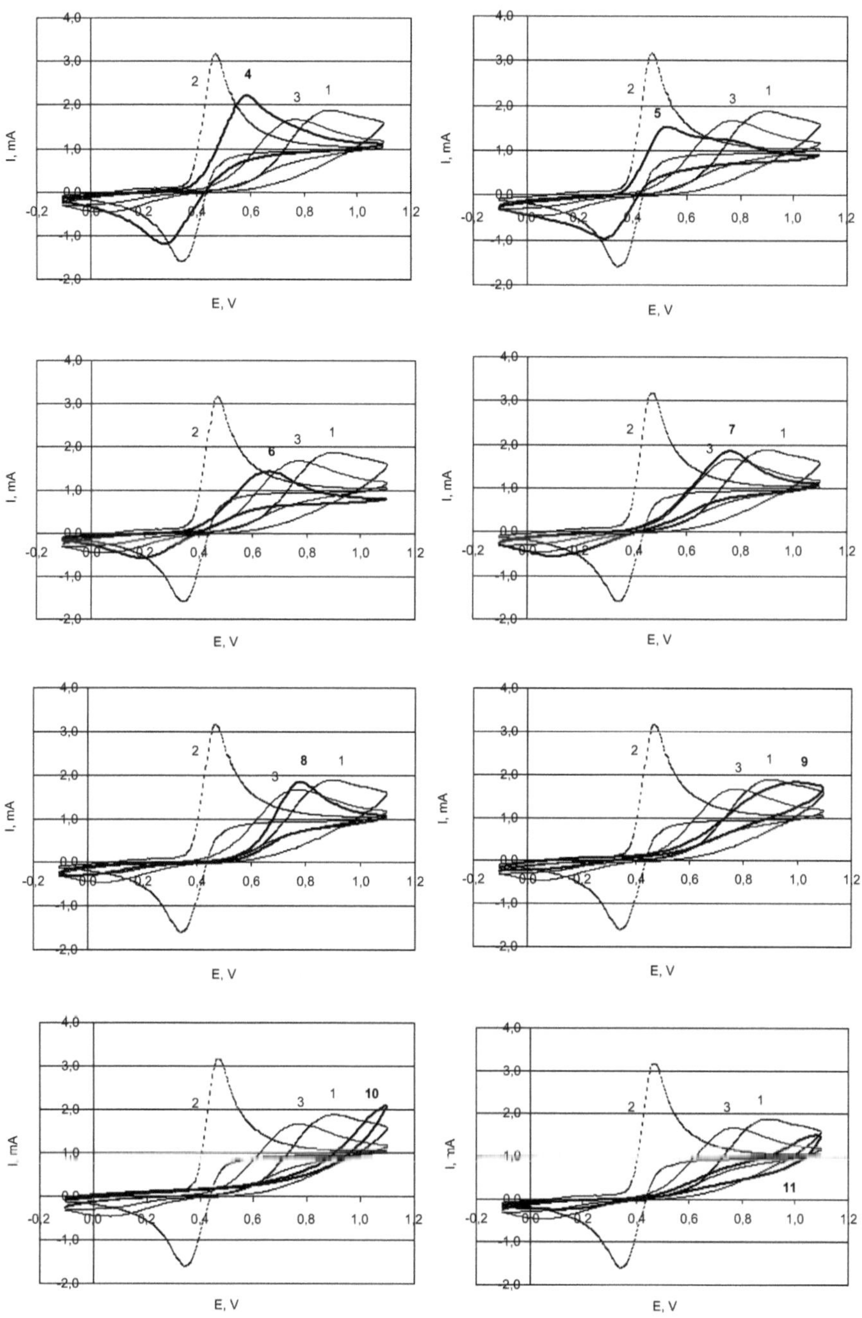

Fig. 6.7 Electrochemical response for polyaniline to hydroquinone/quinone redox in 0.1 M solution of hydroquinone in 0.05 M H_2SO_4: (*1*) bare nickel, (*2*) freshly polyaniline modified nickel, (*3*) reference polyaniline; treated polyaniline with: (*4*) *Ch. merdarium*, (*5*) *Ch. pannorum*, (*6*) *A. radicina*, (*7*) *B. cinerea*, (*8*) *Trichoderma* spp., (*9*) *A. alternata*, (*10*) *Rh. pusillus*, (*11*) *Penicillium* spp. (potential cycling between $E = -0.1$ and 1.1 V, 0.02 Vs^{-1}) vs. Ag/AgCl (From Binkauskiene et al. 2013)

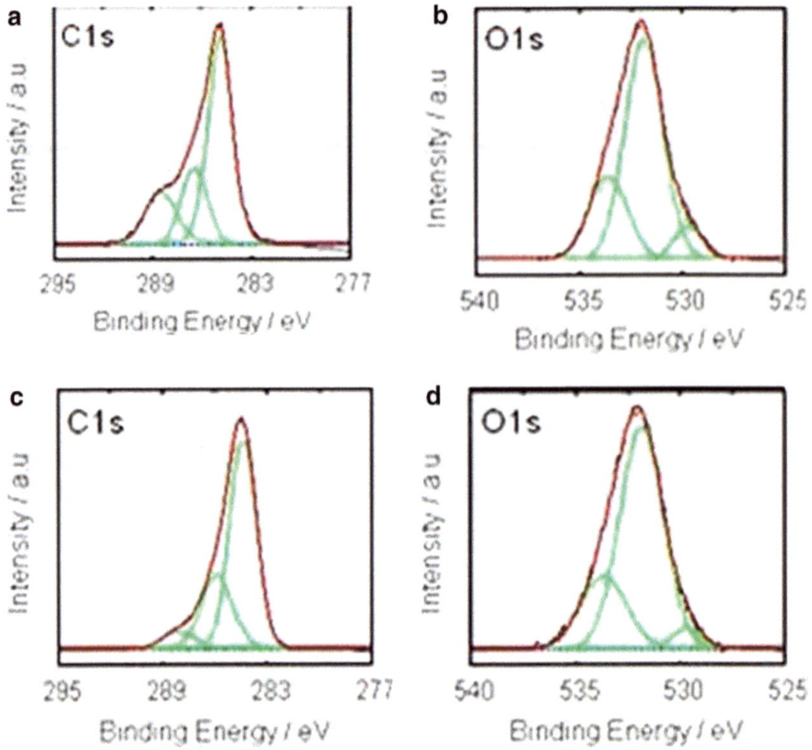

Fig. 6.8 XPS spectra of C1s and O1s of PANI films: (**a**), (**b**) treated with Rh. pusillus and (**c**), (**d**) treated with Ch. merdarium (the biofilms were removed)

Table 6.2 C1s and O1s peak areas (at.%) of biologically-treated PANI samples

	C1s (at.%)			O1s (at.%)		
Sample	C1	C2	C3	O1	O2	O3
Rh. pussilus (after 56 days of exposure)	36.55	13.58	11.96	2.64	23.61	8.98
Ch. merdarium (after 56 days of exposure)	52.49	22.07	5.31	0.62	10.19	3.50

References

Arsov LD, Plieth W, Koβmehl G (1998) Electrochemical and Raman spectroscopic study of polyaniline; influence of the potential on the degradation of polyaniline. J Solid State Electrochem 2:355–361. doi:10.1007/s100080050112

Binkauskiene E, Binkauskas BJ (2009) Electroreduction of hexavalent chromium by polyaniline. Polym Int 58:869–871. doi:10.1002/pi.2601

Binkauskienė E, Jasulaitiene V, Lugauskas A (2008) The influence of *Aspergillus niger* Tiegh. on the electro chemical characterisation of a polyaniline modified Ni electrode. Mater Sci (Medziagotyra) 14:134–137

Binkauskiene E, Jasulaitienė V, Lugauskas A (2009) Effect of *Aspergillus niger* Tiegh. L-10 on the physical and chemical properties of a polyaniline coating in the growth substrate. Synth Met 159:1365–1368. doi:10.1016/j.synthmet.2009.03.011

Binkauskienė E, Lugauskas A, Bukauskas A (2013) The mycological effect on morphological, electrochemical and redox properties of the polyaniline surface. Surf Interface Anal 45:1792–1798. doi:10.1002/sia.5324

Cui CQ, Ong LH, Tan TC, Lee JY (1993) Origin of the difference between potentiostatic and cyclic potential sweep depositions of polyaniline. J Electroanal Chem 346:477–482. doi:10.1016/0022-0728(93)85032-C

Förch R, Schönherr H, Tobias A, Jenkins A (2009) Surface design: applications in bioscience and nanotechnology. Wiley-VCH. doi:10.1002/9783527628599.fmatter

Gray JC, Goddard MR (2012) Gene-flow between niches facilitates local adaptation in sexual populations. Ecol Lett 15:955–962. doi:10.1111/j.1461-0248.2012.01814.x

Inzelt G (2012) Conducting polymers: a new era in electrochemistry. Monographs in electrochemistry. Springer-Verlag, Berlin/Heidelberg. doi:10.1007/978-3-642-27621-7

Inzelt G, Pineri M, Schultze JW, Vorotyntsev MA (2000) Electron and proton conducting polymers: recent developments and prospects. Electrochim Acta 45:2403–2421. doi:10.1016/S0013-4686(00)00329-7

Langer K, Barczynski P, Baksalary K, Filipiak M, Golczak S, Langer SS (2007) A fast and sensitive continuous flow nanobiodetector based on polyaniline nanofibrils. Microchim Acta 159:201–206. doi:10.1007/s00604-007-0767-2

Lugauskas A, Leinartas K, Griguсevičienė A, Sielskienė A, Binkauskienė E (2008) Possibility of micromycetes detected in dust to grow on metals (Al, Fe, Cu, Zn) and on polyaniline-modified Ni. Ecology 54:149–157. doi:10.2478/V10055-008-0023-z

Lugauskas A, Prosyčevas I, Selskienė A, Demčenko I, Narkevičius A, Bučinskienė D, Binkauskienė E (2010) The influence of microscopic fungi on chromated galvanized zinc coatings. Eng 2:979–997. doi:10.4236/eng.2010.212125

Polcaro AM, Mascia M, Palmas S, Vacca A (2010) In: Comninellis C, Chen G (eds) Electrochemistry for the environment. Springer, New York/Dordrecht/Heidelberg/London, pp 205–227. doi:10.1007/978-0387-68318-8

Rajendra Prasad K, Munichandraiah N (2002) Electrocatalytic efficiency of polyaniline by cyclic voltammetry and electrochemical impedance spectroscopy studies. Synth Met 126:61–68. doi:10.1016/S0379-6779(01)00496-9

Ruotolo LAM, Gubulin JC (2003) Reduction of hexavalent chromium using polyaniline films. Effect of film thickness, potential and flow velocity on the reaction rate and polymer stability. J Appl Electrochem 33:1217–1229

Stejskal J, Trchová M, Bober P, Humpolíček P, Kašpárková V, Sapurina I, Shishov MA, Varga M (2015) Conducting polymers: polyaniline. Encyclopedia of polymer. Science and Technology. Wiley, Hoboken, pp 1–44. doi:10.1002/0471440264.pst640

Chapter 7
Synthesis Techniques and Evaluation Methods of Nanoparticles as Fungicides

Youn Su Lee and Ahmed I.S. Ahmed

Abstract The scientists' attention concentrated on developing the materials for addition and desirable properties especially in nano-level because the materials allow improving their properties significantly. Nanostructured materials are involved in all fields of science with great applications in Agriculture. Among the widely used of these different types of nanoparticles in investigations to control a lot of fungal plant pathogens, there are different methods to prepare each type, also many methods and techniques followed to evaluate it as control agents against some plant pathogens. Also engineering of nanoparticles is the backbone of nanotechnology with very fast development and growing. This chapter focused on the methodology of nanoparticles preparation and its evaluation as pesticides in plant pathology researches

1 Classification of Nanoparticle Types

There is growing attention and increasing research activities in the behavior, fate, also environmental outcomes of the nanomaterial's. Information about the nanoparticles (NPs) synthesis and preparation will increase our understanding of the nanomaterials behavior in the environment which is essential to study any effects due to the NPs application in the environment especially with living organisms such microorganisms. Depend on the origin, NPs can be classified into groups (natural and anthropogenic or artificial), while the artificial group can be subdivided into manufactured and accidental (Figs. 7.1 and 7.2). On the other hand, NPs can be divided according to their chemical structures into organic and inorganic (Nowack and Bucheli 2007).

Y.S. Lee
Division of Bioresource Sciences, Kangwon National University,
Chuncheon 200-701, South Korea
e-mail: younslee@kangwon.ac.kr

A.I.S. Ahmed (✉)
Plant Pathology Unit, Plant Protection Department, Desert Research Center,
Cairo 11753, Egypt
e-mail: Ahmed_drc@yahoo.com

R. Prasad (ed.), *Advances and Applications Through Fungal Nanobiotechnology*, Fungal Biology, DOI 10.1007/978-3-319-42990-8_7

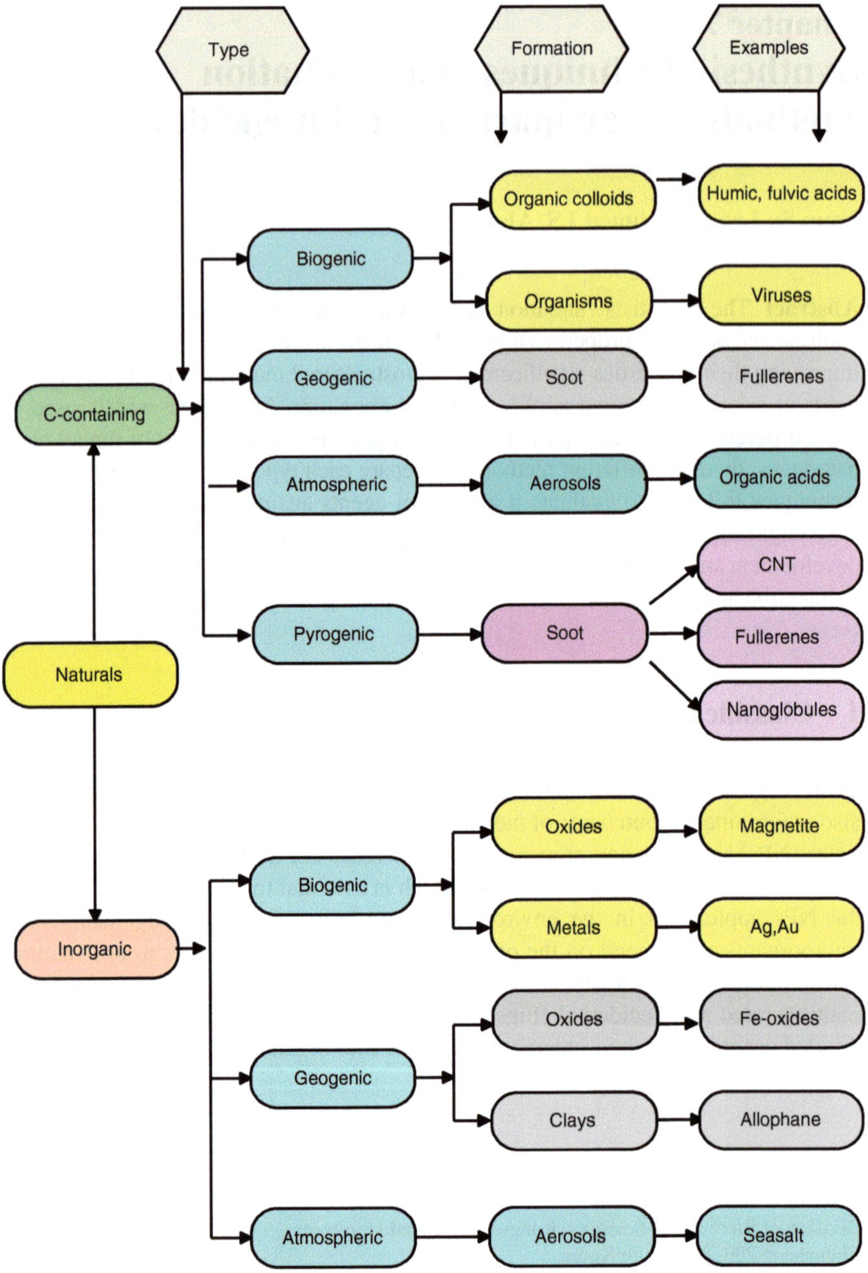

Fig. 7.1 Classification of natural nanoparticles

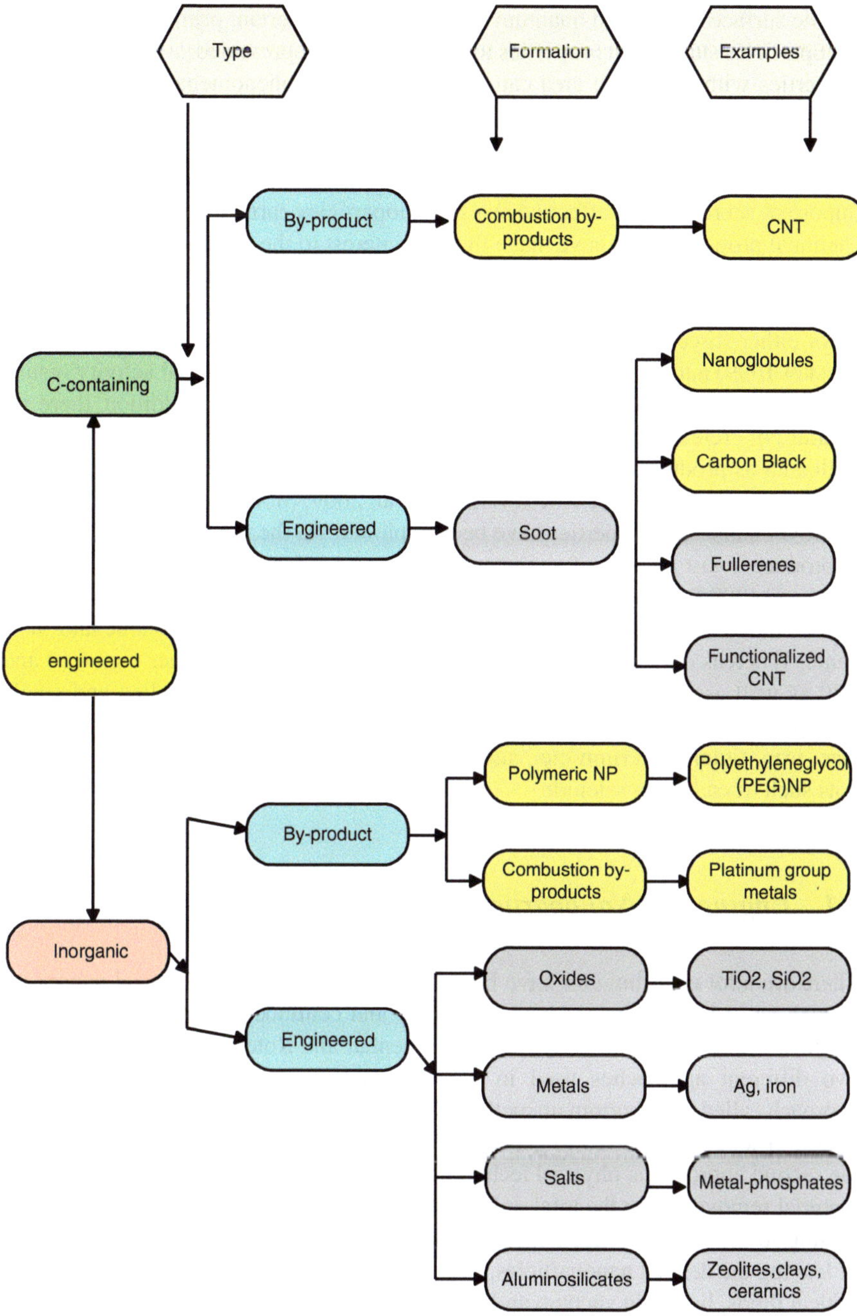

Fig. 7.2 Classification of manufactured nanoparticles

The surface effects and quantum effects are two important primary factors make nanoparticles superior in properties to bulk materials represented, where the surface properties with very high area can enhanced surface phenomena to volume ratio that help the atoms are on or near the surface and more reactive, and the quantum effects due the small dimensions of electrons leading to delocalized on the surface of the NPs as discontinuity behavior (Roduner 2006). These factors are the most important secrets in the effectiveness of nanoparticles earned a lot of the physico-chemical properties that have made them of interest to the scientific community.

Natural NPs are distributed throughout the atmosphere and found in many environments, also microorganism could be prepare the natural nanoparticle through to ways either directly to make the metabolic requirements available (Schüler and Frankel 1999) and (Suzuki et al. 2002) or indirect way of microbial activity results (Banfield et al. 2000; Glasauer et al. 2002; Hansel et al. 2004). In addition, there are natural NPs releasing due to the human activities as a nano particulate matter for millennia as products of some activities such as agriculture, construction, mineral processing and mining activities (Wiesner et al. 2009; Sioutas et al. 2005).

Novel nanoscale properties have been exploited via the modern nanotechnology to produce a vast amount of engineered NPs to serve for special purposes. Ju-Nam and Lead (2008) mentioned that manufactured NPs can be clustered into some of classes based on their core materials, these where classified into organic and inorganic. Whereas organic NPs defined in to groups included fullerenes e.g., C60 and C70 as well as derivatives, and nanotubes of carbon like multi walled and single walled CNTs, on the other hand, inorganic NPs were divided into metal oxides like zinc, iron, titanium cerium etc., and metals mainly silver and gold, also quantum dots such as cadmium selenide.

1.1 Engineered Nanoparticles

There are a lot of techniques have been employed to generate the metal nanoparticles, these techniques have advanced rapidly and continue to improved control the shape and size of the generated particles. Shenhar and Rotello (2003) mentioned to two different approaches used in nanostructures generation included chemical approach called (The bottom up method), where the atoms that composed from diminution of ions are amassed to generate nanostructures and the adverse approach or the second approach is physical technique called the top down method, whereas the material removed from the total or bulk structure through disintegrating or grinding milling so the vapors components proceed from the chemical methods condensation of leaving the desired nanoparticles. These two methods implemented in gas, supercritical fluids, liquid, in vacuum and as solid states. Also many aspects are important in the ability to control it in particles such as (size, shape, distribution, composition and degree of agglomeration). The stabilization of the particles considered one of the important aspects to avoid coalescence and aggregation and it can occur in many different ways (Ju-Nam and Lead 2008). The synthesis of small nanoparticles is more difficult and the capping is important in synthesis process to keep the particles in dispersed status to facing the increase of forces surface energy (Sardar et al. 2009). Top down assembly methods are superior for the possibility of interconnection and integration, while in electronic circuitry. Bottom up assemblage is influential and authoritative in manufacturing identical nanostructures with the precision of atomic, especially in living organisms, both of them can be combined to prepare nanoparticles with specific physicochemical properties (Cigang et al. 2006).

Many types of nanomaterial have been produced for different applications and commercial activities. Metal oxides are important inorganic nanoparticles due to the novel optical, magnetic and electrical properties (Grancharov et al. 2005; Shtykova et al. 2007). Metal oxides NPs used in many applicable activities including sensors, catalysis, biomedical diagnostics, environmental remediation and electronic materials (Kamat 1993; Hoffmann et al. 1995; Oskam 2006; Prasad et al. 2015). The following will highlight the most common NP types (Iron oxide, Titanium oxide, Cerium oxide, Zinc oxide) and metal NPs (Silver, Gold).

1.2 Iron Oxide

There are two most common forms of iron oxide NPs (γ-Fe_2O_3 and magnetite Fe_3O_4) and they have offer a high potential for several biomedical applications due to their super-magnetic properties, this applications including tissue repair; magnetic resonance imaging, drug delivery and hyperthermia (Widder et al. 1978; Arbab et al. 2003; Perez 2007) but the investigation of its effects against microorganisms still limited.

1.3 Titanium Oxide

Titanium dioxide has four naturally crystal forms: brookite, rutile, anatase and TiO_2, it can be synthesized using some important techniques: co-precipitation, sol–gel synthesis process, chemical vapor deposition, reverse micelle synthesis, microemulsion synthesis process and hydrothermal reaction method (Wang et al. 2005; Kim et al. 2006; Ramaswamy et al. 2008; Kawai-Nakamura et al. 2008; Li et al. 2008; Rashidzadeh 2008). It is important that the changing behavior from the amorphous to the anatase phase is influenced by the manufacturing conditions. NPs are exploited by a variety of fields due to their attractive properties as a high refractive index, light absorption/scattering especially with low-cost production of chemical relatively titanium dioxide, (Baldassari et al. 2005; Kim et al. 2007). Titanium oxide has a wide range of application such as pigments, cosmetics, catalysts and photocatalysts (Pratsinis et al. 1996; Chen and Yang 1993; Rao and Dube 1996).

1.4 Zinc Oxide

Zinc oxides in nanoscale have received the scientific attention in recent years due to their properties and its ability to be applicable in many fields (Müller and Weißenrieder 1994; Makino et al. 2002; Wang and Song 2006) and have important role as antimicrobial agents (Jones et al. 2008; Bhuyan et al. 2015). A lot of publications mentioned to several synthesis methods, such as chemical vapor deposition, thermal decomposition, spray pyrolysis, laser ablation, sol-gel method, molecular beam epitaxy, hydrothermal synthesis, etc. have been studied for the synthesis of ZnO structures in addition to the changes of surface using organic compounds have also showed the applications or chances to use new ways of zinc oxide NPs applications (Tarasenko et al. 2010; Guo et al. 2000).

1.5 The Metal NPs

Nanoscale metals can be manufactured in many shapes and sizes dependent on the desired properties which make the particles different from both bulk and atomic structure of materials (Daniel and Astruc 2003). Some important applications exploit many unique nanoscale aspects (Jana et al. 1998; Mirkin et al. 1996), antimicrobial uses to water purifications or wastewater treatment (Aziz et al. 2015). In the different ways employed in manufacturing of NPs metal a lot of methods are applied to tune the size and shape of particles into a specific target and purpose (Lee et al. 2001).

2 Concept of Synthesis and Characterization of Nanoparticles

Nanoparticles have created a high interest in recent years by virtue of their unusual mechanical, electrical, optical and magnetic properties where nano structures having their largest dimension in between ‘0’ to ‘100’ nm (Suman et al. 2010; Prasad 2014; Prasad et al. 2014). These are called nano powders and finding their wide applications in all fields of engineering and biology. Many techniques, including both top-down and bottom-up approaches, have been developed and applied for the synthesis of nanoparticles. There are two ways to manufacturing of nanostructures; such as (bottom-up. and top-down), Top down methods means the breaking down of the bulk material into nano particles. An example of such a technique is high-energy wet ball milling. The other approach is the ‘bottom- up’ that refers to create of a material atom-by-atom, molecule-by-molecule as showed in (Fig. 7.3)

The creation of small size is not all requirements. But also the identical size of all particles must be have a similarity or uniform size distribution that mean the NPs

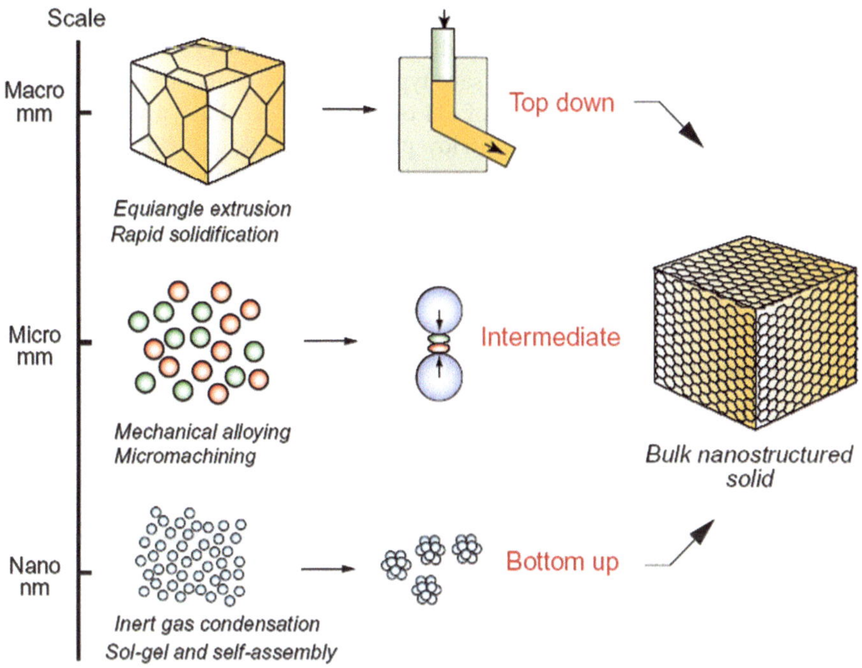

Fig. 7.3 The 'bottom up' and top down' synthesis processes of nanomaterials with the popular techniques (*Source*: http://article.sapub.org/10.5923.j.nn.20130303.06.html)

are mono sized also the identical shape or morphology and identical chemical composition as well as crystal structure that are desired among and within individual particles such as composition and core must be the same and avoiding the agglomeration (Cigang et al. 2006).

This part deals with some of the common processes that have been used in recent years and some laboratory steps to preparation of NPs especially in biological laboratories.

2.1 *Silver Nanoparticles*

Nanoparticles silver among all different types of metal is the most rapidly growing application due and has applications in some of scientific fields and in products (Tolaymat et al. 2010). The manufacturing of silver NPs Well can be achieved through different approaches using numerous capping agents depending on desired applications, the common approaches in publication are photochemical methods such as microwave processing, laser ablation, thermal decomposition of silver oxalate, reduction of organic and inorganic agents and radiation (Lee et al. 2001; Navaladian et al. 2006; Bönnemann and Richards 2001; Pulit et al. 2013).

Also the silver NPs one of most save types where due to its fate and residue in the environment due to the effect of pH on the Ag NPs which size transformed either silver ions or soluble silver with low pH aqueous nitric acid with similar acidity to natural conditions (Elzey and Grassian 2010). Due to the surface properties in preparation of particles the stability and mobility of NPs determined either in aquatic or terrestrial systems also in the interactions process with the living organisms (Navarro et al. 2008)

2.1.1 Preparation of Silver Nanostructures

And colleagues used three methods to synthesize different-sized AgNPs, in typical procedure:

The first method according to Feng et al. (2000):

The spherical AgNPs were prepared by reducing the aqueous $AgNO_3$ and sodium citrate under boiling temperature.

- Heating the 50 ml of 0.001 M $AgNO_3$ to boiling temperature.
- Add 5 ml of 1 % trisodium citrate as drop by drop.
- Heating the solution under continuous stirring at boiling point.
- The reaction was allowed to take place until the color changed to a greenish yellow solution.
- The solution was then cooled to room temperature.
- The AgNPs in this solution were called citrate-AgNPs.

The second method, preparation of spherical silver nanoparticles by reducing aqueous silver nitrate with $NaBH_4$, and the production of nanostructures using this method named borohydride-AgNPs.

- Adding the aqueous solution of $NaBH_4$ (0.002 M).
- Stirring and cooling the solution for 20 min.
- Two milliliters of 0.001 M $AgNO_3$ was dropped into the stirred $NaBH_4$ solution at approximately 1 drop per second, all of the $AgNO_3$ added stop the stirring of the solution.

The third method, AgNPs was synthesized by using $NaBH_4$ and PVP as reducing and stabilizing agents, respectively.

- Add the solution of trisodium citrate (0.5 ml, 30 mM) into flask contained 50 ml deionized water.
- Add $AgNO_3$ aqueous solution (1 ml, 5 mM).
- add the freshly $NaBH_4$ aqueous solution prepared (0.5 ml, 50 mM) quickly,
- Suspension immediately turned into a light yellow color.
- Add the aqueous solution of PVP (0.5 ml, 5 mg/ml, Mw = 40,000) After 30 s.

- The suspension will changed into a dark yellow color after 30 min of reaction proceeding and mark the silver particles using PVP-AgNPs.
- Following the procedure of Torres and coworkers the triangular particles or triangular-AgNPs could be synthesized.
- Preparing of PVP: $AgNO_3$ precursor solution, which including solution of $AgNO_3$ (1 ml, 5 mM) in 50 ml deionized water with the PVP presence (3 ml, 32 mg, Mw=40,000) as the capping agent.
- Add the sodium citrate (3 ml, 30 mM) and hydrogen peroxide to the solution under stirring conditions (400 rpm).
- The reaction took place in the dark.
- After 30 min, the solution will change from yellow to a light blue.
- The reaction was allowed to continue for 5 h.

2.1.2 Biologically Methods to Produce of Silver Nanoparticles

Different organisms have been used for the biosynthesis of nanoparticles like plants, bacteria, yeast and fungi (Pallub et al. 2011; Forough and Farhadi 2010; Chen and Yang 1993; Krishnaraj et al. 2012; Souza et al. 2004) but fungi has advantage over other organisms like simple to handle, eco-friendly, requires very less time, produces the large amount of extracellular enzymes and can uptake the metal ion as compared to physical and chemical method (Prasad et al. 2015).

2.1.3 Method to Biosynthesis of Ag Nanoparticles Using Fungi (Khabat et al. 2011)

- Mixing 10 g of endophytic fungus wet biomass with a 100 ml aqueous solution of 1 mM silver nitrate ($AgNO_3$).
- Placing the mixture at 28 °C for 120 h duration in rotating shaker (100 rpm).
- In this process silver nanoparticles will produce through reduction of the silver ions to metallic silver.
- Using visual inspection of the solution and measuring the UV–Visible spectra of the solution by periodic aliquots sampling (2 ml) of the component to monitoring the reduction of silver ions.
- Record the UV–Vis spectroscopy measurements on a Shimadzu dual beam spectrophotometer with operation at 1 nm resolution.
- Using of Perkin-Elmer LS 50B luminescence spectrophotometer, carry the fluorescence measurements out.
- To perform Fourier transform infrared spectroscopic (FTIR), the films of nanoparticles will produce on substrates by drop-coating the metal nanoparticle solution.

Using this biosynthesis method the silver nanoparticles have been introduced extracellulary in different fungal genus and species e. g., *Trichoderma reesei* (Khabat et al. 2011; Khandelwal et al. 2010), *Aspergillus clavatus* (Verma et al. 2010;

Vigneshwaran et al. 2007); *Cladosporium cladosporioides* (Balaji et al. 2009); Filamentous fungus *Penicillium* sp.*; Fusarium oxysporum* (Durán et al. 2005; Senapati et al. 2005; Souza et al. 2004; Oksanen et al. 2000); *Fusarium semitectum* (Basavaraja et al. 2008); *Verticillium* species (Medentsev and Alimenko 1998), *Aspergillus fumigatus*; and *Neurospora crassa* (Castro-Longoria et al. 2011).

Many plant pathogenic fungi which affected and suppressed by Ag NPs include *Alternaria alternate, Rhizoctonia solani, Botrytes cinerea, Curvularia lunata, Sclerotinia sclerotiorum, Macrophomina phaseolina* using NPs size 10, 50 nm (Krishnaraj et al. 2012), *Cladosporium cladosporoides* with 60 NPs size; *Alternaria alternate; Alternaria brassicicola; Alternaria solani; Botrytis cinerea; Cladosporium cucumerinum; Corynespora cassiicola; Cylindrocarpon destructans; Didymella bryoniae; Fusarium oxysporum* f. sp *cucumerinum.; F. oxysporum* f. sp. *lycopersici; F. oxysporum; Fusarium solani; Fusarium sp; Glomerella cingulate; Monosporascus cannonballus; Pythium aphanidermatum; Pythium spinosum; Stemphylium lycopersici* using 7–25 nm size of Ag NPs (Kim et al. 2012). *Golovinomyces cichoracearum, Sphaerotheca fusca* by 7–25 nm size (Lamsal et al. 2011); *Colletotrichum* spp. also; *Sclerotium cepivorum* by 7–25 nm (Jung et al. 2010); *Sclerotinia sclerotiorum; Rhizoctonia solani, Sclerotinia minor* using 4–8 nm (Lamsal et al. 2011; Min et al. 2009). *Pythium altimum, Mangnoprthe grise, Colletotricum gloeosporiodes, Botrytes cinerea* by 20, 100 nm size (Park et al. 2006); *Fusarium culmorum* by 15, 100 nm (Marek et al. 2013); *Fusarium culmorum* using 5, 10 nm (Marek et al. 2010); *Colletotrichum gloesporioides* by 5, 24 nm (Miguel et al. 2011) and *Aspergillus fumigatus* using 420 nm size (Fig. 7.4) (Khalil Neveen 2013).

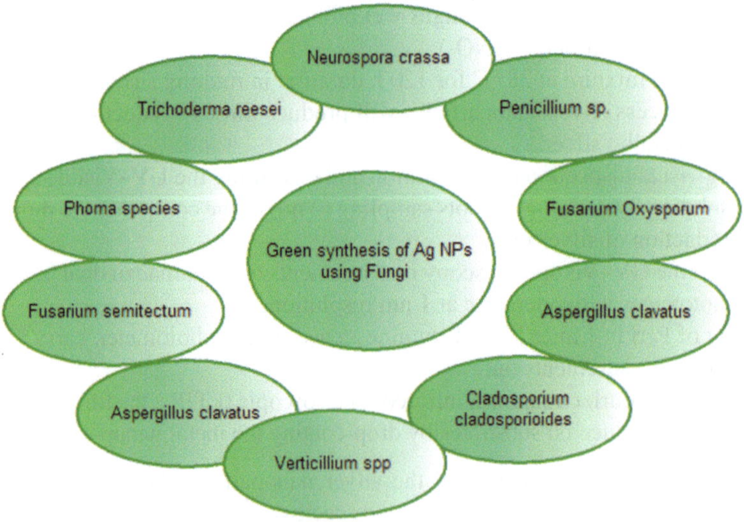

Fig. 7.4 Examples for the fungal species used for synthesis of Ag NPs

3 Sulfur NPs

Sulfur nanoparticles have many practical applications especially as nanoparticles. It was prepared by different method included chemical precipitation, electrochemical method, composing of oil, micro emulsion technique, surfactant, co-surfactant, aqueous phases with the specific compositions and ultrasonic treatment of sulfur-cystine solution. Sulfur nanoparticles are very important application for antibacterial, anticancer, pharmaceuticals, fertilizers, fiber manufacturing, nano carbon tubes modification. For the applications of sulfur in agriculture area, sulfur can be used as fungicide against many plant diseases such as the apple scab disease in the cold conditions, also S used in the culture of grapes, vegetables, strawberry and many cultivated plants (Suleiman et al. 2013). However; S can be considered as a high efficiency pesticide that used in agriculture where it has good effect against a wide range of powdery mildew diseases as well as black spot

3.1 Methods for S-NPs Preparation

According to Suleiman et al. (2013), the S-NPs can be synthesized by different methods as follow:

- Wet chemical precipitation method it used to prepare the S-NPs by dissolving the sodium thiosulpahte in double distilled water and different acid solutions, using different surfactants (TX-100, CTAB, SDBS, and SDS) as the stabilizer effect on the particle size. The anionic surfactant SDBS is higher effective for obtaining a uniform size in both the acid media. The lowest size of particles is (30 nm) were obtained in a certain reactant concentration range using CTAB surfactant.
- S-NPs could be manufactured from H_2S gas by using novel chelates of biodegradable iron in water/organic micro emulsion system. Fe^{3+} malic acid chelate and studied it in w/o micro emulsion including cyclohexane, *n*-hexanol and triton X-100 in oil phase, co surfactant and surfactant for catalytic oxidation of H_2S gas at temperature ambient conditions, neutral pH and pressure, the morphology of sulfur nanoparticles synthesized is nearly undetected in size as an average size of particle 10 nm, and distribution of narrow particle size in range of 5–15 nm, as compared to aqueous surfactant systems.
- Monoclinic sulfur nanoparticles have been prepared via the chemical reaction between sodium polysulfide and hydrochloric acid in a reverse micro emulsions system, with theolin, butanol, and a mixture of Span 80 and Tween 80 (weight ratio 8:1) as the oil phase, co-surfactant and surfactant, respectively. Transparent micro emulsions were obtained by mixing the oil phase; surfactant, co-surfactant, and the aqueous phase in appropriate proportion using an emulsification machine at the room temperature, the sulfur nanoparticles prepared via this method have an average diameter of about 20 nm, a narrow size distribution, uniform spherical shape, and high purity.

- An electrochemical method is used to prepare the sulfur nanoparticles from thiosulfate ion. The particle size of the S-NPs can be adjusted between 35 and 65 nm by adjusting the operation parameters including the initial sodium thiosulfate. in this case , the used of hot alcohol and cold water as solvent/non-solvent system along with 100 ml · min^{-1} flow rate for co-mixing of non-solvent resulted in the formation of S-NPs in a typical size of 250 nm that are fairly homogeneous in shape and have a narrow particle size distribution. To dissolve sulfur, the Dimethyl sulfoxide DMSO was used as the solvent. Solution processing was developed to prepare conductive sulfur/carbon nanocomposites for electrochemical used as in Fig. 7.5.

4 Chitosan and Preparation Nanoparticles

Chitin and chitosan are naturally occurring compounds that have important role in plant diseases control due to its toxicity and inhibition effects on fungal growth and development. Chitin and chitosan are known to have eliciting activities leading to enhance of defense in host plants as a response to microbial attack by the accumulation of pathogen related (PR) proteins, proteinase inhibitors and phytoalexins.

Zhu et al (2012) pointed to the simple and new method to prepare Chitosan nanoparticles in vitro as follow: Using of (low molecular weight, 20–300 cp (centipoise; 1 % in 1 % acetic acid), and degree of deacetylation 75–85 %), TGA (thioglycolic acid ≥98 %), Ellman's reagent (5,5′-dithiobis(nitrobenzoic acid), ≥98 %), and mucin (from porcine stomach, Type III, bound sialic acid 0.5–1.5 %, partially purified powder), St Louis,

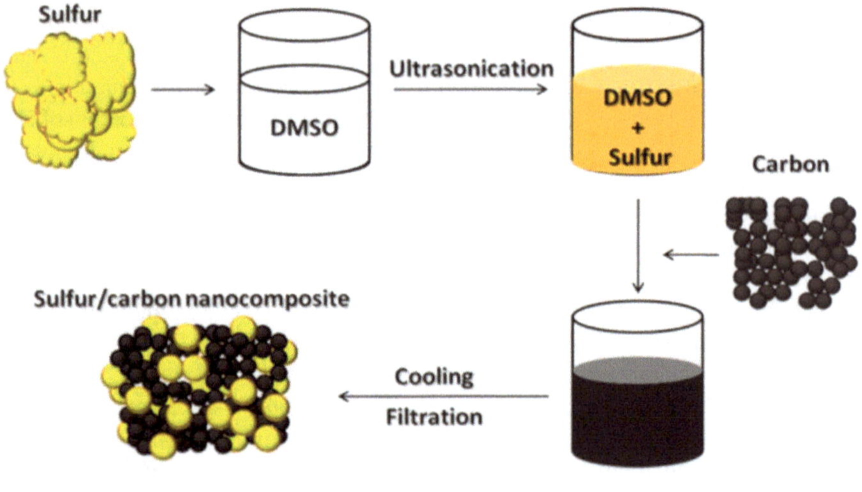

Fig. 7.5 Illustration of the solution processing for the preparation of sulfur/carbon (*Source*: Suleiman et al. 2013)

MO. EDAC·HCl (*N*-(3dimethylaminopropyl)-*N*′-ethylcarbodiimide hydrochloride, 99 %) and NHS (*N*-hydroxysuccinimide, 98 %). FITC (fluorescein isothiocyanate, isomer 1, 95 %); DMF (*N,N*-dimethylformamide) and SA (sodium alginate of low viscosity (0.02 Pa·s) for a 1 % solution at 20 °C). All other chemicals in reagent grade.

5 Synthesis of Thiolated chitosan (TCS)

There is a new method includes two steps.

The First Process

- Added volume of (1 ml of TGA, 3500 mg of EDAC·HCl and 2000 mg of NHS) to 2 ml DMF in flask with overnight constant stirring.
- After completion of the reaction the reactive NHS-ester will produce.

In the Second Process

Hydrate 500 mg of chitosan in 4 ml of 1 M HCl and dissolve it by the demineralized water addition to 2.5 % of chitosan hydrochloride solution.

- Add the reactive NHS-ester drop wise into solution of chitosan hydrochloride with pH maintained at 5 by using of 10 M NaOH.
- Incubating of the mixture under room temperature and overnight continuous stirring.
- The step of TCS isolation using dialyse of the polymer solutions exhaustively in tubing with molecular weight cut off 12 kDa and dialysis tubings as well as membrane of cellulose against 5 mM HCl, then three times against 5 mM HCl including 1 % NaCl, finally, three times against 1 mM HCl under 8 °C in the dark conditions.
- The controls were synthesized in the same way but omitting EDAC·HCl and NHS.
- Lyophilization of the frozen aqueous polymer solutions for each of samples and controls (FreeZone, Labconco, Kansas City, MO), then stored at 4 °C for further use.

5.1 Determination of the Thiol Groups in TCS by Ellman's Method (Zhu et al. 2012)

Using Ellman's reagent, the degree of thiol group substitution in the modified polymers was determined spectrophotometrically.

Ellman's reagent or 5,5′-dithiobis (2-nitrobenzoate) (DTNB) is a symmetric aryl disulfide, which is very sensitive to the reaction of the thiol-disulfide interchange with a free thiol.

- Initially, 5 mg of each of the conjugates and controls dissolved in ultrapure water (PURELAB Classic UVF, ELGA LabWater, High Wycombe, Buckinghamshire, UK) to prepare 2 mg/ml solution.

- 250 μl aliquots were added to 250 μl of 0.5 M phosphate buffer (pH 8.0) and to 500 μl of Ellman's reagent (0.4 mg/ml of DTNB in 0.5 mol/l phosphate buffer, pH 8.0).
- The sample was incubated at room temperature and shielded from light for 3 h.
- The resulting solution was then centrifuged at 18,000 × g for 10 min.
- Transfer 200 μl of the supernatant to a microtitration plate and measure the absorbance at a wavelength of 450 nm with a Multimode Reader (Infinite M200; Tecan Group Ltd., Männedorf, Switzerland).
- A blank control was created with non-modified chitosan.
- A TGA standard curve was established between 0.125 and 2 μmol/ml, precisely following the same protocol used for sample determination.

5.2 Preparation of CS/TCS-SA NPs

According to Zhu et al. (2012) they prepared of CS or TCS with SA nanoparticles as follow:

- CS was dissolved in 1 % acetic acid aqueous solutions at various CS concentrations: 0.5, 1.0, 1.5 mg/ml,
- The pH was adjusted using 1 M NaOH to pH 5.5. In contrast
- TCS was easily soluble in aqueous solutions below pH 5
- Ultrapure water used as the solvent for TCS.
- SA was then dissolved in ultrapure water at concentrations of 0.625 or 1.25 mg/ml, using 1 M HCl to adjust the pH to 5.3.
- All the solutions of the materials were purified through a 0.45 μm microporous membrane filter.
- NPs were obtained by addition of SA solution (2 ml, pH 5.3) to aqueous solutions of polymers CS or TCS (8 ml) under magnetic stirring at room temperature.
- Samples were then visually analyzed for three kinds of phenomena: clear solution, opalescent suspension, and aggregates.
- The opalescent suspension was the objective when preparing the mucoadhesive nano particulate system.
- To optimization of CS/TCS-SA NPs, Nanoparticles were formed by interaction between the negative groups of alginate and the positively charged amino groups of CS or TCS. High concentrations of CS/TCS (1.5 mg/ml) and SA (2.5, 1.25 mg/ml) resulted in the formation of a high concentration of fibrous aggregates.

6 Preparation of Copper Nanoparticles

Copper nanoparticles play an important role in control of some fungal and bacterial plant pathogens due to its antifungal activity of copper nanoparticles against many of crop pathogenic fungi. Copper nanoparticles were synthesized by chemical reduction of Cu^{2+} in the presence of cetyl trimethyl ammonium bromide and isopropyl alcohol.

One of simple methods to synthesis of copper nanoparticles CuNPs when synthesized by reduction of copper (II) nitrate with isopropyl alcohol (IPA) in the presence of the cationic surfactant cetyl trimethylammonium bromide (CTAB) by the following method (Kanheda et al. 2014).

- Prepared of 0.0030 M copper nitrate and 0.09 M of CTAB were in IPA.
- Carry the reaction in clean dry 250 ml Erlenmeyer flasks open to air.
- Add the copper nitrate drop wise to the 0.09 M CTAB/(IPA) solution.
- The reaction mixture stirred vigorously on a magnetic stir plate.
- Turn of the solution violet after the addition of 2 ml of copper nitrate,
- Darker violet after addition of 5 ml copper nitrate and dark violet on further addition, indicating the synthesis of CuNPs.
- Use of IPA as a reducing agent in the synthesis of CuNPs. CTAB molecules catalyze the reduction of Cu^{2+} ions to Cu0 with IPA.
- Act the CTAB also as a capping agent by surrounding the surface of CuNPs.
- The long chain cetyl groups stabilize the CuNPs and prevent aggregation.

Using other method, mentioned to this steps to create the Copper NPs. As follow:

- Take of 5 ml $CuSO_4$ (0.1 M) and 10 ml dodecyl benzene sulfonic acid sodium (DBS) (0.5 M) solution were into 210 ml deionized water in a three necked flask equipped with a spherical condenser.
- Add 30 ml hydrazine (0.5 M) solution, and then drop wise to the above solution when the reaction mixture was heated to 100 °C under constant severe stirring.
- Reflux The reaction solution was for 40 min. to completion of reaction and allowing the mixture to cool to room temperature.
- Separate the reaction liquid using a hydroextractor at 3500 rpm for 30 min,
- Wash it three times with water and ethanol.
- Dry it at 60 °C for 5 h in a vacuum dryer.

6.1 *General Procedure for the Reduction of Nitro Aryl Compounds*

- added the mixture of nitrobenzene (1 mmol), copper nanoparticles (0.1 mmol) and THF/H_2O (v/v = 1:2) into a three-necked flask,
- Dissolve 3 mmol sodium borohydride in 5 ml distilled water were.
- Added dropwise to the above solution under the continuous stirring.
- Store the mixture at 50 °C and monitor it by TLC.
- After the completion of the reaction, allowing the mixture to cool to room temperature and filtration.
- Wash the filtrate with methylene chloride.
- Wash the organic layer successively with brine and water,
- Dry with anhydrous Na_2SO_4,

- Remove the solvent under vacuum.
- Purify the residue by chromatography on silica gel to give desired product.

Many publication studied the effect of CuNPs against pathogenic fungi e.g., *Phoma destructiva, Curvularia lunata; Alternaria alternata; Fusarium oxysporum* using 3–10 nm NPs (Prachi et al. 2014).

7 Iron Oxide NPs

Iron oxides are one of the most important transition metal oxides of technological importance. About Sixteen iron oxides phases in pure status e.g., oxides, hydroxides or oxyhydroxides are known already. Characterization of iron oxide compounds includes most of the trivalent iron form with brilliant colors and low solubility.

These types of iron oxides could be prepared by all known chemical techniques and methods but there are challenging tasks to tailor the nano-size range and morphology for particular application. These oxides employed in many applications like catalysts, pigments, sorbents, coatings, flocculants, gas sensors manufacturing, ion exchangers. Similar investigations have mentioned that NPs of Iron oxide have ability to change the stability, solubility, and aggregation behavior in due to pH changes and concentration of organic matter (Basavaraja et al. 2008).

In the past years, many difficulties faced the production of nano-particles of iron; it was difficult to precipitate iron oxide particles directly in the desired shape and size. So the preparation process was conducted using various precursor particle of iron oxide. Also the preparative method was very sensitive. But now especially with developing of the colloidal chemical synthetic procedures to produce nanoparticles of various materials, the process becomes available. The following will highlight briefly on the methods of preparation:

7.1 Chemical Precipitation

Using the co-precipitation methods, Dove and Kalaniya (2012) and Kandbal et al. (2014) pointed to a simplified method for the preparation of ferrite nanoparticles, the co-precipitation has been carried out only from Fe (II) salt without any composition of Fe (III) and the steps are as follow;

- Using a glass round bottom flask on heating mantle (EXPO make) maintained at 90 °C.
- Using a glass burette of 100 ml fitted with a burette stand for the addition of ammonium hydroxide drop wise.
- Take the deionized distilled water (100 ml) in the bottom flask.
- Add the weighed amount of $FeCl_2.4H_2O$ (3 g) after reaching thermal equilibrium (9 °C ± 10.1) into the flask.

- Load the 100 ml ammonium hydroxide solution with 24.5 g 0.7 mol ammonium hydroxide and the add ammonia solution with the rate 0.007 mol/s followed by stirring for 1 s.
- After the system reached a precipitation state, cooling and settling of the precipitate in the bottom of the flask.
- The supernatant liquid decant carefully into Whatman qualitative filter paper having poor size 22–25 μ without disturbing the precipitate.
- Take the filtrate into centrifuged tubes after centrifugation at 5000 rpm.
- The supernatant will able to pass through the filter paper followed by the transfer of the precipitate.
- Washing of the precipitate with the half of ethanol solution in water.
- Expose the washed precipitate in air for 24 h where it turns into the black magnetic phase.
- Dry the precipitate at 100 °C for 1 h in vacuum.

7.2 *Biological Method for Iron NPs Preparation*

Described the method of nanoparticles preparation from several species of plants as a ecofriendly and green synthesis method, they got the nanoparticles from ten various species of plants that include mango leaves, clove buds, black tea, green tea leaves, coffee seeds, rose leaves, cumin seeds, origano leaves, thymol seeds, curry leaves for reducing ferric chloride ($FeCl_3$) as Fe precursor was purchased from Sigma Chemicals limited. The steps of precursor for the synthesis of iron nanoparticles as follow:

- 0.001 M Ferric Chloride was prepared by using triple distilled water.
- Plant extracts were prepared by taking approximately 25 g leaves/seeds/buds.
- These were of thoroughly washed with sterile distilled water, dried and finely crushed with the help of mortar and pestle by adding 5–10 ml of deionized water gradually.
- The mixture was poured in a flask and heated for 5–10 min at 700 °C before finally decanting it.
- The mixture was then filtered using Whatman No. 1 filter paper.
- Wherever necessary the plant mixture was centrifuged at 5000 rpm for 5 min.
- The supernatant was collected as the plant extract and used for further process.
- During the synthesis of iron nanoparticles both the precursor and the reducing agent were mixed in a clean sterilized flask in 1:1 proportion.
- For the reduction of Fe ions, 5 ml of plant extract was mixed to 5 ml of 0.001 M aqueous of $FeCl_3$ solution with constant stirring at 50–60 °C.
- Characterization UV–vis spectroscopy using light in the visible and adjacent near-UV also near infrared NIR ranges, and with ultraviolet–visible spectroscopy UV–Vis, the results will refer to absorption spectroscopy in the UV–visible spectral area, the perceived color of the chemicals involved will affected by the

absorption in the visible range. In this electro-magnetic spectrum region molecules go through electronic transitions.

8 Synthesis of ZnO Nanoparticles

Zinc, a metallic chemical element with ionic state Zn^{2+} is an essential trace element necessary for plants, microbes and animals (Park et al. 2009; Sangeetha et al. 2012). Zinc Oxide (ZnO) has been widely used in a lot of fields such as near-UV emission, transparent conductor, gas sensors, and piezoelectric application (Bhuyan et al. 2015). ZnO nanostructures can be synthesized on a scale at economical and low costs using simple chemical components as solutions such as chemical precipitation, Sol–gel synthesis, and solvothermal/hydrothermal reaction. Aneesh et al. (2007) could prepare the ZnO NPs using hydrothermal technique and they mentioned to this methods as a favorable alternative synthetic techniques because of the low temperature and easy to particle size control. In their paper described the hydrothermal process which has several influences over other processes for instance simple equipment using, economical cost, friendly environment and less risks. Also this technique has been successfully employed to prepare luminescent materials and ZnO in nanoscale and the properties of particles which produced from this technique can be controlled via process of the hydrothermal by adjusting the temperature of reaction, concentration and time of precursors.

8.1 The Preparation of ZnO NPs via Hydrothermal Technique

To synthesize the ZnO nanoparticles using hydrothermal technique as Aneesh et al. (2007) described the steps as follow:

- Prepare the stock solutions of Zn $(CH_3COO)_2.2H_2O$ of (0.1 M) in 50 ml methanol under stirring condition.
- Making of the stock solution (25 ml of NaOH) varying 0.2–0.5 M solution, and prepare it in methanol which added under stirring to get the pH in range of (8–11).
- Transfer the solutions into Teflon lined sealed stainless steel autoclaves and maintained at temperature ranged from 100 to 200 °C for 6 and 12 h under autogenously pressure.
- After complete of reaction, wash the resulting white solid with methanol.
- Filtration and drying the sample using laboratory oven at 60 °C.
- Characterization of the samples for their structure by X-ray diffraction (Rigaku D max-C) with Cu Kα radiation.

- Examine the samples using the transmission electron microscopy (TEM), high resolution transmission electron microscopy (HRTEM) and selected area electron diffraction SAED with a JEOL JEM-3100 F TEM operating at 200 kV.
- Preparing of the sample by placing a drop of the ZnO suspension using methanol onto a standard carbon coated copper grid.
- Drying of grids before micrographs recording.
- Estimation of the optical band gap from the UV-vis-NIR diffuse reflectance spectroscopic (UV-vis-NIR DRS) and studied under wavelength range from 190 nm to 1200 nm with spectrophotometer. Also there are many publication highlighted the effect of ZnO NPs on phytopathogenic fungi like *Fusarium oxysporum and Penicilium expansum* using NPs size 70 nm (Yehia and Ahmed 2013); *Alternaria alternate, Fusarium oxysporum; Rhizopus stolonifer; Mucor plumbeus* with 30 and 50 nm (Wani and Shah 2012); *Botrytes cinerea, Penicillium expansum* using 70 nm and *Bipolaris sorokiniana*, *Magnaporthe grisea* by 20, 30 nm NPs size (Jones et al. 2008).

9 TiO_2 Nanoparticles

TiO_2 is one of the promising materials high photochemical stability with low cost and small sizes in many applications such as pigments, adsorbents and catalytic supports. Much research has been mentioned upon the reduction of TiO_2 particle size. Its common compound, titanium dioxide, is a popular photo-catalyst, and is used in manufacture of pigments (Chen and Yang 1993; Raliya and Tarafdar 2014). There are some important the methods to prepare this type of NPs, Vijayalakshmi and Rajendran (2012) employed two techniques to manufacturing TiO_2 as follow:

9.1 Sol–Gel Route

- The sol-gel synthesized TiO_2 was obtained from Titanium (IV) isopropoxide (TTIP) was dissolved in absolute ethanol
- Distilled water was added to the solution in terms of a molar ratio of Ti: $H_2O = 1:4$.
- Nitric acid was used to adjust the pH and for restrain the hydrolysis process of the solution.
- The solution was vigorously stirred for 30 min in order to form sols.
- After aging for 24 h, the sols were transformed into gels.
- The gels were dried under 120 °C for 2 h to evaporate water and organic material to the maximum extent.
- The dry gel was sintered at 450 °C for 2 h were subsequently carried out to obtain desired TiO_2 nano-crystalline.

9.2 Hydrothermal Method

- Analytical grade titanium tetrachloride was adopted as the source material and sodium hydroxide as mineralizer.
- An aqueous solution of titanium was obtained by mixing one molar stoichiometric ratio of TTIP in 50 ml of distilled water.
- The solution 2–3 mol of NaOH with stirring at several minutes, resulting in a white colloidal sol. The final volume was adjusted to 90 ml using distilled water.
- 90 ml sol was transferred to a 100 ml Teflon lined auto clave vessel.
- The sealed vessel was heated to 240 °C for 12 h and the resultant precipitate was dried at 450 °C for 2 h to obtain TiO_2 nanoparticles.

10 Preparation of Magnesium Oxide Nanoparticles

Magnesium is an alkaline earth metal with its ionic form Mg^{2+}, essential to all living cells, where it plays a crucial role in employ of important biological polyphosphate like ATP, RNA and DNA (Raliya and Tarafdar 2014). Magnesium oxide (MgO) is an important type of basic oxide that has many applications. Therefore, nanoscale MgO has been extensively used in catalysis, toxic waste remediation, and refractory materials industries based on its versatile properties (Camtakan et al. 2011; Rodriguez 2007; Klabunde 2001). As experimental process which employed in synthesis of MgO, Camtakan et al. 2011 describe the steps as follow:

- MgO nanoparticles have been prepared via hydroxide precipitation from aqueous solutions followed by thermal decomposition of the hydroxide.
- The samples were obtained by precipitation at a controlled temperature, ($MgCl_2$) of a magnesium salt solution of concentration 1 mol/l by addition of an alkaline solution (NaOH) of 2 mol/l concentration.
- Both reactants were simultaneously added to 100 ml cylindrical container in a bath at 80 °C for 2 h.
- Vigorous stirring was applied during the addition of the reactants, as well as during the aging of the precipitate in the mother liquor.
- The suspension was allowed to age at the synthesis temperature for 2 h, then subsequently at room temperature for 1 day.
- The solid phase obtained was recovered by using a Whatman filter paper No: 44 and washed twice with deionized water and absolute alcohol, then air-dried at 60 °C for 4 h.
- The samples were submitted to a hydrothermal treatment: the precipitation was placed in a closed Teflon bottle in an oven pre-heated at 2508 °C during 1 h and 370 °C for 2 h and 435 °C for 3 h according to Ding et al. (2001).
- A stock solution of uranium (1000 mg U L21) was prepared by dissolving 2.1 g of uranyl nitrate hexahydrate (UO_2 $(NO_3)_2.6H_2O$) in 1 l of a 1 % v/v HNO_3, solution. Serial dilutions of the stock solution were made to prepare working standards (10–55 lg U ml 21): all standard solutions were acidified with nitric acid to pH 2.0.

- The initial pH of the working solutions was adjusted by addition of HNO_3 or Na_2CO_3.
- The buffer solutions (pH 4, 7, and 9) to calibrate the pH-meter.
- Batch Sorption experiment were carried in a thermostatic shaker bath,
- MgO (0.01 g), which has 18.7 nm average particle sizes, was added to 10 ml solution containing various uranium concentrations at different temperatures for various contact time.
- The suspension was filtered by using a Whatman filter paper No: 44.
- A simple and sensitive spectrophotometric method was used in the experiments to determine uranium in solution.
- A simple and sensitive spectrophotometric method based on colored complexes with 1,2 Pyridyl Azo Resorcinol (PAR) in aqueous medium was used for determination of uranium.
- The concentration in the solution was determined with UV–vis spectrophotometer by measuring absorbance at 510 nm for uranium.

To determine the limit of detection, the suggested procedure requires taking 10 measurements at 10 different locations on a blank sample; the limit of detection given as;

$$LOD = 3S / b$$

where, (s) the standard deviation of the spectroscopic signals of the 10 measurements on the blank sample and (b) is the slope of the linear calibration curve Ding et al. (2001). Camtakan et al. (2011) also calculate the detection limit of the method according to this equation. And they mentioned to the detection limit of this method is 0.02 lg U ml 21 and range of uranium concentration is between 0.08 and 16 lg U ml 21 according to Onishi (1989). The influence of specific process parameters such as initial uranium concentration, pH of the solution, contact time, and temperature was determined by calculating uranium (VI) sorption by nanoparticle MgO and changing a parameter and keeping other parameters constant. The percentage sorption of uranium from aqueous solution was computed as follows:

$$\text{Sorption }\% = \frac{\text{Cint} - \text{Cfin}}{\text{Cint}} \times 100$$

where C_{int} and C_{fin} are the initial and final uranium concentration, respectively.

11 To Desorption Experiments

- MgO nanoparticles loaded with uranium at optimum removal conditions was used for the desorption studies.
- The U-loaded MgO nanoparticles filtered and dried before use in desorption experiments.

- Loaded sorbent was contacted with desorption reagents using different concentrations for various contact times at 25 °C in the thermostated shaker.
- The desorption ratio was calculated from the amount of uranium adsorbed on the sorbents and the final uranium concentration in the desorption medium, using the following equation,

$$\text{Desorption ratio}\left(\%\right)=\frac{\text{amount of metal ion desorbed}}{\text{amount of metal ion desorbed}}\times 100$$

12 Biosynthesis of Zinc, Magnesium and Titanium Nanoparticle

Biological methods for nanoparticle synthesis would help circumvent many of the detrimental features by enabling synthesis at mild pH, pressure, temperature and at a substantially lower cost (Raliya and Tarafdar 2012; Lakkakula et al. 2009).

Nanoparticles of Zn, Mg and Ti have been introduced by using 14 different isolates from fungi belonged to (*Aspergillus flavus, Aspergillus terreus, Aspergillus tubingensis, Aspergillus niger, Rhizoctonia bataticola, Aspergillus fumigatus and Aspergillus oryzae*) Raliya and Tarafdar (2012) as the steps follow:

- Each fungus was grown in 250 ml Erlenmeyer flask containing 100 ml modified malt extract, glucose, yeast extract, peptone (MGYP) medium, containing 0.3 % malt extract, 1 % sucrose, 0.3 % yeast extract, and 0.5 % peptone.
- Grow the culture with continuous shaking using a rotary shaker (150 rpm) at 28 °C for 72 h. After 72 h, this step after adjusting the pH of medium to 6.8.
- Separate the fungal balls of mycelia from the culture broth using centrifuge (4000 rpm) at 4 °C for 10 min and then the fungal mycelia were washed thrice with sterile distilled water.
- Harvesting of fungal biomass by re-suspend it in 100 ml sterile Milli-Q-water using 250 ml Erlenmeyer flask and again kept on shaker (150 rpm) at 28 °C for 62 h.
- After incubation, the cell-free filtrate was obtained by spreading the fungal biomass by filtration using membrane filter.
- Using cell-free filtrate, salt solution of precursor salts ZnO, $ZnSO_4$, $ZnCl_2$, $ZnNO_3$ for Zn, MgO, $MgSO_4$, MgC_{12}, $MgNO_3$ for Mg, and TiO_2 rutile, TiO_2 anatase for Ti, was prepared in various concentrations ranging from 1 M to 0.01 mM in Erlenmeyer flasks, 0.1 mM concentration used for all conditions unless specified it.
- The pH ranges between 4.0 and 8.0 were tested, covering both acid and alkaline range.
- The entire mixture was put into shaker (150 rpm) at various temperatures of reaction mixture of precursor metal salt and extracellular enzyme was analyzed

from 20 to 40 °C with an increment of one and two degree temperature, and the reaction time allowed from 0 to 120 h was set up and observation taken at regular intervals of time.
- The biotransformation was collected periodically and monitored for characterization.

13 The Methodology for Nanoparticle as Control Agents Against Fungal Plant Pathogens

13.1 Preparation of Spore Suspension in Different Concentration of Nanoparticles to Evaluate Their Effect on the Spore Germination of Fungi

- Prepare the fungal inoculates on potato dextrose agar (PDA) media (a common microbial media for culturing fungus) at 28 °C in Petri plates.
- Prepare spore suspension of each isolate of fungi containing at least 20–30 spores per microscopic field was from 10 days old fungal culture.
- One drop about 0.1 ml of spore suspension was put in a cavity glass slide containing a drop (about 0.1 ml) of different concentration of nanoparticles.
- Keep these slides in moist chamber by putting two folds of filter paper in both sides of Petri-plates.
- Incubated these Petri plates at 24 ± 2 °C for 24 h.
- Replicate each treatment four or five times to confirm the obtained results.
- Recorded the percent spore germination using formula as:

$$\text{percent spore germination} = \frac{\text{No. of spores germation}}{\text{Total No. of spores examined}} \times 100$$

13.2 Assay for Sclortium Forming Phytopathogenic Fungi

- Freshing the sclerotia by incubating the fungal plates at room temperature without any treatment.
- Washing the sclerotia (2–4 weeks old) with sterilized distilled water and rinsed with 70 % alcohol.
- For testing the sclerotial germination growth, one sclerotium placed in the center of MEA medium supplemented with either concentrations of the nanoparticles or an equal volume of water.
- Incubate the plates at 24 °C, and use for the measurement of sclerotial germination growth.
- Determined the sclerotial germination rate by measuring the diameter of mycelial colonies.

13.3 Field Assay Method to Evaluate Nanoparticle Effects on Foliar Diseases in Green House Condition

- Infection of host plant with the disease naturally.
- Use nanoparticle at different concentrations,
- Use the aerial spray method to apply nanoparticles around the shoot portion of the whole plants 3–4 weeks before the outbreak of the disease and after disease occurrence.
- Distilled water was used as a control.
- Calculate the disease index by counting the numbers of infected leaves out of 100 - 150 leaves among the treated plants.

Wani and Shah (2012) mentioned to the antifungal effect of magnesium oxide nanoparticles *on some pathogens like Alternaria alternate, Fusarium oxysporum; Rhizopus stolonifer; Mucor plumbeus* and reported that the highest effect happened when they used the 0 and 50 nm nanoparticles size.

References

Aneesh PM, Vanaja KA, Jayaraj MK (2007) Synthesis of ZnO nanoparticles by hydrothermal method. Proc Spie 663966390J:1–9

Arbab AS, Bashaw LA, Miller BR, Jordan EK, Lewis BK, Kalish H, Frank JA (2003) Characterization of biophysical and metabolic properties of cells labeled with super paramagnetic iron oxide nanoparticles and transfection agent for cellular MR imaging. Radiology 229:838–846

Aziz N, Faraz M, Pandey R, Sakir M, Fatma T, Varma A, Barman I, Prasad R (2015) Facile algae-derived route to biogenic silver nanoparticles: synthesis, antibacterial and photocatalytic properties. Langmuir 31:11605–11612

Balaji DS, Basavaraja S, Deshpande R, Bedre MD, Prabhakar BK, Venkataraman A (2009) Extracellular biosynthesis of functionalized silver nanoparticles by strains of *Cladosporium cladosporioides* fungus. Colloids Surf B Biointerfaces 68(1):88–92

Baldassari S, Komarneni S, Mariani E, Villa C (2005) Microwave hydrothermal process for the synthesis of rutile. Mater Res Bull 40:2014–2020

Banfield JF, Welch SA, Zhang H, Ebert TT, Penn RL (2000) Aggregation-based crystal growth and microstructure development in natural Iron oxyhydroxide biomineralization Products. Science 289:751–754

Basavaraja S, Balaji SD, Lagashetty A, Rajasab AH, Venkataraman A (2008) Extracellular biosynthesis of silver nanoparticles using the fungus *Fusarium semitectum*. Mater Res Bull 43(5):1164–1170

Bhuyan T, Mishra K, Khanuja M, Prasad R, Varma A (2015) Biosynthesis of zinc oxide nanoparticles from *Azadirachta indica* for antibacterial and photocatalytic applications. Mater Sci Semicon Process 32:55–61

Bönnemann H, Richards RM (2001) Nanoscopic metal particles synthetic methods and potential applications. Eur J Inorg Chem 2001:2455–2480

Camtakan Z, Erenturk S, Yusan S (2011) Magnesium oxide nanoparticles: Preparation, characterization, and uranium sorption properties. Environ Progr Sustain Energy 31(4): doi:10.1002/ep

Castro-Longoria E, Vilchis-Nestor AR, Avalos-Borja M (2011) Biosynthesis of silver, gold and bimetallic nanoparticles using the filamentous fungus *Neurospora crassa*. Colloids Surf B Biointerfaces 83(1):42–48

Chen JP, Yang RT (1993) Selective catalytic reduction of NO with NH_3 on SO_4-^{2}/TiO_2 super acid catalyst. J Catal 139:277–288

Cigang Xu, Harm van Zalinge , John LP, Andrew Glidle, Jonathan M Cooper, David R S Cumming, Wolfgang Haiss, Jian Lin Yao, David J Schiffrin, Maria Proupin-Perez, Richard Cosstick, Richard J Nichols (2006) A combined top-down bottom-up approach for introducing nanoparticle networks into nanoelectrode gaps. Nanotechnology 17:3333

Daniel MC, Astruc D (2003) Gold nanoparticles: assembly, supramolecular chemistry, quantum-size-related properties, and applications toward biology, catalysis, and nanotechnology. Chem Rev 104:293–346

Ding Y, Zhang G, Wu H, Hai B, Wang L, Qian Y (2001) Nanoscale magnesium hydroxide and magnesium oxide powders: control over size, shape and structure via hydrothermal synthesis. Chem Mater 3:435–440

Dove PN, Kalaniya RY (2012) Synthesis and magnetic properties of Fe_3O_4, Fe_2O_3 core / shell nanoparticles. J Ind Chem Soc 89:853–856

Durán N, Marcato PD, Alves OL, Souza GIH, Esposito E (2005) Mechanistic aspects of biosynthesis of silver nanoparticles by several *Fusarium oxysporum* strains. J Nanobiotechnol 3:8. doi:10.1186/1477-3155-3-8

Elzey S, Grassian V (2010) Agglomeration, isolation and dissolution of commercially manufactured silver nanoparticles in aqueous environments. J Nanopart Res 12:1945–1958

Feng QL, Wu J, Chen GO, Cui FZ, Kim TN, Kim JO (2000) A mechanistic study of the antibacterial effect of silver ions on *Escherichia coli* and *Staphylococcus aureus*. J Biomed Mater Res 52:662–668

Forough M, Farhadi K (2010) Biological and green synthesis of silver nanoparticles. Turk J Eng Enviro Sci 34:1–7

Glasauer S, Langley S, Beveridge TJ (2002) Intracellular iron minerals in a dissimilatory iron-reducing bacterium. Science 295:117–119

Grancharov SG, Zeng H, Sun S, Wang SX, O'brien S, Murray CB, Kirtley JR, Held GA (2005) Bio-functionalization of monodisperse magnetic nanoparticles and their use as biomolecular labels in a magnetic tunnel junction based sensor. J Phys Chem 109:13030–13035

Guo L, Yang S, Yang C, Yu P, Wang J, Ge W, Wong GKL (2000) Synthesis and characterization of poly (vinylpyrrolidone)-modified zinc oxide nanoparticles. Chem Mater 12:2268–2274

Hansel CM, Benner SG, Nico P, Fendorf S (2004) Structural constraints of ferric (hydr) oxides on dissimilatory iron reduction and the fate of Fe (II). Geochim Cosmochim Acta 68:3217–3229

Hoffmann MR, Martin ST, Choi W, Bahnemann DW (1995) Environmental applications of semiconductor photocatalysis. Chem Rev 95(1):69–96

Jana NR, Sau TK, Pal T (1998) Growing small silver particle as redox catalyst. J Phys Chem 103:115–121

Jones N, Ray B, Ranjit KT, Manna AC (2008) Antibacterial activity of ZnO nanoparticle suspensions on a broad spectrum of microorganisms. FEMS Microbiol Lett 279:71–76

Ju-Nam Y, Lead JR (2008) Manufactured nanoparticles: an overview of their chemistry interactions and potential environmental implications. Sci Total Environ 400:396–414

Jung JH, Kim S, Min JS, Kim YJ, Lamsal K, Kim KS, Lee YS (2010) The effect of nano-silver liquid against the white rot of the green onion caused by *Sclerotium cepivorum*. Micobiology 38(1):39–45

Kamat PV (1993) Photochemistry on nonreactive and reactive (semiconductor) surfaces. Chem Rev 93:267–300

Kandbal ND, Shah N, Oshali RL, Joshi R, Prasad J (2014) Co-Precipitation methods of synthesis and characterization of iron oxide nanoparticles. J Sci Ind Res 73

Kanheda P, Birlaa S, Gaikwada S, Gadea A, Seabrab B, Rubilarc O, Durane N, Mahendra R (2014) In vitro antifungal efficacy of copper nanoparticles against selected crop pathogenic fungi. Mater Lett 115(15):13–17

Kawai-Nakamura A, Sato T, Sue K, Tanaka S, Saitoh K, Aida K, Hiaki T (2008) Rapid and continuous hydrothermal synthesis of metal and metal oxide nanoparticles with a microtube-reactor at 523 K and 30 MPa. Mater Lett 62:3471–3473

Khabat V, Mansoori GA, Karimi S (2011) Biosynthesis of silver nanoparticles by fungus *Trichoderma reesei* (A route for large-scale production of AgNPs). Insciences J 1(1):65–79
Khalil Neveen M (2013) Biogenic silver nanoparticles by *Aspergillus terreus* as a powerful nanoweapon against *Aspergillus fumigatus*. Afr J Microbiol Res 7(50):5645–5651
Khandelwal N, Singh A, Jain D, Upadhyay MK, Verma HN (2010) Green synthesis of silver nanoparticles using *Argimone mexicana* leaf extract and evaluation of their antimicrobial activities. Dig J Nanomater Biostruct 5:483–489
Kim JH, Lee GD, Park SS, Hong SS (2006) Hydrothermal synthesis of titanium dioxides using acidic and basic peptizing agents and their photocatalytic activity on the decomposition of orange II. In: Hyun-Ku Rhee I-SN, Jong moon P (eds) Studies in surface science and catalysis. Elsevier, Amsterdam
Kim SY, Lim TH, Chang TS, Shin CH (2007) Photocatalysis of methylene blue on titanium dioxide nanoparticles synthesized by modified sol-hydrothermal process of $TiCl_4$. Catal Lett 117:112–118
Kim SW, Jung JH, Lamsal K, Kim YS, Min JS, Lee YS (2012) Antifungal effects of silver nanoparticles (AgNPs) against various plant pathogenic fungi. Mycobiology 40(1):53–58
Klabunde KJ (2001) Nanoscale materials in chemistry. Wiley, Canada
Krishnaraj C, Ramachandran RK, Mohan P, Kalaichelvan T (2012) Optimization for rapid synthesis of silver nanoparticles and its effect on phytopathogenic fungi. Spectrochem Acta Part A 93:95–99
Lakkakula JR, Kolekar NS, Mendhulkar VD, Kashid SB (2009) Phytosynthesis of silver nanoparticle using *gliricidia sepium* (Jacq.). Curr Nanosci 5:117–122
Lamsal K, Kim SW, Jung JH, Kim YS, Kim SK, Lee YS (2011) Inhibition effects of silver nanoparticles against powdery mildews on cucumber and pumpkin. Mycobiology 39(1): 26–32
Lee I, Han SW, Kim K (2001) Simultaneous preparation of SERS-active metal colloids and plates by laser ablation. J Raman Spectrosc 32:947–952
Li H, Duan X, Liu G, Jia X, Liu X (2008) Morphology controllable synthesis of TiO_2 by a facile hydrothermal process. Mater Lett 62:4035–4037
Makino T, Chia CH, Segawa Y, Kawasaki M, Ohtomo A, Tamura K, Matsumoto Y, Koinuma H (2002) High-throughput optical characterization for the development of a ZnO-based ultraviolet semiconductor-laser. Appl Surf Sci 189:277–283
Marek JK, Magdalena K, Gorczyca A (2010) The effect of silver nanoparticles on phytopathogenic spores of *Fusarium culmorum*. Can J Microbiol 56:247–253
Marek JK, Anna G, Rasmus JNF (2013) The effect of nanosilver on pigments production by *Fusarium culmorum* (W.G.Sm.) Sacc. Pol J Microbiol 62(4):365–372
Medentsev AG, Alimenko VK (1998) Naphthoquinone metabolites of the fungi. Photochemistry 47:935–959
Miguel AA, Eduardo SM, Lesli O, Georgina C, Esther S (2011) Synthesis and characterization of silver nanoparticles: effect on phytopathogen *Colletotrichum gloesporioides*. J Nanopart Res 13:2525–2532
Min JS, Kim KS, Kim SW, Jung JH, Lamsal K, Kim S P, Jung MY, Lee YS (2009) Effect of colloidal silver nanoparticles on sclortium forming phytopathogenic fungi. Plant Pathol J 25(4): 376–380
Mirkin CA, Letsinger RL, Mucic RC, Storhoff JJ (1996) A DNA based method for rationally assembling nanoparticles into macroscopic materials. Nature 382:607–609
Müller J, Weißenrieder S (1994) ZnO-thin film chemical sensors. Fresenius J Anal Chem 349:380–384
Navaladian S, Viswanathan B, Viswanath R, Varadarajan T (2006) Thermal decomposition as route for silver nanoparticles. Nanoscale Res Lett 2:44–48
Navarro E, Baun A, Behra R, Hartmann N, Filser J, Miao AJ, Quigg A, Santschi P, Sigg L (2008) Environmental behavior and ecotoxicity of engineered nanoparticles to algae, plants, and fungi. Ecotoxicology 17:372–386

Nowack B, Bucheli TD (2007) Occurrence, behavior and effects of nanoparticles in the environment. Environ Pollut 150:5–22

Oksanen T, Pere J, Paavilainen L, Buchert J, Viikari L (2000) Treatment of recycled kraft pulps with *Trichoderma reesei* hemicellulases and cellulases. J Biotechnol 78(1):39–44

Onishi H (1989) Photometric determination of trace metals (4th Edition), Part IIB. Wiley, Canada

Oskam G (2006) Metal oxide nanoparticles: synthesis, characterization and application. J Sol-Gel Sci Technol 37:161–164

Pallub S, Arun C, Sidhartha sanker G (2011) Induction of apoptosis in cancer cells at low silver nanoparticles concentrations using chitosan nano carrier. Appl Mater Interface 3:218–228

Park H, Kim SH, Kim HJ, Choi SH (2006) Anew composition of nano silica-silver for control of various plant diseases. Plant Pathol J 23(3):295–302

Park S, Lee JH, Kim HS, Park HJ, Lee JC (2009) Effect of ZnO nanopowder dispersion on photocatalytic reactions for the removal of Ag ions from aqueous solution. J Electro Ceram 22:105–109

Perez JM (2007) Iron oxide nanoparticles: hidden talent. Nat Nano 2:535–536

Prachi K, Birlaa S, Swapnil G, Aniket G, Amedea B, Seabrab OR, Durane N, Mahendra R (2014) In vitro antifungal efficacy of copper nanoparticles against selected crop pathogenic fungi. Mater Lett 115:13–17

Prasad R (2014) Synthesis of silver nanoparticles in photosynthetic plants. J Nanoparticles, Article ID: 963961. doi:10.1155/2014/963961

Prasad R, Kumar V, Prasad KS (2014) Nanotechnology in sustainable agriculture: present concerns and future aspects. Afr J Biotechnol 13(6):705–713

Prasad R, Pandey R, Barman I (2015) Engineering tailored nanoparticles with microbes: quo vadis. WIREs Nanomed Nanobiotechnol. doi:10.1002/wnan.1363

Pratsinis SE, Zhu W, Vemury S (1996) The role of gas mixing in flame synthesis of titania powders. Powder Technol 86:87–93

Pulit J, Banach M, Szczygłowska R, Bryk M (2013) Nanosilver against fungi; silver nanoparticles as an effective biocidal factor. Acta Biochim Pol 60:795–798

Raliya R, Tarafdar JC (2012) Novel approach for silver nanoparticles synthesis using *Aspergillus terreus* CZR-1: mechanism perspective. J Bionanosci 6:12–16

Raliya R, Tarafdar JC (2014) Biosynthesis and characterization of zinc, magnesium and titanium nanoparticles: an eco-friendly approach. Int Nano Lett 4:93

Ramaswamy V, Jagtap NB, Vijayanand S, Bhange DS, Awati PS (2008) Photocatalytic decomposition of methylene blue on nanocrystalline titania prepared by different methods. Mater Res Bull 43:1145–1152

Rao NN, Dube S (1996) Photocatalytic degradation of mixed surfactants and some commercial soap/detergent products using suspended TiO_2 catalysts. J Mol Catal A Chem 104:L197–L199

Rashidzadeh M (2008) Synthesis of high-thermal stable titanium dioxide nanoparticles. Int J Photoenergy Article ID 245981:4. doi:10.1155/2008/245981

Rodriguez JA (2007) Synthesis properties and applications of oxide nanomaterials. Wiley, Canada, p 6

Roduner E (2006) Size matters: why nanomaterials are different. Chem Soc Rev 35:583–592

Sangeetha G, Rajeshwari S, Venckatesh R (2012) Green synthesized ZnO nanoparticles against bacterial and fungal pathogens. Prog Nat Sci Mater Int 22(6):693–700

Sardar R, Funston AM, Mulvaney P, Murray RW (2009) Gold nanoparticles: past, present, and future. Langmuir 25:13840–13851

Schüler D, Frankel RB (1999) Bacterial magnetosomes: microbiology, bio mineralization and biotechnological applications. Appl Microbiol Biotechnol 52:464–473

Senapati S, Ahmad A, Khan MI, Sastry M, Kumar R (2005) Extracellular biosynthesis of bimetallic Au–Ag alloy nanoparticles. Small 1(5):517–520

Shenhar R, Rotello VM (2003) Nanoparticles: scaffolds and building blocks. Accounts Chem Res 36:549–561

Shtykova EV, Huang X, Remmes N, Baxter D, Stein B, dragnea B, Svergun DI, Bronstein LM (2007) Structure and properties of iron oxide nanoparticles encapsulated by phospholipids with poly (ethylene glycol) tails. J Phys Chem C 111:18078–18086

Sioutas C, Delfino RJ, Singh M (2005) Exposure assessment for atmospheric ultrafine particles (UFPs) and implications in epidemiologic research. Environ Health Perspect 113

Souza GIH, Marcato PD, Duran N, Esposito E (2004) Utilization of *Fusarium oxysporum* in the biosynthesis of silver nanoparticles and its antibacterial activities. In: IX National Meeting of Environmental Microbiology Curtiba, PR (Brazil) Abstract p 25

Suleiman M, Ali AA, Hussein A, Hammouti B, Hadda TB, Warad I (2013) Sulfur nanoparticles: synthesis, characterizations and their applications. J Mater Environ Sci 4(6):1029–1033

Suman, Prasad R, Jain VK, Varma A (2010) Role of nanomaterials in symbiotic fungus growth enhancement. Curr Sci 99:1189–91

Suzuki Y, Kelly SD, Kemner KM, Banfield JF (2002) Radionuclide contamination: nanometre-size products of uranium bioreduction. Nature 419(134):134

Tarasenko N, Nevar A, Nedelko M (2010) Properties of zinc-oxide nanoparticles synthesized by electrical-discharge technique in liquids. Phys Status Solidi 207:2319–2322

Tolaymat TM, El badawy A M, Genaidy A, Scheckel KG, Luxton TP, Suidan M (2010) An evidence-based environmental perspective of manufactured silver nanoparticle in syntheses and applications: a systematic review and critical appraisal of peer-reviewed scientific papers. Sci Total Environ 408:999–1006

Verma VC, Kharwar RN, Gange AC (2010) Biosynthesis of antimicrobial silver nanoparticles by the endophytic fungus *Aspergillus clavatus*. Nanomedicine 5(1):33–40

Vigneshwaran N, Ashtaputrea NM, Varadarajana PV, Nachanea RP, Paralikara KM, Balasubramanyaa RH (2007) Biological synthesis of silver nanoparticles using the fungus *Aspergillus flavus*. Mater Lett 61(6):1413–1418

Vijayalakshmi R, Rajendran V (2012) Synthesis and characterization of nano-TiO_2 via different methods. Arch Appl Sci Res 4(2):1183–1190

Wang ZL, Song J (2006) Piezoelectric nano generators based on zinc oxide nanowire arrays. Science 312:242–246

Wang WN, Lenggoro IW, Terashi Y, Kim TO, Okuyama K (2005) One-step synthesis of titanium oxide nanoparticles by spray pyrolysis of organic precursors. Mater Sci Eng B 123:194–202

Wani AH, Shah MA (2012) A unique and profound effect of MgO and ZnO nanoparticles on some plant pathogenic fungi. J Appl Pharmaceutical Sci 02(03):40–44

Widder KJ, Senyel AE, Scarpelli GD (1978) Magnetic microspheres: a model system of site specific drug delivery in vivo. Proc Soc Exp Biol Med. 158:141–146

Wiesner MR, Lowry GV, Jones KL, Hochella JMF, Giulio DI, Casman RTE, Bernhardt ES (2009) Decreasing uncertainties in assessing environmental exposure, risk, and ecological implications of nanomaterial. Environ Sci Technol 43:6458–6462

Yehia RS, Ahmed OF (2013) In vitro study of the antifungal efficacy of zinc oxide nanoparticles against *Fusarium oxysporum* and *Penicilium expansum*. Afr J Microbiol Res 7(19):1917–1923

Zhu X, Su M, Tang S, Wang L, Liang X, Meng F, Hong Y, Xu Z (2012) Synthesis of thiolated chitosan and preparation nanoparticles with sodium alginate for ocular drug delivery. Molecular 18:1973–1982

Chapter 8
Plant Fungal Disease Management Using Nanobiotechnology as a Tool

Sandra Pérez Álvarez, Norma Elena Leyva López, Jesús Méndez Lozano, Edgar Antonio Rodríguez Negrete, and María Elena Santos Cervantes

Abstract Fungal diseases cause significant economic agricultural losses around the world and their control have been limited to chemical fungicide in an irrational manner. The majority of the fungal pathogenic species belongs to the Ascomycetes (genera: *Alternaria, Fusarium, Verticillium*) and Basidiomycetes (genera: *Sclerotium, Rhizoctonia*). Nanobiotechnology as a novel tool could improve actual delivering techniques to management common plant fungal diseases; for example, using chemicals nanoparticles to tag specific sites at the cellular levels. Nowadays, applications of nanoparticles that provide better efficacy for the control of plant diseases are nanoforms of carbon, silver, silica, alumino-silicates and chitosan. To understand the possible benefits of employing nanobiotechnology to agriculture, it is necessary to analyze the penetration and transport of nanoparticles in plants. Some of the current advances, challenges and potential of nanobiotechnology in fungal diseases management are discussed in this chapter.

1 Nanoparticles in Plant Diseases Management

Today farmers around the world use chemicals such as pesticides, fungicides and herbicides as the fastest way to control pests and diseases even when it is expensive. Abuse in the use of these products has caused many problems such as: adverse effects on human health, problems with pollinating insects and domestic animals, development in pest and diseases resistance, and entering this material into the soil and water affect direct and indirect the environment (Chowdappa and Gowda 2013).

The use of nanotechnology to avoid chemicals can be a suitable solution for this problem. Nanoparticles can provide diagnostic tools for early disease detection and they can be used into the part of plant that was attacked by disease or pest (Sharon et al. 2010). Nanoparticles are stable and biodegradable; they can be

S.P. Álvarez (✉) • N.E.L. López • J.M. Lozano • E.A.R. Negrete • M.E.S. Cervantes
Instituto Politécnico Nacional, CIIDIR-IPN, Unidad Sinaloa, Departamento de Biotecnología Agrícola, Blvd. Juan de Dios Bátiz Paredes No 250. Guasave, Sinaloa 81101, Mexico
e-mail: perezalvarezsandra2015@gmail.com

R. Prasad (ed.), *Advances and Applications Through Fungal Nanobiotechnology*, Fungal Biology, DOI 10.1007/978-3-319-42990-8_8

employed in production of nanocapsules for delivery of pesticides, fungicides, herbicides, and some other agrochemicals (Johnston 2010; Chen and Yada 2011; Prasad 2014; Prasad et al. 2014). Some of the nanoparticles that are widely used to control plant diseases are nanoforms of carbon, silver, silica and alumina-silicates (Saurabh et al. 2015).

1.1 Carbon Nanoparticles

Carbon based nanomaterials have a great application in agriculture not only for diseases control but also for the enhancing growth effect in plants (Patel et al. 2014). In Brazil for the improvement of natural fibers, as coconuts and sisal, carbon nanofibers are used and from this they are making nanoparticles that contain pesticides and control their release (Patel et al. 2014). Carbon nanomaterials can have different form like hollow spheres, ellipsoids (fullerenes), or cylindrical [nanotubes such as single-walled carbon nanotube (SWCNT) and multi-walled carbon nanotubes (MWCNT)] (Saurabh et al. 2015).

The effect of CNT as enhance growth in tomato seeds was proved by Khodakovsky et al. (2000) and because they found that enhanced growth was due to increased water uptake caused by penetration of the CNT could be possible to use CNT as a vehicle to deliver desired molecules into the seeds during germination that can protect them from diseases (Patel et al. 2014).

1.2 Silver Nanoparticles (SNPs)

Silver has antimicrobial activity both in ionic or nanoparticle forms. The great antimicrobial effect of silver mainly in unicellular microorganisms is believed to be brought about by enzyme inactivation (Kim et al. 1998). Application of silver in management of fungal plant diseases has been tested for two fungal pathogens of cereals viz. *Bipolaris sorokiniana* (Sacc.) (spot blotch of wheat) and *Magnaporthe grisea* (T.T. Hebert). M.E. Barr (rice blast) resulted in vitro that silver, both in ionic and nanoparticle forms, inhibited colony growth of both the pathogens but *M. grisea* was comparatively more sensitive to silver application; in vivo the test with perennial ryegrass (*Lolium perenne* L.) showed that silver ions and nanoparticles brought significant reduction in disease severity when applied three hours prior to pathogen inoculation (Jo et al. 2009).

The effect of SNPs on the growth of sclerotium-forming species *Rhizoctonia solani* J.G. Kühn, *Sclerotinia sclerotiorum* (Lib) de Bary and *Scerotinia minor* Jagger was analyzed and revealed that SNPs effectively inhibit the hyphal growth in a dose-dependent manner. Additional, the microscopic observation of hyphae exposed to SNPs showed severe damage resulting in the separation of layers of hyphal wall and collapse of fungal hyphae (Min et al. 2009).

The effective usage of SNPs instead of commercial fungicides has been proved by Lamsal et al. (2011). They evaluated the effect of this nanoparticle against six

Colletotrichum species associated with pepper anthracnose under different culture conditions and concluded that, concentration of 100 ppm inhibited the growth of fungal hyphae as well as conidial germination in vitro when compared to the control. SNPs also showed significantly high inhibition of fungi in field conditions when applied on the plants before disease outbreak (Lamsal et al. 2011).

1.3 Silica Nanoparticles

Mesoporous silica nanoparticles (MSN) can deliver DNA and chemicals into plants for targeted delivery into plant cells (Wang et al. 2002). These are silica (SiO_2) nanoparticles with normally arranged pores that increase the surface area of the nanoparticles (Metha and Mitra 2011). MSN system such as honeycomb was used to deliver DNA and chemicals into plant cells and intact leaves (tobacco and maize). Results indicated that the DNA-coated Type-II MSN can serve as an efficient delivery system for protoplasts and make the DNA accessible to transcription machinery, leading to the transgene expression; also the study showed an application of silica nanoparticles in target-specific delivery of proteins, nucleotides and chemicals in plant biotechnology (Torney et al. 2007).

Nanosized silica-silver particles were applied in field condition to control powdery mildew diseases of cucurbits and 100 % control was achieved after three weeks by Park et al. (2006). Suriyaprabha et al. (2014) studied resistance against phytopathogens such as *Fusarium oxysporum* (Schltdl) and *Aspergillus niger* (P.E.L. van Tieghem) in maize treated with nanosilica (20-40 nm) compared with that of bulk silica and results indicate that nanosilica-treated plant shows a higher expression of phenolic compounds and a lower expression of stress-responsive enzymes against both fungi, also significantly higher resistance in maize treated with nanosilica was found, so they concluded that silica nanoparticles can be used as an alternative potent antifungal agent.

1.4 Alumino-Silicates Nanoparticles

At present alumino-silicates nanoparticles are used for the formulation of pesticides because aluminum-silicate nanotubes can be sprayed on plant surfaces and then they could be easily picked up in insect hairs and these pathogens actively groom and consume pesticide-filled nanotubes. These pesticides are biologically more active and relatively environmentally safe (Sharon et al. 2010; Misra et al. 2013; Patel et al. 2014).

1.5 Chitosan Nanoparticles

Chitosan as nanoparticles has some application in biology because it is biodegradable and non-toxic (Chowdappa and Gowda 2013). These nanoparticles are very effective against plant pathogens like *Fusarium solani* so they can be formulated and applied as a natural antifungal agent (Ing et al. 2012). Depending on size of chitosan nanoparticles and molecular weight of chitosan products they could be used as antibacterial too because antimicrobial activity of chitosan nanoparticles depends on its zeta potential, which plays a significant role in binding with negatively charged microbial membrane (Chen et al. 2010).

Chitosan is considered as a chelated agent because it is possible to chelate various organic and inorganic compounds, being possible improving the stability, solubility and biocidal activity of chelated fungicides or other pesticides (Sudheesh et al. 2013). As an excellent example, copper (Cu) compounds are wide used for their antifungal nature and several times this metal has been used with chitosan for antibacterial and antifungal activities (Sharp 2013). Chitosan also can induce plant defenses in many crops, such as tomato (Benhamou et al. 1994), cucumber (Ben-shalom et al. 2003), strawberry fruits (El Ghaouth et al. 1992) and some others.

Soil improvement with chitosan can control Fusarium wilts (caused by *Fusarium oxysporum* Schlecht) (Rabea et al. 2003) and gray molds (*Botrytis* sp.) (Ben-shalom et al. 2003; Aziz et al. 2006) in a number of crops. Chitosan nanoparticles based technology has a promising future with value in crop productivity in sustained and eco-friendly ways (Kashyap et al. 2015).

2 Fungal Diseases Scenario

Phytopathogenic fungi are one of the main infectious agents in plants, producing alterations during developmental stages including, growth, flowering, maturity, post-harvest, gaining nutrients from the plants they invade and, therefore, resulting in huge economic damage (Pusztahelyi et al. 2015). The study of fungal plant pathogens had a great development in the past decade due to the knowledge of their genomic sequence data and resources for functional genomics analysis, including transcriptomics, proteomics, and metabolomics. All these studies together with targeted mutagenesis or transgenic have shown the complex mechanisms involving pathogenesis and avoidance (Walter et al. 2010).

Many fungi can produce a type of microscopic structure such as a spore that is necessary for their dispersal and/or survival under adverse conditions (Stajich et al. 2009). However, some fungi also produce fungal tissue called sclerotia that are persistent structures and help them to survive severe conditions such as freezing temperatures, desiccation, microbial attack, or the long-term absence of a host (Coley-Smith and Cooke 1971). Reports of sclerotium formation in species from 85 fungal genera in at least 20 orders of Basidiomycota and Ascomycota have been documented by Smith et al. (2015).

2.1 *Basidiomycetes, genera: Sclerotium, Rhizoctonia*

Basidiomycota (basidiomycetes) make up 32 % of the described fungi and include most wood-decaying species, as well as pathogens and mutualistic symbionts (Riley et al. 2014). The genus *Sclerotium* and *Rhizoctonia* belongs to Basidiomycetes class.

Sclerotia have been identified in many fungal lineages; sclerotium formation is primarily recognized as a key life history trait in several necrotrophic plant pathogens (e.g. *Sclerotium rolfsii, S. sclerotiorum, Rhizoctonia solani*). For example, *S. sclerotiorum* and *S. rolfsii* each attack >400 plant species, including major crops such as peanuts, potatoes, and soybeans (Jenkins and Averre 1986; Cintas and Webster 2001).

Rhizoctonia solani causes a wide range of commercially significant plant diseases. It is one of the fungi responsible for Bronw patch (a turfgrass disease), damping off in seedlings, as well as black scurf of potatoes, bare patch of cereals, root rot of sugar beet, belly rot of cucumber, sheath blight of rice, and many other pathogenic conditions (Akira 1987). This pathogen is a species complex with several anastomosis groups (AGs) that differ morphologically and phylogenetically with respect to host range and pathogenicity, susceptibility to different fungicides and geographic distribution (González et al. 2006).

In 2014 the first reported of *Rhizoctonia solani* anastomosis group AG-4 HG-I from the Lao PDR was done by Ireland et al. It was isolated from gai lan (*Brassica oleracea* var. alboglabra) affected by collar rot, seedling death, root rot and stunting of older plants from the Paksong area of Champasak province. The anastomosis group was confirmed by sequencing and Koch's postulates were fulfilled. Besides infecting vegetables in the Brassicaceae, *R. solani* AG-4 is known to have a wide host range, infecting crops in the Chenopodiaceae, Fabaceae and Solanaceae (Ogoshi 1987) and Cucurbitaceae (Kuramae et al. 2003). Limited but potentially suitable rotation vegetable crops should be tested for susceptibility to AG-4 HG-I in the Alliaceae, Apiaceae and Asteraceae (Ireland et al. 2014).

Fungi in genus Sclerotium form sclerotia and sterile mycelia but no spores (Saccardo 1899). These fungi are known to cause several diseases such as damping-off of seedlings, stem canker, crown blight, root, crown, bulb, tuber and fruit rots. Sclerotial diseases affect a wide variety of plants like vegetables, legumes, cereals, forage plants and weeds. Sclerotium includes more than 40 plant-pathogenic species (Farr et al. 2008).

Sclerotium rolfsii is a soil-borne pathogen that can cause many diseases in numerous crops worldwide in the tropics, subtropics, and other warm temperate regions. It has an extensive host range and at least 500 species in 100 families are susceptible. Crops such as cocoyam, tomato, lentil, turmeric, groundnut, betel vine and many others are drastically infected by this fungus resulting in significant crop loss. *S. rolfsii* by the formation of sclerotia can maintain continuity of generation under adverse situation (Vivek et al. 2014). Chemical control is the most effective method against this pathogen but it is expensive, toxic and can destroy benefice microorganism. The leaf extract of *Pimenta dioica* completely inhibited the mycelial growth of *S. rolfsii*, also leaf extract of *Aglaia roxburghiana* and bark extract of *Persea macrantha* can be used as natural fungicides to control this pathogen (Vivek et al. 2014).

Sclerotinia disease that affect lettuce, among some others crops, caused by the fungus *Sclerotinia sclerotiorum* is a major problem worldwide with losses of up to 50 % (Young et al. 2001). Infection can be initiated by ascospores, which are released from apothecia produced through carpogenic germination of soil borne sclerotia near the soil surface. Sclerotinia disease control is mainly through fungicides resulting in a lack of resistant in lettuce varieties. The target of fungicides in this case should be the airborne ascospores, but accurate spray timing to achieve good control and avoid unnecessary applications is difficult and is a challenge in all crops affected by *S. sclerotiorum*. A nice solution to this big problem may be the developing of the forecasting models based on the key biotic and abiotic factors governing the development of Sclerotinia disease in order to try and predict periods of risk. This approach has been attempted for several crops including bean, carrot and oilseed rape (Koch et al. 2007; Foster et al. 2011).

A model proposed by Clarkson et al. (2014) was validated by a further series of independent controlled environment experiments where both Relative Humidity and temperature were varied and generally simulated the pattern of disease development well.

White mould is a disease caused by *S. sclerotiorum* and occurs in most of the common bean producing regions of the world, causing significant crop damage. For example, in Florida *S. sclerotiorum* caused an estimated 5–10 % annual losses (Purdy 1979), while reductions of 40–70 % in dry bean yields grown under irrigation are commonly observed in Brazil (Nasser et al. 1995). In 2013 this disease was reported by the first time in Cuba where a disease incidence was found between 21 and 85 % (Martínez-de la Parte et al. 2013).

2.2 Ascomycetes, genera: *Alternaria*, *Fusarium*, *Verticillium*

Ascomycota is a division or phylum that the members are commonly known as the sac fungi or ascomycetes. They are the largest phylum of Fungi, with over 64,000 species. The defining feature of this fungal group is the “ascus”, meaning “sac” or “wineskin”, a microscopic sexual structure in which nonmotile spores, called ascospores, are formed. However, some species of the Ascomycota are asexual, meaning that they do not have a sexual cycle and thus do not form asci or ascospores (Lutzoni et al. 2004).

The cosmopolitan fungal genus *Alternaria* contains several saprophytic and pathogenic species. Based on phylogenetic and morphological studies, the genus is divided into 26 sections. Alternaria section contains most of the small-spored of Alternaria species with concatenated conidia, including important plant, human and postharvest pathogens. Species within section Alternaria have been mostly described based on morphology and/or host-specificity, yet molecular variation between them is minimal. Woudenberg et al. (2015) investigated how the described morphospecies within section Alternaria are supported by molecular data, and they sequenced the whole-genome of nine Alternaria morphospecies supplemented with transcriptome sequence of 12 Alternaria morphospecies as well as multi-gene sequence of 168 Alternaria isolates. Results showed that the assembled genomes ranged in size from 33.3 to 35.2 Mb within section Alternaria and from 32.0 to 39.1 Mb for all Alternaria genomes. The number of repetitive sequences differed significantly between the different Alternaria genomes; ranging from 1.4 to 16.5 %. The repeat content within section Alternaria was relatively low with only 1.4–2.7 % of repeats. Whole-genome alignments revealed 96.7–98.2 % genome identity between section Alternaria isolates, compared to 85.1–89.3 % genome identity for isolates from other sections to the *A. alternata* reference genome (Woudenberg et al. 2015). These results provide unique fixed nucleotides that will help plant pathologists and medical mycologists to choose which genes to sequence for quick and accurate identification of their species of interest.

Alternaria alternata is known as the cause of leaf spot and other diseases in over 100 host species of plants (Rotem 1994), but also as postharvest disease in various crops (Coates and Johnson 1997). Some other examples of plant pathogens in section Alternaria include *A. longipes*, the causal agent of brown spot of tobacco, *A. mali*, the causal agent of Alternaria blotch of apple, *A. gaisen*, the causal agent of black spot of Japanese pear, *A. arborescens*, the causal agent of stem canker of tomato and *A. solani*, the causal agent of early blight of potato and tomato. The first descriptions of the *A. alternata, A. tenuissima, A. cheiranthi* and *A. brassicicola* species-groups, based on sporulation patterns, were made by Simmons (1995). More recent molecular-based studies revealed that Alternaria species cluster in several distinct species clades, now referred to as sections (Lawrence et al. 2013; Woudenberg et al. 2013), which do not always correlate with the species-groups that were delineated based on morphological characteristics.

Many *Fusarium* species are globally distributed and are economically important as producers of toxic secondary metabolites and infective agents of plants, animals and humans (Leslie and Summerell 2006). Some of the most characteristics examples include *Fusarium poae*, *F. verticillioides* and members of the *F. solani* species complex (FSSC), *F. oxysporum* species complex (FOSC) and the *F. graminearum* species complex (FGSC) (Streit et al. 2012). Although most cultivated plants are host to one or more pathogens in this genus (Leslie and Summerell 2006).

Five novel Fusarium species (*F. fracticaudum, F. marasasianum, F. parvisorum, F. pininemorale* and *F. sororula*) were described by Herron et al. (2015). Three of the species identified (*F. marasasianum, F. parvisorum* and *F. sororula*) are aggressive pathogens of *Pinus* spp. in Colombia. This study allow to diagnose new diseases and to improve quarantine measures to stop the spread of these pathogen to others areas (Herron et al. 2015).

F. proliferatum was reported in 2015 in Canada as a causal agent of soybean root rot being most aggressive of the four Fusarium species identified (*F. avenaceum, F. culmorum, F. oxysporum and F. proliferatum*), causing the greatest root rot severity and reduction of seedling emergence. Identification of *F. proliferatum* was confirmed by PCR analysis with the *F. proliferatum*-specific primer set CLPRO1/CLPRO2. Amplicons of the target fragments (partial calmodulin (cld) gene, 526 bp) were obtained only from DNA of isolates tentatively identified as *F. proliferatum*, and sequencing of the amplicon showed it shared 100 % identity with the cld gene sequences of *F. proliferatum* in GenBank (Chang et al. 2015).

The fungal genus *Verticillium* has several pathogenic species that are distributed worldwide in temperate and subtropical regions (Fradin and Thomma 2006). *V. dahliae* and *V. albatrum* are well described, as they cause severe wilting diseases on many economically important crops. The host-range of the near-diploid fungus *V. longisporum*, named after its characteristic spores, is limited to Brassicaceae, where it causes severe crop loss on oilseed rape (Klosterman et al. 2011). As *V. longisporum* infection does not cause pleiotropic wilting symptoms, it might serve as an interesting model for studying long distance signaling between infected roots and shoots (Iven et al. 2012).

Verticillium dahliae Kleb. is a phytopathogenic fungus that causes a severes wilt disease in many crops, including several economically important ones (Fradin and Thomma 2006). Many molecules purified from *V. dahliae* culture fluids have behaved both as defense elicitors and (or) pathogenicity factors (Yao et al. 2011). Previously was reported the isolation of a secreted protein elicitor PevD1 from *V. dahliae*, which could trigger hypersensitive response (HR) and induce systemic acquired resistant (SAR) to tobacco mosaic virus (TMV) in tobacco (Wang et al. 2012).

2.3 Fungal Postharvest Diseases

One of the most important losses in agricultural production, involving the greatest costs on the farm economy are because postharvest diseases. It is estimated that worldwide between 10 and 40 % losses of agricultural production occur postharvest. Losses are more severe in developing than developed nations of the world. Several species of fungi participate in postharvest deterioration and rots of tubers and agro-produce. These include species of *Aspergillus, Fusarium, Colletotrichum,* amongst some others (Enyiukwu et al. 2014). Shukla et al. (2012) reported that in developing countries, the greatest losses in yam tubers and cereals are fungi induced.

Tomato (*Solanum lycopersicum* Mill) is the world's most processed crop with the magnitude of microbes-induced postharvest losses in the fresh fruits estimated between 25 and 80 % (Ijato et al. 2011). Postharvest losses in tomato, according to Arya (2010), are about 1–20 % in USA and in India about 10–40 %. Losses of up to 20–30 % occur in onion (*Allium cepa* L.) due to black mould caused by *Aspergillus niger* Van Tieghem (Gupta et al. 2012).

Aspergillus flavus Link. is a fungal pathogen that can colonized maize (*Zea mays* L.) and peanut (*Arachis hypogaea* L.) resulting in the contamination of kernels with carcinogenic mycotoxins known as aflatoxins leading to economic losses and potential health threats to humans. By utilizing biomarkers, breeding programs can be optimized to select not only for aflatoxin resistance but also for associated abiotic stress tolerance (Fountain et al. 2015). Aflatoxins not only have been associated with numerous diseases and disorders in humans and livestock, but also have a negative economic impact due to loss of crop value (Fountain et al. 2014; Wild and Gong 2010).

The genus Colletotrichum has a number of plant pathogens of major importance. This fungus causes diseases in a variety of woody and herbaceous plants and it is distributed primarily in tropical and subtropical climates. Several fruit production is especially affected, such as strawberry, mango, citrus, avocado and banana. Colletotrichum species cause devastating disease of coffee berries in Africa, and also they can seriously affect cereals including maize, sugar cane and sorghum. The genus was recently voted the eighth most important group of plant pathogenic fungi in the world, based on perceived scientific and economic importance (Dean et al. 2012).

Pepper (*Capsicum* spp.) is an important crop for many countries that is affected by anthracnose (*Colletotrichum* spp.). This disease affect fruits at pre- and post-harvest stage, leads to severe economic losses in tropical and subtropical areas and it is one of the main barriers to pepper production (Kim et al. 2008; Xia et al. 2011). Pepper anthracnose is caused by several Colletotrichum spp., including *C. acutatum* (teleomorph *Glomerella acutata*), *C. gloeosporioides* (teleomorph *Glomerella cingulata*), *C. capsici* (a synonym of *C. dematium*), and *C. coccodes* (Park and Kim 1992). *C. acutatum* and *C. gloeosporioides* are the most destructive and widely distributed (Sarath Babu et al. 2011). These fungi attack fruit in this crop at both the green and the red fruit stages, and can affect also leaves and stems. Typical anthracnose symptoms on pepper fruit are sunken necrotic tissues, with concentric rings of acervuli. These fruit blemishes lead to unmarketability (Than et al. 2008).

The main sources of resistance to anthracnose diseases have been identified in two pepper species, *Capsicum baccatum* L. and *C. chinense* Jacq., by Asian Vegetables Research and Development Center (AVRDC) in 1999, and researchers have used these sources to study the inheritance of anthracnose resistance (Kim et al. 2010; Lee et al. 2010). Genetic analyses of segregating populations showed that the resistance inheritance pattern varied depending on the Colletotrichum species and isolate, the resistance source, and also the fruit maturation stage (ChunYing et al. 2015).

3 Plant Fungal Pathogens in Biosynthesis of Nanoparticles

Nanotechnology offers some green and eco-friendly solutions for plant disease management and also can be used as bio-manufacturing units, which will provide an added benefit in being easy to use, as compared to other. Some fungal species are non-pathogenic and combined with the simplicity of production and handling will improve the mass production of nanoparticles. Mycosynthesis of gold, silver, gold-silver alloy, selenium, tellurium, platinum, palladium, silica, titania, zirconia, quantum dots, usnic acid, magnetite, cadmium sulphide and uraninite nanoparticles has also been reported by various researchers (Narayanan and Sakthivel 2010). Application of nanotechnology in plant pathology is still in the early stages. For example, nanofungicides, nanopesticides and nanoherbicides are being used extensively in agriculture practices (Alghuthaymi et al. 2015; Prasad et al. 2014).

The metal nanoparticles mycosynthesis is the use of fungi in nanotechnology for the synthesis of nanoparticles because filamentous fungi can grow on readily available and inexpensive substrates, and also they have the ability to produce a wide range of commercially interesting metabolites that have attracted considerable interest to exploit them as production microorganisms in biotechnology (Sastry et al. 2010; Dhillon et al. 2012).

3.1 Biosynthesis of Silver Nanoparticles (SNPs)

Extracellular biosynthesis of SNPs by *Fusarium solani* (USM-3799), a phytopathogen causing disease in onion, when challenged with 1 mM silver nitrate ($AgNO_3$) was reported by Ingle et al. (2009). They study the formation of nanoparticles and characterized them by visual observation followed by UV–Vis spectrophotometric analysis, which showed a peak at about 420 nm, a specific for SNPs. Further analysis carried out by Fourier Transform Infrared Spectroscopy (FTIR), provides evidence for the presence of proteins as capping agent, which helps in increasing the stability of this synthesized nanoparticles and the formation of SNPs was confirmed by Transmission Electron Microscopy (TEM).

Dhandhukia et al. (2012) reported a rapid and extracellular synthesis of SNPs using a plant pathogenic fungus *F. oxysporum* f.sp. cubense (Foc) by the incubation of Foc mycelium with silver nitrate solution produced SNPs in 90 min. SNPs were characterized by UV–Vis spectroscopy, FTIR and TEM. The particles synthesized were in range of 10–100 nm, capped by proteins and possess antimicrobial activity against *Pseudomonas* sp. The pathogenic fungus *F. oxysporum* f.sp. cubense (Foc) was isolated from wilt infected banana plants. Some further studies are necessary to know if the bacteria can develop resistance against nanoparticles and the toxicity for human before their use.

A research focused on extracellular synthesis of SNPs using cell free culture supernatant of strain GP-23 of *Bacillus* species was confirmed through 16S rRNA sequence analysis. The nanoparticles were characterized by FTIR, XRD (X-ray diffraction) and they were found to be spherical in shape with size in the range of 7–21 nm, also they were stable in aqueous solution for 5 months period of storage at room temperature under dark condition. The biosynthesized SNPs exhibited strong antifungal activity against plant pathogenic fungus, *F. oxysporum* at the concentration of 8 μg ml^{-1} so, this result suggest that the synthesized SNPs act as an effective antifungal agent/fungicide (Gopinath and Velusamy 2013).

Sarsar et al. (2015) synthesized SNPs by using the filtrate extract of novel fungal strain *Penicillium atramentosum* KM and they found as evident from the FTIR that the protein components of fungal extract caused the reduction of silver nitrate, the experiment also showed that SNPs had a characteristic UV–visible peak at 420 nm with an average size of 5–25 nm. The XRD (X-ray diffraction) record exhibited the characteristic peaks of 111, 200, 220 and 311 nanoparticles signifying that these nanoparticles were crystalline in nature. The synthesized SNPs showed antimicrobial activity against bacterial strains.

3.2 Biosynthesis of Gold Nanoparticles

The bioreduction of aqueous $AuCl_4^-$ ions by the fungus *Verticillium* sp. has been demonstrated by Mukherjee et al. (2001). They obtained the reduction of the noble metal ions on the surface of the mycelia as well as on the cytoplasmic membrane leading to the formation of gold nanoparticles of fairly well defined dimensions and good monodispersity. These nanoparticles were bound to the surface of the fungal cells and can be used for other applications including catalysis and as precursors for synthesis of coatings for electronic applications.

Gardea-Torresdey et al. (2002) have demonstrated the synthesis of gold nanoparticles within live alfalfa plants by gold uptake from solid media where the Au (III) ions are reduced to Au (0) by the plant and then the atoms are absorbed into the plant where the nucleation and growth of gold nanoparticles takes place. The method can be used to decontaminating soil polluted with heavy metal ions.

Geranium leaves (*Pelargonium graveolens*) and its endophytic fungus (*Colletotrichum* sp.) can be used for the extra-cellular synthesis of gold nanoparticles. The biogenic gold nanoparticles synthesized using the fungus were essentially spherical in shape while the particles grown using the leaves exhibited a variety of shapes that included rods, flat sheets and triangles, this shape variability are not clear at this stage, the possibility of achieving nanoparticle shape control in a host leaf–fungus system is potentially stimulating (Shankar et al. 2003).

Gold and gold-silver nanoparticles (Au and Au-Ag) have synthesized by Sawle et al. (2008) as crystallized and spherical-shaped particles that were stabilized using a fungus, *F. semitectum* (non-pathogenic saprobe fungus) in an aqueous system. The result was a colloidal suspension highly stable for many weeks that were characterized by the surface plasmon resonance (SPR) peaks using a UV–vis spectrophotometer, and the structure, morphology and size were determined by FTIR, XRD, and TEM. The biosyntheses of metallic and bimetallic nanoparticles are important to an increase the efficiency of biosynthetic procedures using environment-benign resources as an alternative to chemical synthesis protocols.

3.3 Biosynthesis of Cadmium Sulphide Nanoparticles

In the past decade it has been found that the ascomycetous fungus *F. oxysporum* has several advantages for nanoparticles production because it has sulphoxide reductases, which are active in the medium in the presence of appropriate metal salts known to mediate the reduction of sulphate ions and can thus generate cadmium sulphide (CdS) nanoparticles extracellularly (Ahmad et al. 2002).

El-Shanshoury et al. (2012) described a rapid and low cost biosynthesis of using culture supernatants of *Escherichia coli* ATCC 8739, *Bacillus subtilis* ATCC 6633 and *Lactobacillus acidophilus* DSMZ 20079 T. They found that nanoparticles were formed within 24 h and them UV–vis spectroscopy study revealed the build-up of absorption bands at 419.5, 381.5 and 362.5 nm for *E. coli* ATCC 8739, *B. subtilis* ATCC 6633 and

L. acidophilus DSMZ 20079 T, respectively for assisted synthesis of CdS nanoparticles. XRD, TEM and fluorescence spectral analyses were used to determine the formation of CdS nanoparticles. This investigation result may help in the development of an easy and eco-friendly route for the synthesis of CdS nanoparticles.

Cadmium sulphide nanoparticles are considered very promising in applied chemistry, bioscience and medicine so stable luminescent CdS nanocrystals can be synthesized by incubating *P. ostreatus* mycelium with inorganic CdS and sodium sulphide, TEM revealed that the obtained quantum dots were of a spherical shape and predominantly from 4 to 5 nm in size and the electron diffraction pattern confirmed the wurtzite crystalline structure of the synthesized CdS quantum dots. The results from this research confirmed for the first time that the system based on basiodiomycota fungi could be considered promising for synthesizing semiconductor quantum dots (Borovaya et al. 2015).

4 Advances, Challenges and Potential of Nanobiotechnology in Fungal Diseases Management

Sustainable agriculture relies heavily upon high inputs of pesticides to protect crops against pathogens and pests. To preserve biodiversity, it is becoming to reconsider the strategies used for disease management by alternate approaches such nanotechnology. Nanofungicides based on nano sized inorganic molecules have demonstrated their capability to tackle plant fungal pathogens and hold promise in the future. However, current research is focused in the development of nanofungicides based on biopolymers and hybrid nano products designed to obtain cheaper, more reliable and eco-friendly products (Prasad et al. 2014).

4.1 Towards Eco-friendly Nanofungicides

Fungal plant diseases are generally managed with the application of chemical fungicides, but it induces non-specific effects destroying both beneficial and pathogenic organisms. Such ecological disorders open the route to undesirable health and environmental risks (Manczinger et al. 2002). Recent applications of nanotechnology are focused in the development of more unhazardous (greener) pesticides with fewer effects to human and animals, avoiding or rationalizing the use of synthetic fungicides. Plant and microorganisms secondary metabolites have been used in the formulation of nanoparticles increasing the effectiveness of active ingredients used to reduce the spread of plant diseases since they represent a rich source of bioactive chemicals, biodegradable and non-polluting molecules. Most of synthetic methods for the NPs production are based in the use highly contaminant organic solvents; however the use of eco-friendly organisms is compatible with the green chemistry principles (Raveendran et al. 2003; Prasad 2014; Prasad et al. 2015).

4.1.1 Inorganic-Based Nanofungicides

Different types of nanomaterials based in metals including titanium, palladium, aluminum copper, zinc, gold, iron and silver have been developed, but silver SNPs have proved to be most effective molecule as it exhibit potent antifungical efficacy (Guo et al. 2003). Besides chemical nanoparticles synthesis, current methods use biofactories by exploiting the capability of organisms (plants and microorganisms) to hyperaccumulate metals and transform them in nanoparticles in order to obtain inorganic molecules with desirable nano characteristics. Nowadays the research in the field is focused mainly in the biosynthesis of CuO NPs (690 papers), Au and ZnO NPs (4640 papers) and SNPs (7699 papers) (Thakkar et al. 2010).

The antimicrobial activity of titanium is well recognized and several studies have suggested that applying titanium dioxide to crops can suppress fungal pathogens development (Norman and Chen 2013). The main advantage of titanium formulations is its lower ecological and toxicological risks compared with other metal-based nanoformulations. It is generally believed that titanium microorganism disinfection relies on the interaction between microorganisms and reactive oxygen species (ROS) generated by photocatalysis upon its activation from proper light illumination. This capability is very desirable for pathogen inactivation because it causes decomposition of micotoxins produced by fungi. The disinfection of wheat phytopathogen *Fusarium graminearum* macroconidia was demonstrated using a visible-light activated titanium oxide nanoparticle photocatalysis (Zhang et al. 2013). In other study, the biocidal effectiveness of zinc titanium oxide ($ZnTiO_3$) nanopowder against the fungus *Aspergillus niger* was assessed. Strong inhibition of pathogen growth was observed in plate assays (Ruffolo et al. 2010).

Zinc oxide (ZnO) is a non-toxic chemically stable compound exhibiting antimicrobial and photocatalytic characteristics (Bhuyan et al. 2015). ZnO nanoparticles were synthetized from weed plant *Parthenium hysterophorus* by an inexpensive, ecofriendly and simple method. Nanoparticles displayed an antifungal activity against *Aspergillus flavus* and *Aspergillus niger* by using a concentration of 25 μg ml^{-1} of NPs in a size particle dependent manner (Rajiv and Venckatesh 2013). A promising application of Zn NPs resides in its capability to synergistically act with pesticides to produce residue-free antifungal products. Oxidative stress caused by ZnO NPs was enhanced by pesticide thiram that results in a synergistic antifungal effect against plant pathogen *Phytohtora capsici*, following photocatalytic degradation of pesticide avoiding negative influence of pesticide residue on human health and environment (Xue et al. 2014).

Copper is a microelement required for important plant physiological processes as protein regulation, photosynthetic electron transport, mitochondrial respiration, oxidative stress response and acts as cofactor of different enzymatic reactions. Copper-based nanoparticles are of great interest because of low cost, availability and antimicrobial properties (Shobha et al. 2014). The Cu antimicrobial activity reside in its capability to induce the production on hydroxyl radical that subsequently bind with DNA molecules and lead a disorder of helical structure by cross-linking with and between the nucleic acid strand and damage essential proteins by binding to sulfhydryl amino and carboxyl groups of amino acids.

CuNPs synthesized by chemical reduction of Cu^{2+} in the presence of cetyl trimethyl ammonium bromide and isopropyl alcohol demonstrated significant antifungal activity against plant pathogenic fungi: *Phoma destuctiva*, *Curvularia lunata*, *A. alternata*, and *F. oxisporum* (Kanhed et al. 2014).

Nowadays silver is accepted as agrochemical replacement eliminating unwanted microorganisms in plants, soils and hydroponic systems. Applications of SNPs in seed/seedlings or in soil produces control of the phytopathogen, but also stimulate plant grow by known mechanisms so far (Patel et al. 2014).

Nanosilver particles disrupt cellular transport systems causing rapid accumulation of toxic silver ions interrupting general metabolism and respiration by reacting with molecules. Silver ions also induce the production of ROS, which are detrimental to cells causing damage to lipids, proteins and nucleic acids (Hwang et al. 2008; Prasad 2014; Aziz et al. 2015). Most of fungal pathogens exhibit high inhibition effect to SNPs in a dose dependent manner. DNA loses its ability to replicate, resulting in a decreased expression of ribosomal RNA subunits and enzymes essential to ATP production. Furthermore, it has been hypothesized that silver ions affects the function of membrane-interacting proteins, and in consequence respiratory chain is inhibited. SNPs nano colloidal solutions were evaluated as antifungal compounds against eighteen different plant pathogenic fungi. Results revealed pathogen sensitivity to colloidal SNPs in a dose dependent manner (optimal results at 100 ppm of SNPs in PDA media) (Kim et al. 2012).

In other study, SNPs were biosynthesized by using white radish (*Raphanus sativus*). SNPs showed wide-spectrum antifungal activity against *F. solani*, *F. oxisporum*, *F. graminearum* and *Penicillium expansum* (Safaa et al. 2015).

4.1.2 Polymer-Based and Hybrid Nanofungicides

Polymer-based nanoformulations seem to have great potential according their greater efficacy compared to commercial formulations e.g. slow release, protection against degradation, and low solubility of the AI), which makes them suitable for a large number on applications. The types of polymers considered for nanopesticides consisting mainly of polysaccharides (chitosan, alginates and starch), polyesters (poly-ε-caprolactone and polyethylene glycol). Recently there has been an increase in the use of biodegradable materials or biological origin such corn oil, beeswax, lecithin or cashew gum (Nguyen et al. 2012; Abreu et al. 2012). Chitosan is a natural and versatile biopolymer deriver by partial deacetylation on chitin, derived from cell wall of fungi and insects. This polymer possesses a wide-spectrum of antifungal activities and capability to enhance plant innate defenses (Amborabé et al. 2008). An antifungal dispersion system was prepared by oleoyl-chitosan (O-chitosan) nanoparticles chemically synthetized, and evaluated against several plant pathogenic fungi. Mycellium growth experiment demonstrated that *Nigrospora sphaerica*, *Botryosphaeria dothidea*, *Nigrospora oryzae* and *A. tenuissima* were chitosan-sensitive (Xing et al. 2016). An interesting alternative for chitosan isolation is the use of fungus-derived cell walls as polymer source. Chitosan was obtained from plant pathogen *F. oxysporum* and used to challenge the same pathogen. Treated tomato plants showed delay in wilt disease symptom expression and severity and also enhanced of yield (Sathiyabama and Einstein 2015).

4.2 Nanobiofungicides

The name of nanofungicides is used to describe any fungicide formulation that, (1) intentionally includes entities in the nanometer size range up to 100 nm, (2) is designated with a "nano" prefix (e.g., nanohybrid, nanocomposite), and/or (3) is claimed to have novel properties associated with the small size (Abd-Elsalam and Alghuthaymi 2015).

In this entire chapter the use of metal nanoparticles to control plant diseases has been explain. Some important examples are:

- A development of an eco-friendly fungicide that use nanomaterials to liberate its pathogen killing properties only when it is inside the targeted fungal pathogen (Choudhury et al. 2010);
- Copper nanoparticles dissolved in water have been used since 1930s as a fungicide for controlling grapes and fruit trees diseases (Hatschek 1931);
- Chitosan and Cu-chitosan nanoparticles have uniform size and stability, which may contribute to their higher antifungal activity against *A. alternata*, *Macrophomina phaseolina* and *Rhizoctonia* in in vitro studies (Saharan et al. 2013).
- Cu-chitosan nanoparticles also proved to have maximum inhibition rate of spore germination of *A. alternata*. Compared to chitosan and Cu-chitosan nanoparticles, the chitosan-saponin nanoparticles were found poor in antifungal activity (Saharan et al. 2013).

Nanoformulations seems to be a friendly environmental way to control plant fungal diseases but it is important to study the toxicity of nanoparticles that can be dangerous for men and some others organism (Banik and Sharma 2011). For example nanosized silica-silver particles were applied under field condition to control of cucurbit powdery mildew, and 100 % control was achieve after 25 days (30), in this case the nanoparticles were phytotoxic only at a very high dose of 3200 ppm when tested in cucumber and pansy plants.

The application of glucan nanoparticles (GNPs) like β glucan has been study as a resistant activator in turmeric for control of rhizome rot disease (Anusuya and Sathiyabama 2015). The Soil borne Oomycete *Pythium aphanidermatum* is the causal agent of rhizome rot disease, one of the most serious threats to turmeric crops. At present, effective fungicides are not available and the effectiveness of β-D-glucan nanoparticles (GNPs) have been proved by Anusuya and Sathiyabama (2015), found that β-D GNPS significantly reduced the rot incidence offering 77 % protection to *Curcuma longa* (syn *C. domestica*).

Nanofungicides can be new types of biohybrids nanocide materials that will be used as a new environment friendly antimicrobial against different fungal pathogenic organisms of plant (Abd-Elsalam and Alghuthaymi 2015).

A series of cuprous oxide (Cu_2O) nanocrystals with different structures were synthesized by reductive reaction by Qiang (2015) and nanocrystals were obtained in an aqueous mixture of $CuSO_4$, fructose, NaOH and templates at 45–75 °C. The author found an inhibition rate of 90 % towards *Colletotrichum capsici* (Syd.) E.J. Butler and Bisby (causal agent of anthracnose of chili), as soon as 70 mg nano-Cu_2O was added into 100 ml culture medium (Qiang 2015).

4.3 Nanobiosensors for Fungal Plant Pathogen Detection

Nanosensors for plant pathogen and pesticide detection are other area in agriculture that can benefit from nanotechnology. One application of the nanosensors is the enzyme immobilization using nanomaterials (Kim et al. 2006).

Application of nano sensors in agriculture is still at the basic research level. Using these sensors farmers will monitor environmental conditions closely for plant growth and protection. These detection systems can contribute to increased productivity and decreased the use of agrochemicals (e.g. antibiotics, pesticides, nutrients) by early intervention (Ghormade et al. 2011).

4.3.1 Nanobiosensors for Pesticide Detection

Some studies were realized for development of analytical tools to determine pesticide residue. In parallel with typical chromatography, immunochemical assays based on bio-molecules were employed as an alternative for pesticide measurement by virtue of its high selectivity, sensitivity, and reliability as well as its rapidity (Gabaldon et al. 1999).

Several uni-molecular and array type of nanomaterial based biosensors are being developed for detection of pesticides for axample, gold NPs (30 nm) based dipstick competitive immuno-assay with sensitivity of 27 ng ml^{-1} was developed to detect organochlorine pesticide such as DDT (Lisa et al. 2009) because gold NPs have the property of agglomeration associated with color production and this can be used for pesticide detection.

Vinayaka et al. (2009) to detect 2,4-dichlorophenoxyacetic acid (2, 4-D) (herbicide) up to 250 pg l^{-1} used cadmium telluride quantum dots (CdTeQDs), semiconductor fluorescent NPs, in a fluoroimmunoassay.

A nucleic acid sensor was fabricated by Kaushik et al. (2009) via immobilization of single standard calf thymus deoxyribose nucleic acid onto a chitosan nanobiocomposite film containing iron oxide NPs (Fe_3O_4, 22 nm), deposited onto indium-tin-oxide coated glass surface for pyrethroid, cypermethrin and permethrin, detection Acetylcholinesterase coated iron oxide magnetic NPs (30 nm) bound carbon nanotubes and zirconium oxide NPs (31.5 nm) composite on the screen-printed electrode surface for detection of dimethoate, an organophosphorus pesticide was developed by Gan et al. (2010).

For an easy monitoring of the environment the development of multianalytic array sensors, microfluidic devices and cantilever arrays are necessary.

4.3.2 Nanobiosensors for Microorganism Detection

At present several methods for plant pathogens detection are available, but some of them are time consuming like traditional culture based methods (Fletcher et al. 2006), others are specific of reactions, such as the antibody staining specificity of ELISA (Enzyme-Linked ImmunoSorbent Assay) (Uddin et al. 2003), the high cost of nucleic-acid based polymerase chain reaction methods such as restriction fragment length polymorphism, DNA fingerprinting and amplification of the internal transcribed spacer region from rRNA gene increase specificity of identification (Gao et al. 2004; Doorn et al. 2007). The detection technology based in NPs is a novel microbial detection that starts to revolutionize agriculture. Silica-based NPs (60 nm) were filled with a fluorescent dye and conjugated to an antibody specific to a surface antigen of the microbe of interest (Zhao et al. 2004), the method is sensible for the detection of plant pathogens.

The identification of Trichoderma species world-wide is tedious and can lead to some mistake based on micro-morphological descriptions for this reason an electrochemical DNA biosensor was successfully developed based on ionic liquid, ZnO nanoparticles and a chitosan nanocomposite membrane on a modified gold electrode (AuE). A single-stranded DNA probe was immobilized on this electrode and methylene blue (MB) was used as the hybridization indicator to monitor the hybridization reaction of the target DNA. Under optimal conditions using differential pulse voltammetry (DPV), the target DNA sequences were detectable at different concentration ranges such as, 1.0×10^{-18}, and the detectable limit was 1.0×10^{-19} mol l^{-1}. The developed DNA biosensor enables the study of hybridization with crude DNA fragments and the results of this study confirm that this DNA biosensor provides a fast, sensitive and convenient way for the species level identification of *Trichoderma harzianum* RIFAI (Siddiquee et al. 2014).

5 Conclusions and Future

Nanobiotechnology great potential in agriculture specifically in plant protection has been demonstrated. It is gradually evolving from theory to application regime not only in agriculture but in medicine, biology, chemistry and so on. Nanoparticles are used to help the agricultural industry combat viruses, fungi, bacteria and some other crop pathogens increasing the efficiency of fungicides, pesticides and herbicides not only in the way of application but also in form (nanomaterials, nanotubes, mesoporous).

Diminish environmental hazard is another main contribution of nanobiotechnology to human life. In agriculture the use of nanomaterials for delivery of pesticides and fertilizers allow to reduce the dosage and ensure a controlled slow delivery. Nanobiotechnology need to be incorporate increasingly in the agriculture system of the development countries with further research studies and practical application in the fields.

References

Abd-Elsalam KA, Alghuthaymi MA (2015) Nanobiofungicides: are they the next-generation of fungicides? J Nanotechnol Mater Sci 2(1):1–3

Abreu F, Oliveira EF, Paula HCB, de Paula RCM (2012) Chitosan/cashewgumnanogels for essential oil encapsulation. Carbohyd Polymer 89(4):1277–1282

Ahmad A, Mukherjee P, Mandal D et al (2002) Enzyme-mediated extracellular synthesis of CdS nanoparticles by the fungus, *Fusarium oxysporum*. J Am Chem Soc 124:12108–12109

Akira O (1987) Ecology and pathogenicity of anastomosis and intraspecific groups of *Rhizoctonia solani* Kuhn. Annu Rev Phytopathol 25:125–143

Alghuthaymi MA, Almoammar H, Rai M, Said-Galiev E, Abd-Elsalam KA (2015) Myconanoparticles: synthesis and their role in phytopathogens management. Biotech Biotec Eq 29(2):221–236

Amborabé BE, Bonmort J, Fleurat-Lessard P, Roblin G (2008) Early events induced by chitosan on plant cells. J Exp Bot 59:2317–2324

Anusuya S, Sathiyabama M (2015) Foliar application of β–D-glucan nanoparticles to control rhizome rot disease of turmeric. International J Biol Macromol 72:1205–1212

Arya A (2010) Recent advances in the management of plant pathogens: botanicals in the fungal pest management. In: Arya A, Perello AE (eds) Management of fungal plant pathogens. CAB International, UK, pp 1–25

Aziz A, Trotel-Aziz P, Dhuicq L, Jeandet P, Couderchet M, Vernet G (2006) Chitosan Oligomers and copper sulfate induce grapevine defense reactions and resistance to gray mold and downy mildew. Phytopathology 96(11):1188–1194

Aziz N, Faraz M, Pandey R, Sakir M, Fatma T, Varma A, Barman I, Prasad R (2015) Facile algae-derived route to biogenic silver nanoparticles: synthesis, antibacterial and photocatalytic properties. Langmuir 31:11605–11612. doi:10.1021/acs.langmuir.5b03081

Banik S, Sharma P (2011) Plant pathology in the era of nanotechnology. Indian Phytopathol 64(2):120–127

Benhamou N, Lafontaine PJ, Nicole M (1994) Induction of systemic resistance to Fusarium crown and root rot in tomato plants by seed treatment with Chitosan. Phytopatol 84(12):1432–1444

Ben-shalom N, Ardi R, Pinto R, Aki C, Fallik E (2003) Controlling gray mould caused by Botrytis cinerea in cucumber plants by means of chitosan. Crop Prot 22:285–290

Bhuyan T, Mishra K, Khanuja M, Prasad R, Varma A (2015) Biosynthesis of zinc oxide nanoparticles from *Azadirachta indica* for antibacterial and photocatalytic applications. Mat Sci Semicon Proc 32:55–61

Borovaya M, Pirko Y, Krupodorova T, Naumenko A, Blume Y, Yemets A (2015) Biosynthesis of cadmium sulphide quantum dots by using *Pleurotus ostreatus* (Jacq.) *P. Kumm*. Biotech Biotec Eq 29(6):1156–1163

Chang KF, Hwang SF, Conner RL, Ahmed HU, Zhou Q (2015) First report of Fusarium proliferatum causing root rot in soybean (Glycine max L.) in Canada. Crop Prot 67:52–58

Chen H, Yada R (2011) Nanotechnologies in agriculture: new tools for sustainable development. Trends Food Sci Tech 22:585–594

Chen L, Kung SK, Chen HH, Lin SB (2010) Evaluation of zeta potential difference as an indicator for antibacterial strength of low molecular weight chitosan. Carbohyd Polym 82:913–919

Choudhury SR, Nair KK, Kumar R et al (2010) Nanosulfur: a potent fungicide against food pathogen, *Aspergillus niger*. AIP Conf Proc 1276:154–157
Chowdappa P, Gowda S (2013) Nanotechnology in crop protection: status and scope. Pest Manag Hortic Ecosyst 19(2):131–151
ChunYing S, Sheng ML, Hai Z, Alain P, Hao LW, Bao Xi ZH (2015) Resistances to anthracnose (*Colletotrichum acutatum*) of Capsicum mature green and ripe fruit are controlled by a major dominant cluster of QTLs on chromosome P5. Sci Hortic 181:81–88
Cintas NA, Webster RK (2001) Effects of rice straw management on *Sclerotium oryzae* inoculum, stem rot severity, and yield of rice in California. Plant Dis 85(11):1140e1144
Clarkson JP, Fawcett L, Anthony SG, Young C (2014) A model for *Sclerotinia sclerotiorum* infection and disease development in lettuce, based on the effects of temperature, relative humidity and ascospore density. PLoS One 9(4):e94049–e94061
Coates L, Johnson G (1997) Postharvest diseases of fruit and vegetables. In: Brown JF, Ogle HJ (eds) Plant pathogens and plant diseases. Rockvale Publications, Armidale, pp 533–548
Coley-Smith JR, Cooke RC (1971) Survival and germination of fungal sclerotia. Annu Rev Phytopathol 9:65–92
Dean R, Van Kan JAL, Pretorius ZA, Hammond-Kosack KE, Di Pietro A et al (2012) The top 10 fungal pathogens in molecular plant pathology. Mol Plant Pathol 13:414–430
Dhandhukia PC, Patel M, Thakker JN (2012) Biosynthesis of silver nanoparticles using a plant pathogenic fungus. *Fusarium oxysporum* f. Sp. Cubense. J Pure Appl Sci 20:10–14
Dhillon GS, Brar SK, Kaur S, Verma M (2012) Green approach for nanoparticle biosynthesis by fungi. Curr Trends Appl 32:49–73
Doorn RA, Szemes M, Bonants P, Kowalchuk GA, Salles JF, Ortenberg E et al (2007) Quantitative multiplex detection of plant pathogens using a novel ligation probe- based system coupled with universal, high-throughput real-time PCR on Open- Arrays™. BMC Genomics 8:276
El Ghaouth A, Arul J, Grenier J, Asselin A (1992) Antifungal activity of chitosan on two postharvest pathogen of strawberry fruits. Phytopathology 82:398–402
El-Shanshoury R, Elsilk SE, Ebeid ME (2012) Rapid biosynthesis of cadmium sulfide (CdS) nanoparticles using culture supernatants of *Escherichia coli* ATCC 8739, *Bacillus subtilis* ATCC 6633 and *Lactobacillus acidophilus* DSMZ 20079 T. African J Biotech 11(31):7957–7965
Enyiukwu DN, Awurum AN, Nwaner JA (2014) Efficacy of plant-derived pesticides in the control of myco-induced postharvest rots of tubers and agricultural products: A review. Net J Agr Sci 2(1):30–46
Farr DF, Rossman AY, Palm ME, McCray EB (2008) Fungal databases. Systematic Botany & Mycology Laboratory, ARS, USDA. http://nt.ars-grin.gov/fungaldatabases
Fletcher J, Bender C, Budowle B, Cobb WT, Gold SE, Ishimaru CA et al (2006) Plant pathogenforensics: capabilities, needs, and recommendations. Microbiol Mol Biol 7:450–471
Foster AJ, Kora C, McDonald MR, Boland GJ (2011) Development and validation of a disease forecast model for Sclerotinia rot of carrot. Canadian Journal of Plant Pathol 33:187–201
Fountain JC, Scully BT, Ni X, Kemerait RC, Lee RD, Chen ZY, Guo BZ (2014) Environmental influences on maize-*Aspergillus flavus* interactions and aflatoxin production. Front Microbiol 5:1–7
Fountain JC, Khera P, Yang L, Nayak SN, Scully BT, Lee RD, Chen ZY, Kemerait RC, Varshney RK, Guo B (2015) Resistance to *Aspergillus flavus* in maize and peanut: Molecular biology, breeding, environmental stress, and future perspectives. The Crop J 3:229–237
Fradin EF, Thomma BP (2006) Physiology and molecular aspects of Verticillium wilt diseases caused by *V. dahlia* and V. albo-atrum. Mol Plant Pathol 7(2):71–86
Gabaldon JA, Maquieira A, Puchades R (1999) Current trends in immunoassay based kits for pesticide analysis. Crit Rev Food Sci Nutr 39:519–538
Gan N, Yang X, Xie D, Wu Y, Wen WA (2010) Disposable organophosphorus pesticides enzyme biosensor based on magnetic composite nanoparticles modified screen printed carbon electrode. Sensors 10:625–638
Gao X, Jackson TA, Lambert KM, Li S, Hartman GL, Niblack TL (2004) Detection and quantification of *Fusarium solani* f. sp. glycines with real-time quantitative polymerase chain reaction. Plant Dis 88:1372–1380

Gardea-Torresdey JL, Parsons JG, Gomez E, Peralta-Videa J, Troiani HE, Santiago P, Jose-Yacaman M (2002) Formation and growth of Au nanoparticles inside live alfalfa plants. Nano Lett 2:397

Ghormade V, Deshpande MV, Paknikar KM (2011) Perspectives for nano-biotechnology enabled protection and nutrition of plants. Biotech Adv 29(6):792–803

González D, Cubeta MA, Vilgalys R (2006) Phylogenetic utility of indels within ribosomal DNA and β-tubulin sequences from fungi in the *Rhizoctonia solani* species complex. Mol Phylogenet Evol 40:459–470

Gopinath V, Velusamy P (2013) Extracellular biosynthesis of silver nanoparticles using *Bacillus* sp. GP-23 and evaluation of their antifungal activity towards *Fusarium oxysporum*. Spectrochim Acta Part A:Mol Biomol Spectrosc 106:170–174

Guo WZ, Li JJ, Wang YA, Peng XG (2003) Conjugation chemistry and bioapplications of semiconductor box nanocrystals prepared via dendrimer bridging. Chem Mater 15:3125–3133

Gupta R, Khokhar MK, Lai R (2012) Management of black mould of onion. J Plant Pathol Microbiol 3(5):3

Hatschek E (1931) Inventor. Electro Chem Processes Ltd assignee Brouisol. British patent no 392:556

Herron DA, Wingfield MJ, Wingfield BD, Rodas CA, Marincowitz S, Steenkamp ET (2015) Novel taxa in the Fusarium fujikuroi species complex from *Pinus* spp. Stud Mycol 80:131–150

Hwang E, Lee JH, Chae YJ, Kim YS, Kim BC, Sang BI, Gu MB (2008) Analysis of the toxic mode of action of silver nanoparticles using stress-specific bioluminescent bacteria. Small 4(746):750

Ijato JY, Adebiyi AO, Ijadnola JÁ (2011) Antifungal effects of four tropical plants aqueous and ethanolic extracts on postharvest rot tomato (*Lycopersicon esculenta*) in Ado-Ekiti, Nigeria. New York Sci J 4(1):64–68

Ing LY, Zin NM, Sarwar A, Katas H (2012) Antifungal activity of chitosan nanoparticles and correlation with their physical properties. Int J Biomat 2012:1–9

Ingle A, Rai M, Gade A, Bawaskar M (2009) *Fusarium solani*: a novel biological agent for the extracellular synthesis of silver nanoparticles. J Nanoparticle Res 11(8):20179–22085

Ireland KB, Weir BS, Phantavong S, Phitsanoukane P, Vongvichid K, Vilavong S, Tesoriero LA, Burgess LW (2014) First report of *Rhizoctonia solani* anastomosis group AG-4 HG-I in the Lao PDR. Australasian Plant Dis Notes 10:152

Iven T, König S, Singh S, Braus-Stromeyer SA, Bischoff M, Tietze LF, Braus GH, Lipka V, Feussner I, Dröge-Laser W (2012) Transcriptional activation and production of tryptophan-derived secondary metabolites in Arabidopsis roots contributes to the defense against the fungal vascular pathogen *Verticillium longisporum*. Mol Plant 5(6):1389–1402

Jenkins SF, Averre CW (1986) Problems and progress in integrated control of southern blight of vegetables. Plant Dis 70(7):614e619

Jo Y-K, Kim BH, Jung G (2009) Antifungal activity of silver ions and nanoparticles on phytopathogenic fungi. Plant Dis 93:1037–1043

Johnston CT (2010) Probing the nanoscale architecture of clay minerals. Clay Miner 45:245–279

Kanhed P, Birla S, Gaikwad S, Gade A, Seabra AB, Duran N, Rai M (2014) In vitro antifungal efficacy of copper nanoparticles against selected crop pathogenic fungi. Mat Lett 115:13–17

Kashyap PL, Xiang X, Heiden P (2015) Chitosan nanoparticle based delivery systems for sustainable agriculture. Int J Bio Macromol 77:36–51

Kaushik A, Solanki PR, Ansarib AA, Malhotra BD, Ahmad S (2009) Iron oxide-chitosan hybrid nanobiocomposite based nucleic acid sensor for pyrethroid detection. Biochem J 46:132–140

Khodakovsky A, Schroder P, Sweldens W (2000) In Sig graph. Comput Graphics P 271-278

Kim TN, Feng QL, Kim JO, Wu J, Wang H, Chen GC, Cui FZ (1998) Antimicrobial effects of metal ions (Ag^{+}, Cu^{2+}, Zn^{2+}) in hydroxyapatite. J Mater Sci Mater in Med 9:129–134

Kim J, Jia H, Wang P (2006) Challenges in biocatalysis for enzyme-based biofuel cells. Biotechnol Adv 24:296–308

Kim SH, Yoon JB, Do JW, Park HG (2008) A major recessive gene associated with anthracnose resistance to *Colletotrichum capsici* in chili pepper (*Capsicum annuum* L.). Breeding Sci 58:137–141

Kim S, Kim KT, Kim DH, Yang EY, Cho MC, Jamal A, Chae Y, Pae DH, Oh DG, Hwang JK (2010) Identification of quantitative trait loci associated with anthracnose resistance in chili pepper (*Capsicum* spp.). Korean J Hortic Sci 28:1014–1024

Kim SW, Jung JH, Lamsal K, Kim YS, Min JS, Lee YS (2012) Antifungal effects of silver nanoparticles (AgNPs) against various plant pathogenic fungi. Kor Soc Mycol Mycobiol 40(53):58

Klosterman SJ, Subbarao K, Kang S, Veronese P, God SE, Thomma BHJ et al (2011) Comparative genomics yields insights into niche adaptation of plant vascular wilt pathogens. PLoS Pathogen 7, e1002137

Koch S, Dunker S, Kleinhenz B, Roehrig M, von Tiedemann A (2007) Crop loss-related forecasting model for Sclerotinia stem rot in winter oilseed rape. Phytopathology 97:1186–1194

Kuramae E, Buzeto A, Ciampi M, Souza N (2003) Identification of *Rhizoctonia solani* AG 1-IB in Lettuce, AG 4 HG-I in tomato and melon, and AG 4 HG-III in Broccoli and Spinach, in Brazil. Eur J Plant Pathol 109:391–395

Lamsal K, Kim SW, Jung JH, Kim YS, Kim KS, Lee YS (2011) Application of silver nanoparticles for the control of *Colletotrichum* species in vitro and pepper anthracnose disease in field. Mycobiology 39:194–199

Lawrence DP, Gannibal PB, Peever TL, Pryor BM (2013) The sections of Alternaria: formalizing species-groups concepts. Mycologia 105:530–546

Lee J, Hong J, Do J, Yoon J, Lee JD, Hong JH, Do JW, Yoon JB (2010) Identification of QTLs for resistance to anthracnose to two *Colletotrichum* species in pepper. J Crop Sci Biotechnol 13:227–233

Leslie JF, Summerell BA (2006) The Fusarium laboratory manual. Blackwell, Ames

Lisa M, Chouhan RS, Vinayaka AC, Manonmani HK, Thakur MS (2009) Gold nanoparticles based dipstick immuno-assay for the rapid detection of dichlorodiphenyltrichlor-oethane: an organochlorine pesticide. Biosens Bioelectron 25:224–227

Lutzoni F, Kauff F, Cox CJ, Mclaughlin D, Celio G, Dentinger B, Padamsee M, Hibbett D, James TY, Baloch E, Grube M et al (2004) Assembling the fungal tree of life: progress, classification, and evolution of subcellular traits. Am J Bot 91(10):1446–1480

Manczinger L, Antal Z, Kredics L (2002) Ecophysiology and breeding of mycoparasitic *Trichoderma* strains (a review). Acta Microbiol Immunol Hung 49:1–14

Martínez-de la Parte E, Trujillo M, Cantillo-Pérez T, García D (2013) First report of white mould of beans caused by *Sclerotinia sclerotiorum* in Cuba. New Dis Rep 27:1–5

Metha KC, Mitra MA (2011) Plant pathology in the era of nanotechnology. Indian Phytopathol 64(2):120–127

Min JS, Kim KS, Kim SW, Jung JH, Lamsal K, Kim SB, Jung M, Lee YS (2009) Effects of colloidal silver nanoparticles on sclerotium-forming phytopathogenic fungi. J Plant Pathol 25:376–380

Misra AN, Misra M, Singh R (2013) Nanotechnology in agriculture and food industry. Int J Pure Appl Sci Technol 16(2):1–9

Mukherjee P, Ahmad A, Mandal D, Senapati S, Sainkar SR, Khan MI et al (2001) Bioreduction of $AuCl_4^-$ ions by the fungus. Verticillium sp and surface trapping of the gold nanoparticles formed. Angew Chem Int Ed 40(19):3585–3588

Narayanan KB, Sakthivel N (2010) Biological synthesis of metal nanoparticles by microbes. Adv Colloid Interface Sci 156:1

Nasser LCB, Sutton JC, Boland GJ, James TW (1995) Influence of crop residues and soil moisture on *Sclerotinia sclerotiorum* from the Cerrados region in Brazil. Can J Plant Pathol 17:353–365

Nguyen HM, Hwang IC, Park JW, Park HJ (2012) Photoprotection for deltamethrin using chitosan-coated beeswax solid lipid nanoparticles. Pest Manag Sci 68(7):1062–1068

Norman D, Chen J (2013) Effect of foliar application of titanium dioxide on bacterial blight of *Geranium* and *Xanthomonas* leaf spot of poinsettia. Hort Sci 48:189–192

Ogoshi A (1987) Ecology and pathogenicity of anastomosis and intraspecific groups of *Rhizoctonia solani* kuhn. Annu Rev Phytopathol 25:125–143

Park K, Kim C (1992) Identification, distribution and etiological characteristics of anthracnose fungi of red pepper in Korea. Kor J Plant Pathol 8:61–69

Park HP, Kim SH, Kim HJ, Choi HS (2006) A new composition of nanosized silica-silver for control of various plant diseases. Plant Pathol J 22(3):295–302

Prasad R (2014) Synthesis of silver nanoparticles in photosynthetic plants. J Nanoparticles, Article ID: 963961. doi:10.1155/2014/963961

Prasad R, Kumar V, Prasad KS (2014) Nanotechnology in sustainable agriculture: present concerns and future aspects. Afr J Biotechnol 13(6):705–713

Prasad R, Pandey R, Barman I (2015) Engineering tailored nanoparticles with microbes: quo vadis. WIREs Nanomed Nanobiotechnol. doi:10.1002/wnan.1363

Patel N, Desai P, Patel N, Jha A, Gautam HK (2014) Agronanotechnology for plant fungal disease management: a review. Int J Curr Microbiol Appl Sci 3(10):71–84

Purdy LH (1979) *Sclerotinia sclerotiorum*: history, diseases and symptomatology, host range, geographic distribution, and impact. Phytopathology 69:875–880

Pusztahelyi T, Holb IJ, Pócsi I (2015) Secondary metabolites in fungus-plant interaction. Front Plant Sci 6:1–23

Qiang W (2015) Synthesis of Cu_2O nanocrystals and their agricultural application. Bulg Chem Commun 47(3):929–936

Rabea EI, Badawy M, Stevens C, Smagghe G, Steurbaut W (2003) Chitosan as antimicrobial agent: applications and mode of action. Biomacromol 4(6):1457–1465

Rajiv SR, Venckatesh R (2013) Bio-Fabrication of zinc oxide nanoparticles using leaf extract of *Parthenium hysterophorus* L. and its size-dependent antifungal activity against plant fungal pathogens. Spectrochim Acta Part A Mol Biomol Spectrosc 112:384–387

Raveendran P, Fu J, Wallen SL (2003) Completely green synthesis and stabilization of metal nanoparticles. J Ame Chem Soc 125(46):13940–13941

Riley R, Salamov AA, Brown DW, Nagy LG, Floudas D, Held BW, Levasseur A, Lombard V, Morin E, Otillar R et al (2014) Extensive sampling of basidiomycete genomes demonstrates inadequacy of the white-rot/brown-rot paradigm for wood decay fungi. Proc Natl Acad Sci U S A 111(27):9923–9928

Rotem J (1994) The genus Alternaria. Biology, epidemiology and pathogenicity. APS, St. Paul

Ruffolo SA, La Russa MF, Malagodi M. Rossi CO, Palermo AM, Crisci GM (2010) ZnO and ZnTiO3 nanopowders for antimicrobial stone coating. Appl Phys A 100(3):829–834

Saccardo PA (1899) Saccardo's Sylloge Fungorum XIV. Edwards JW (ed) Edwards Bros Inc., Ann Arbor, pp 1141–1154

Safaa M, NaeimaA YMH, Nafady NA (2015) Application of biosynthesized silver nanoparticles for the control of land snail *Eobania vermiculata* and some plant pathogenic fungi. J Nanomaterials 2015:1–10

Saharan V, Mehrotraa A, Khatik R, Rawal P, Sharma SS, Pal A (2013) Synthesis of chitosan based nanoparticles and their in vitro evaluation against phytopathogenic fungi. Int J Biol Macromol 62:677–683

Sarath Babu B, Pandravada S, Prasada RR, Anitha K, Chakrabarty S, Varaprasad K (2011) Global sources of pepper genetic resources against arthropods, nematodes and pathogens. Crop Prot 30:389–400

Sarsar V, Selwal MK, Selwal KK (2015) Biofabrication, characterization and antibacterial efficacy of extracellular silver nanoparticles using novel fungal strain of *Penicillium atramentosum* KM. J Saudi Chem Soc 19:682–688

Sastry RK, Rashmi HB, Rao NH, Ilyas SM (2010) Integrating nanotechnology (NT) into agri-food systems research in India: a conceptual framework. Technol Forecast Soc Change 77:639–648

Sathiyabama M, Einstein RC (2015) Fungal cell wall polymer based nanoparticles in protection of tomato plants from wilt disease caused by *Fusarium oxysporum* f.sp. *lycopersici*. Carbohydr Polym 133:400–407

Saurabh S, Bijendra KS, Yadav SM, Gupta AK (2015) Applications of nanotechnology in agricultural and their role in disease management. J Nanosci Nanotechnol 5:1–5

Sawle DB, Salimath B, Deshpande R, Bedre MD, Prabhakar BK, Venkataraman A (2008) Biosynthesis and stabilization of Au and Au–Ag alloy nanoparticles by fungus, *Fusarium semitectum*. Sci Technol Adv Mater 9(3):1–6

Shankar SS, Ahmad A, Parischa R, Sastry M (2003) Bioreduction of chloroaurate ions by geranium leaves and its endophytic fungus yields gold nanoparticles of different shapes. J Mater Chem 13:1822–1826

Sharon M, Choudhary AK, Kumar R (2010) Nanotechnology in agricultural diseases and food safety. J Phytol 2(4):83–92

Sharp RG (2013) A review of the applications of chitin and its derivatives in agriculture to modify plant-microbial interactions and improve crop yields. Agronomy 3(4):757–793

Shobha G, Moses V, Ananda S (2014) Biological synthesis of copper nanoparticles and its impact – a review. Int J Pharmaceut Sci Invention 3(8):28–38

Shukla AM, Yadav RS, Shashi SK, Dikshit A (2012) Use of plant metabolites as an effective source for the management of postharvest fungal pest: A review. Inter J Curr Discov Innov 1(1):33–45

Siddiquee S, Rovina K, Yusof NA, Rodrigues KF, Suryani S (2014) Nanoparticle-enhanced electrochemical biosensor with DNA immobilization and hybridization of *Trichoderma harzianum* gene. Bio Sensing Res 2:16–22

Simmons EG (1995) Alternaria themes and variations (112–144). Mycotaxon 55:55–163

Smith ME, Henkel TW, Rollins JA (2015) How many fungi make sclerotia? Fungal Ecol 13:211–220

Stajich JE, Berbee ML, Blackwell M, Hibbett DS, James TY, Spatafora JW, Taylor JW (2009) The fungi. Curr Biol 19(18):R840–R845

Streit E, Schatzmayr G, Tassis P, Tzika E, Marin D, Taranu I, Tabuc C, Nicolau A, Aprodu I, Puel O, Oswald IP (2012) Current situation of mycotoxin contamination and co-occurrence in animal feed-focus on Europe. Toxins 4:788–809

Sudheesh SK, Mishra AK, Arotiba OA, Mamba BB (2013) Chitosan-based nanomaterials: a state-of-the-art review. Inter J Biol Macromol 59:46–58

Suriyaprabha R, Karunakaran G, Kavitha K, Yuvakkumar R, Rajendarn V, Kannan N (2014) Application of silica nanoparticles in maize to enhance fungal resistance. IET Nanobiotechnol 8(3):133–137

Thakkar KN, Mhatre SS, Parikh RY (2010) Biological synthesis of metallic nanoparticles. Nanomedicine 6:257–262

Than PP, Jeewon R, Hyde KD, Pongsupasamit S, Mongkolporn O, Taylor PWJ (2008) Characterization and pathogenicity of *Colletotrichum* species associated with anthracnose on chilli (*Capsicum* spp.) in Thailand. Plant Pathol 57:562–572

Torney F, Trewyn BG, Lin VSY, Wang K (2007) Mesoporous silica nanoparticles deliver DNA and chemicals into plants. Nat Nanotechnol 2:295–300

Uddin W, Viji G, Schumann GL, Boyd SH (2003) Detection of *Pyricularia grisea* causing gray leaf spot of perennial ryegrass turf by a rapid immuno-recognition assay. Plant Dis 87:772–778

Vinayaka AC, Basheer S, Thakur MS (2009) Bioconjugation of CdTe quantum dot for the detection of 2,4-dichlorophenoxyacetic acid by competitive fluoroimmunoassay based biosensor. Biosens Bioelectron 24(6):1615–1620

Vivek MN, Manasa M, Kambar Y, Nawaz AS, Vinayaka KS, Kekuda TR (2014) Antifungal activity of some plants of Western Ghats of Karnataka against *Sclerotium rolfsii*. Indian J Adv Plant Res 1(4):29–33

Walter S, Nicholson P, Doohan FM (2010) Action and reaction of host and pathogen during Fusarium head blight disease. New Phytol 185:54–66

Wang YA, Li JJ, Chen HY, Peng XG (2002) Stabilization of inorganic nanocrystals by organic dendrons. J Am Chem Soc 124(2293):2298

Wang B, Yang X, Zeng H, Liu H, Zhou T, Tan B et al (2012) The purification and characterization of a novel hypersensitive-like response-inducing elicitor from *Verticillium dahliae* that induces resistance responses in tobacco. Appl Environ Microbiol 93(1):191–201

Wild CP, Gong YY (2010) Mycotoxins and human disease: a largely ignored global health issue. Carcinogenesis 31:71–82

Woudenberg JHC, Groenewald JZ, Binder M, Groenwald JZ, Binder M, Crous PW (2013) Alternaria redefined. Stud Mycol 75:171–212

Woudenberg JHC, Seidl MF, Groenewald JZ, de Vries M, Stielow JB, Thomma BPHJ, Crous PV (2015) Alternaria section Alternaria: species, formae speciales or pathotypes? Stud Mycol 82:1–21

Xia H, Wang XL, Zhu HJ, Gao BD (2011) First report of anthracnose caused by *Glomerella acutata* on chili pepper in China. Plant Dis 95:219

Xing K, Shen X, Zhu X, Ju X, Miao X, Tian J, Feng Z, Peng X, Jiang J, Qin S (2016) Synthesis and in vitro antifungal efficacy of oleoyl-chitosan nanoparticles against plant pathogenic fungi. Int J Biol Macromol 82:830–836

Xue J, Luo Z, Li P, Ding Y, Cui Y, Wu Q (2014) A residue-free green synergistic antifungal nanotechnology for pesticide thiram by ZnO nanoparticles. Sci Rep 4:5408

Yao Z, Rashid KY, Adam LR, Daayf F (2011) *Verticillium dahliae's* VdNEP acts both as a plant defence elicitor and a pathogenicity factor in the interaction with *Helianthus annuus*. Can J Plant Pathol 33(3):375–388

Young CS, Smith JA, Watling M, Clarkson JP, Whipps JM (2001) Environmental conditions influencing apothecial production and lettuce infection by *Sclerotinia sclerotiorum* in field conditions. In: Young C, Hughes K (eds) Proceedings of the 6th international Sclerotinia workshop, York, pp 181–182

Zhang J, Liu Y, Zhang X, Shang JK (2013) Antifungal activity and mechanism of palladium-modified nitrogen-doped titanium oxide photocatalyst on agricultural pathogenic fungi *Fusarium graminearum*. ACS Appl Mater Inter 5(21):10953–10959

Zhao X, Hilliard LR, Mechery SJ, Wang Y, Bagwe RP, Jin SA (2004) Rapid bioassay for single bacterial cell quantitation using bioconjugated nanoparticles. Proc Natl Acad Sci 101: 15027–15032

Chapter 9
Antifungal Products by Fungi in Food Nano-Packaging

Arash Babaei, Mohamad Javad Akbarian Mymand, and Samaneh Faraji Kafshgari

Abstract Nowadays, one of the challenges encountered by food packaging industry is a high demand for fresh, healthy and with minimum processed foods. Selecting materials with their appropriate packaging method can be helpful to maintain product quality. Today, due to reductions in petroleum products and petrochemicals, rapid global warming and environmental pollution, bioactive polymers have attracted much attention in bioactive packaging. Nanotechnology has a broad application in all stages of production, processing, storage, packaging and transport of agricultural products. Arrival of nanotechnology to the agricultural and food industry, ensures the increasing of products and their quality and protects environment and earth resources. Nowadays, most of the agricultural and food products, due to the inappropriate storage and packaging conditions, their sale have decreased and it can hurt the economy. Nanotechnology as a powerful instrument, can help us to improve the economy. Antimicrobial packaging is one of types of active packaging with many applications. Active packaging is the packaging system that achieved by adding active components in the packaging system. Antimicrobial packaging is a system that kills or inhibits spoilage and pathogenic microorganisms that are in foods. The antimicrobial packaging achieved by adding antimicrobial agents in the packaging system or using antimicrobial polymers. When an antimicrobial activity in the packaging system establishes the packaging system limits or prevents microbial growth.

A. Babaei (✉) • M.J.A. Mymand
Biology Department, Faculty of Sciences, Malayer University, Malayer, Iran
e-mail: a.babaei@sheffield.ac.uk

S.F. Kafshgari
Department of Food Science and Technology, Agricultural Sciences and Natural Resource university of Gorgan, Gorgan, Iran

R. Prasad (ed.), *Advances and Applications Through Fungal Nanobiotechnology*, Fungal Biology, DOI 10.1007/978-3-319-42990-8_9

1 Introduction

Increasing the shelf life of food products is very important. Nowadays, packaging is essential. Food packaging is a system that prepare food for transport, distribution, storage and retailing (Shin and Selke 2014). Packaging has four main functions including food safety, protection, convenience and communication.

Food packaging is heart of modern food industry. Without packaging and efficiently commercially processed food could not be handled. The World Packaging Organization (WPO) estimates that more than 25 % of food is wasted due to poor packaging (Olsmats and Wallteg 2009). In addition to the primary roles of packaging in protection, secondary functions affected by packaging, such as participation in the sale and increasing of it.

In fact, the main role of food packaging is protection and safe transportation of food products up to the consumer time. During the distribution of food products, their quality may affect by biological, chemical and physical factors. Thus, food packaging increases shelf life and preserves the quality and safety of food products.

Expressed that; the essential requirements for producing packaging are included:

1. Much production
2. Acceptable and effective packing material
3. Appropriate shape and structure
4. Easy to use
5. Considerations related to disposal

Therefore, in accordance with the conditions listed, design of packaging and its development, is not included industrial design, innovation and marketing just, but included environmental and engineering sciences. Food packaging is included of simple protection methods, provide facilities, marketing at the point of purchase (POP), reduce material requirements, safety, Protection against picky and environmental aspects (Table 9.1).

Table 9.1 Time period of performances and implications in food industry

Time period	Performances and implications
1960s	Convenience, marketing at the point of purchase
1970s	Weight reduction, Reduce resource consumption and saving energy
1980s	Safety
1990s	Environmental effects
2000s	Safety and security

2 Antimicrobial Packaging

Antimicrobial packaging is one of types of active packaging with many applications (Floros et al. 1997). Active packaging is the packaging system that achieved by adding active components in the packaging system (Han 2003a, b). Antimicrobial packaging is a system that kills or inhibits spoilage and pathogenic microorganisms that are in foods.

The antimicrobial packaging achieved by adding antimicrobial agents in the packaging system or using antimicrobial polymers. When an antimicrobial activity in the packaging system establishes the packaging system limits or prevents microbial growth.

The goals of conventional food packaging (Fig. 9.1) are included:

1. Shelf-life extension
2. Quality maintenance
3. Safety assurance

There are various methods to achieve these goals, antimicrobial packaging is specifically designed that can control microorganisms and provide the aims of packaging.

Some products, are not sensitive to microbial spoilage or contamination and do not need the antimicrobial packaging system. However, most foods are perishable and susceptible to contamination.

Therefore, the goals of an antimicrobial packaging system (Fig. 9.2) are:

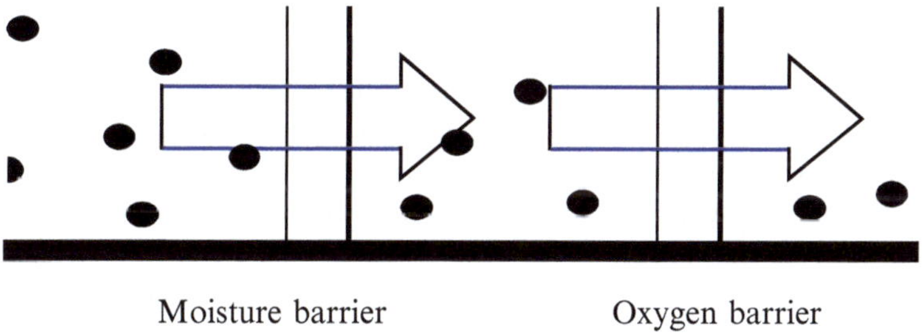

Fig. 9.1 Conventional food packaging

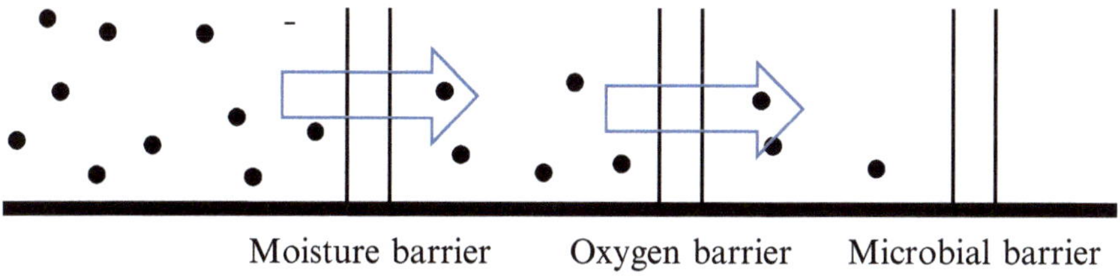

Fig. 9.2 Antimicrobial food packaging

1. Safety assurance
2. Quality maintenance
3. Shelf-life extension

Which these goals are reversed the primary aims of conventional packaging systems. Nowadays, food security is an important issue and antimicrobial packaging could play a role in food security assurance.

Antimicrobial agents with different activities effect on microorganisms. For antimicrobial agent affect against spoilage and pathogenic microorganisms. There is no 'Magic Bullet'. This is due to the features antimicrobial mechanisms and various physiologies of the microorganisms. For select specific antimicrobial agents categorization of microorganisms is very helpful. Categories may consist of oxygen requirement (aerobes and anaerobes), cell wall composition (Gram positive and Gram-negative), growth-stage (spores and vegetative cells), optimal growth temperature (thermophile, mesophile and psychrophile) and acid/osmosis resistance.

The characteristic antimicrobial function of the antimicrobial agent besides the microbial characteristics is important to understand the limits of the activity. Some antimicrobial agents inhibit essential metabolic (or reproductive genetic) pathways of microorganisms while some others alter cell membrane/wall structure (Aziz et al. 2015). For example, lysozyme destroys cell walls without the inhibition of metabolic pathways and results in physical cleavages of cell wall, while lactoferrin and EDTA acts as coupling agents of essential cationic ions and charged polymers.

Two major functions of microbial inhibition are microbicidal and microbiostatic effects. In the case of microbiostatic effects, to prevent regrowth of target microorganisms in during the storage period or shelf-life, the packaging system has to possess the active function of above maintaining by the minimal inhibitory concentration.

Traditional preservation methods sometimes consist of antimicrobial packaging concepts, which include sausage casings of cured/salted/smoked meats, smoked pottery/oak barrels for fermentation, and bran-filled pickle jars.

Hurdle technology is the basic principle of these traditional preservation methods and antimicrobial packaging. Another hurdle to prevent the degradation of the total quality of packaged foods in antimicrobial packaging is the extra antimicrobial function of the packaging system. The microbial hurdle may not contribute to the protection function from physical damage. However, it provides tremendous protection against microorganisms, which has never been achieved by conventional moisture and oxygen barrier packaging materials.

Antimicrobial functions which are achieved by adding antimicrobial agents in the packaging system or using antimicrobial polymeric materials show generally three types of modalities;

1. release
2. absorption
3. Immobilization.

3 Antimicrobial Nano-Packaging

The advent of Nanotechnology in food packaging industry, creates functional solutions in relation to increasing the shelf life of food. In fact, the advent of this technology is duo high ratio surface to volume of the particles with nanometer dimensions. This ratio is in direct relationship to the radius of the spherical nanoparticles. By reducing the particle size in the nanometer range, the strength of surface-active substances and the reaction of the material with the surrounding environment due to the increase of surface-active sites will increase significantly. In general, the active packaging systems have more effective features than barrier properties and are obtained by the addition of active ingredients and components in the packaging system. This type of packaging shows proper reaction by changing the properties of the package to internal and external environmental changes and, thus are important in maintaining freshness of fresh foods (Suppakul et al. 2003). Antimicrobial active packaging made of metal nano composites, is a new generation of Nano-structural packaging that are producing a direct combination of metal nanoparticles with polymers.

The antimicrobial properties of silver and zinc oxide compounds have known from last years and have many applications in disinfect medical equipment, water purification and healing, creams, lotions and antibacterial creams (Prasad 2014; Prasad et al. 2014, 2015). The antimicrobial mechanism of metal nanoparticles that are obtained of this metal is still unclear. According to the studies of the researchers, this mechanism may be through inducing oxidative stress in the cell membrane of microorganisms duo releasing of reactive oxygen species or releasing ions from the particle surface and binding to cell membranes and destroying it.

4 Antimicrobial Agents

Various antimicrobial agents can be used in food and food packaging system to create a new antimicrobial packaging system. Generally they can be classified into three groups: chemical antimicrobial agents, natural agents and probiotics.

4.1 Chemical Antimicrobial Agents

All packaging material that used for food preservation, must be food grade additives. Chemical antimicrobial agents can be blended with food ingredients, food additives and used into the headspace. Antimicrobial agents are in contact with food products and consumed by them. Therefore, chemical antimicrobial agents should be control as food ingredients. In the case of non-edible chemicals, bonding of antimicrobial agents in food packaging materials (immobilization) is the only way to enter chemical agents in the food packaging system. The immobilized antimicrobial agents cannot migrate from packaging materials to foods, while the blended agents can migrate.

Various organic acids are most common chemical antimicrobial agents used by researchers. Organic acids are widely used as chemical antimicrobial agents because their performance is understood well and the use of them is economically. Most organic acids include fatty acids that are chemical agents naturally and had been used past years. Now, most of them are produced by chemical synthesis or chemically modify natural acids. Organic acids have certain sensitivity to microorganisms. For example, Sorbic acid, Sorbets, is very strong fungicide agents, while their antimicrobial activities are not effective. Therefore, in order to have effective antimicrobial agents, the correct choice of organic acids is essential. A mixture of organic acids has wider spectrum antimicrobial and stronger activity antimicrobial than a single organic acid.

Fungicides are also antimicrobial agents. Since fungicides are not permitted as a direct food preservative, they cannot be blended to food compounds or incorporated into the food contact packaging materials as food contact substances. Therefore, when non-edible antimicrobial agents such as fungicides are used, the design of special antimicrobial food packaging systems is essential.

Food disinfectants are food contact materials or food contact surface disinfectants. Food disinfectants residues in food at specific limit is permissible. Therefore, the use of food disinfectants has many advantages than the use of other non-edible antimicrobial agents such as fungicides.

4.2 Natural Antimicrobial Agents

Plant extracts, spices, enzymes and bacteriocins are natural antimicrobial agents. Due to customer demand for food free of chemical preservatives, nowadays food manufacturers use natural antimicrobial agents in order to sterilize or increase shelf life of food. Plant extracts and spices are included several natural composition that has a broad antimicrobial spectrum of numerous antimicrobial. Besides antimicrobial activity, other advantages of them include antioxidant activity and their effect as an alternative to drugs. However, the performance and kinetics of them are unknown and their chemical potential is still under discussion. In addition, they make flavored problems.

The specificity of enzyme activity should be considered, because the antimicrobial activity is very sensitive to the environment and precursors. For example, lysozyme activity can be strongly influenced by temperature and pH value. In most cases, lysozyme is not effective against Gram-negative bacteria. This is due to the complex structure, cell wall of Gram-negative bacteria and specificity of lysozyme to the peptidoglycan.

Different bacteriocin such as niacin, producing lacticin, propionicin etc. can be entered into food or food packaging systems to inhibit the growth of spoilage and pathogenic microorganisms (Daeschel 1989). Extracted bacteriocins are generally low molecular weight peptides that can be used in different ways. However, it is essential to determine their resistance to thermal treatment and the pH. In the fermented food products, live bacteria that produce bacteriocins can be used as probiotics in food packaging to achieve antimicrobial performance.

4.3 Probiotics

Various microorganisms such as lactic acid bacteria, produce bacteriocins and non-peptide growth inhibitor chemicals, such as Reuterin. This natural antimicrobial is able to inhibit the growth of other bacteria. As a result, the use of probiotics can be controlled effectively competing undesirable microorganisms. Many traditional fermented food products containing antimicrobial probiotics.

Many researches and developments in probiotic antimicrobial performance have been done for the preservation of fermented food. Currently, only limited researches on the use of probiotics to designing antimicrobial packaging is being done. With the development of new technology to provide live probiotics, the use of probiotics as an antimicrobial source to create antimicrobial food packaging duo safety and high performance, will become more common in the future (Table 9.2).

Table 9.2 Some of antifungal agents and packaging systems

Antimicrobials	Packaging materials	Foods	Organisms	References
Organic acids				
Benzoic acids	Ionomer	Culture media	*Penicillium spp., Aspergillus niger*	
Benzoicand sorbic acids	PE-co-met-acrylates	Culture media	*Aspergillus niger*, *Penicillium* spp.	
Sorbates	LDPE	Culture media	*S. cerevisiae*	Han and Floros (1997)
Sorbic anhydride	PE	Culture media	*S. cerevisiae*, Moulds	
Fungicides				
Benomyl	Ionomer	Culture media	Moulds	Halek and Garg (1988)
Imazalil	PE	Cheese	Moulds	
Natural extract				
Herb extract, Ag-Zirconium	LDPE	Lettuce, cucumber	*E. coli*, *S. aureus*, *L. Mesenteroides*, *S. cerevisiae*	An et al. (1998)
Oxygen absorber				
Ageless	Sachet	Bread	Moulds	

5 Antimicrobial Mechanisms

An antimicrobial agent has inhibitory activity and specific mechanisms against every microorganism. Therefore, choice of antimicrobial agents depends on their performance against a target microorganism. No magical antimicrobial agent to act effectively against all spoilage and pathogenic microorganisms does not exist, because the antimicrobial agents have different activities and influence on differing microorganisms. This is due to their especially antimicrobial mechanisms and differences physiology of microorganisms.

Simple classification of microorganisms can be useful for especially antimicrobial agent. This classification can be done on the basis of oxygen requirement (aerobic or anaerobic), the composition of the cell wall (Gram-positive and Gram-negative), stage of growth phase (spore or vegetative cell), optimal growth temperature (thermophilic, mesophilic or psychrophilic) or resistant to acid/osmosis. In addition to microbial properties, antimicrobial agent properties, such as range of activity are also important in order for perception of the performance of microorganism. For example, some antimicrobial agent inhibits to essential metabolic pathway (or genetic replication) of microorganisms, while the other change the structure of the membrane/cell wall. Two main functions of inhibition microbial or germicidal effects and microbial inhibitory.

5.1 *Germicidal*

It is expected that antimicrobial packaging systems, will kill pathogenic and spoilage bacteria, because these systems are removed every microorganism from the packaging systems/food. However, in practice it is very difficult to remove all microorganisms, but when the concentration of microbicide for the short time slightly above the minimum inhibitory concentration (MIC), the microbicide antimicrobial system will be able to kill the target microorganisms.

Using other treatments to improve the antimicrobial activity of packaging systems, such as refrigerating or cooling, antimicrobial efficiency will increase. Although generally when the system factors are able to provide the requirements together, refrigerating is not necessary. Refrigerating can be very effective in non-target microbial growth inhibition (unwanted). If the initial concentration of the germicidal antimicrobial systems is less than MIC of the target microorganisms, and concentrations never rise, the antimicrobial agent may show an inhibitory effect, instead of germicidal. So keep concentration of antimicrobial over the MIC for a certain critical time to eliminate target microorganisms is very important. If the package is sealed with solder, packaged foods contain any live microorganisms, even when the concentration due to migration or loss of the factor after the critical time reduce under the MIC.

5.2 *Microbial Inhibitory*

Microbial inhibitory agents are able to inhibit the growth of microorganisms at over the special critical concentrations (MIC). However, when the concentration is below the critical level, or when the antimicrobial agent through leakage, opening or in any other way is removed from the packaging system, under pressure microorganisms will be able to grow or spores germinate. So keep the concentration of the antimicrobial agent over the MIC the entire shelf-life of packaged food product is necessary. Chemical indicators that show density or microbial growth in microbial inhibitory antimicrobial packaging systems will be very useful and this is the same intelligent packaging concept (Han 2003a, b).

6 The Performance and Volatility

In most solid or semi-solid packed food, microorganisms grow on the surface of food in the beginning (Brody et al. 2001). Therefore, antimicrobial activity should be applied to surface. The antimicrobial activity may occur depending on the method of putting up antimicrobial agent in packaging materials, headspace or in the atmosphere inside the package. Therefore, methods of incorporating and techniques of transferring in the design of effective antimicrobial packaging systems are important.

As examples of the methods of incorporating microbial agents, microbial agents have been floated in packaging materials before the final extrusion (Han and Floros 1997; Nam et al. 2002), dissolved into coating solvents, added to edible coating materials and mixed in the mold/filler material such as paper and pasteboard (Nadarajah et al. 2002).

Gaseous antimicrobial agents can also be added to the package atmosphere (Lanciotti et al. 2003; Krause et al. 2003). Chemical immobilization, bands chemical agents into the packaging structures because the Regulations do not allow agents to immigrate in foods (Appendini 1996; Appendini and Hotchkiss 1997; Miller et al. 1985; Halek and Garg 1988).

Immobilized antimicrobial agents inhibit microbial growth on surfaces in contact packaged products. Chemical stabilizers, chemical agents into the packaging structures are covalent band because the Regulations do not allow immigration agents in foods. Antimicrobial agents stabilized microbial growth on surfaces in contact inhibition provides packaging products.

6.1 *Non-volatile Immigration*

Mass transfer of non-volatile antimicrobial done with diffusion migration. Non-volatile agents primarily places in the packaging materials are among package and surface of food. So, solubility coefficient (Separation coefficient) and diffusion coefficient (Distribution coefficient) of antimicrobial agent in food in order to maintain its concentration over effective levels of MIC during predicted shelf life, are very important features.

If the agent enters the packing material initially, must be leave the packaging material and before releasing of food core, dissolved in food. So important character constants in the mass transfer profile are coefficient distribution of the agent in the packaging material, solubility (or separation coefficient) of the agent in the food and dispersion coefficient of the agent in food.

In food packaging antimicrobial agent system, having kinetics of mass transfer in accordance microbial growth kinetics in order to provide effective antimicrobial activity is important. The non-volatile immigration system requires complete contact between packaging materials and surface of food. The food should have a continuous matrix without voids, holes, air spaces and have heterogeneous particles. Single ingredient solid foods, semi-solid (soft solid) and liquid products are good examples of products that can use non-volatile immigration system antimicrobial packaging.

6.2 *Volatile Immigration*

Many researchers have claimed that having complete contact of the antimicrobial active agent with surface food in order to facilitate the migration of active agent—in order to have maximum performance- is essential (Vermeiren et al. 2002; Suppakul et al. 2003). Although, if the volatile antimicrobial agents is used, it is not necessary.

In order to keep the surface concentration on higher levels of specific MIC, control headspace gas concentration is very important, because the concentration of volatile agent in headspace is equilibrium on the concentration of surface food and substances in the package. Initially, the volatile agent is placed in packing material, whether it is film, container, sachet or tray. After packing food, volatile agent is evaporated into the headspace, achieved to the surface food and be absorbed by the food.

The use of volatile antimicrobial agents has several advantages. This system can effectively use for very porous foods, powdered and crushed foods and foods contain particles without specific shape, such as, small fruits, vegetables.

6.3 *Non-immigration and Absorption*

The non-immigration system uses of non-immigrant antimicrobial polymers that as the result antimicrobial agent does not migrate out of the polymer duo its covalent connection to the polymer background (Steven and Hotchkiss 2003). In addition, antimicrobial agents, other bioactive agents (such as enzymes, proteins and other organic compounds) are able to connect to the polymer through the covalent cross links. Since these agents are not mobile, their activity is limited only to the contact level. This restriction is higher in solid or semi-solid food.

As a food packaging system, this system has unique advantages in terms of marketing and regulations. Since bioactive agents are non-immigrant, this system needs very little connector factors. This may reduce the cost of all packaging systems that the use of very expensive antimicrobial agents.

Non-immigration system can be containing agents that not allowed to use as food ingredients or food additives. After verification of non- immigration, the packaging material can contain any substance in food contact. The marketing aspect, this system is very attractive because the food has any chemical antimicrobial agent in entire of shelf life. However, in contrast to these advantages, it can have a very limited range of antimicrobial agents and use of it limited to a few types of foods.

7 Physical Characteristics of Packaging Material

Physical and mechanical characteristics of the packaging materials affected by antimicrobial agents are applied. If the antimicrobial agent is compatible with packing materials, plenty of agents can be impregnated into the packaging materials without any disruption in its physical and chemical integrity. Therefore, additional antimicrobial agent that cannot mix with the packing materials will reduce the mechanical integrity and physical strength. Polymer morphology studies in anticipation of the possible loss of physical integrity duo adding an antimicrobial agent to the polymer packaging.

Antimicrobial agents with small size can be mixed with polymers and placed on the polymeric structure without any significant interference with polymer-polymer fittings in amorphous region. If a large amount of an antimicrobial agent is mixed into the packaging material, prepared space by the amorphous region will be filled and mixed agent begins to interfere with polymer-polymer fittings in crystalline region. However, after adding low levels of the antimicrobial agent, physical integrity does not injure, but optical properties may be changed. For example, reducing the transparency or discoloration of packing materials may be occurred.

8 Controlled Release Technology

Design of antimicrobial packaging system requires balancing technical factors of the controlled release and kinetics of microbial growth. When the mass transfer of antimicrobial agent faster than the growth of the target microorganisms, before the ending of the storage period, loaded antimicrobial agent will be diluted to lesser than effective critical concentration (MIC) and packaging system will lose their antimicrobial activity because the packed food in comparison amount of packaging material and antimicrobial agents is almost unlimited. As a result, microorganisms, beginning to grow after depletion of antimicrobial agent. By contrast, when the rate of migration is very low to maintain concentration in higher levels of MIC, microorganisms can constantly grow before the antimicrobial agent is released. Thus, the releasing rate of the antimicrobial agent from packaging material for food, especially for communication between the mass transfer kinetic and microbial growth target rate, should be controlled.

Maintain of surface concentration depends on the releasing rate of packing materials (diffusion coefficient of packaging material) and food migration rate (diffusion coefficient). Since the flow of the releasing of packaging material is reduced due to reduce of amounts of antimicrobials in packaging materials over time, regard to the profile of its rapid losing, the time of maintaining surface concentration on higher levels of MIC is estimated accuracy (Fig. 9.3).

When the dissolution of antimicrobial agents in packed food is too low, the antimicrobial concentration of food contact surface is the highest. When the rate of releasing is lower than unlimited free diffusion, which occurs at high solubility in the slow releasing systems, the time to maintain of concentrated in higher than MIC, generally longer than the free diffusion systems. If antimicrobial agents in polymer packaging materials covered, dispersion coefficient of agent in polymer background will control releasing rate (Fig. 9.4).

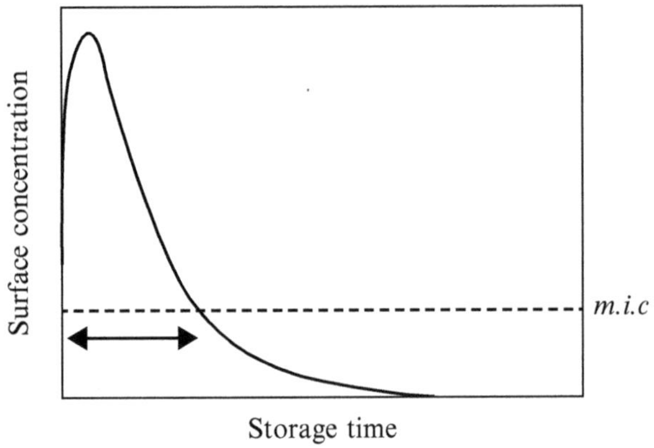

Fig. 9.3 Unconstrained free diffusion system

Fig. 9.4 Monolithic system

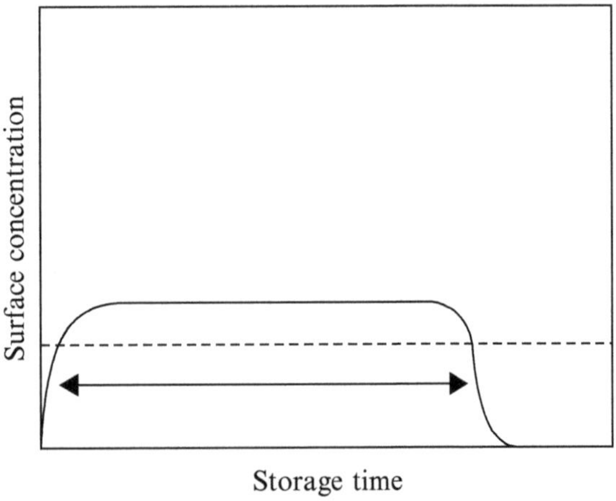

Figure 9.4 shows longer time of maintaining concentration over MIC that creates by membrane systems. This system has a permeable membrane that controls the rate of releasing.

9 Antifungal Agent

This idea that a microorganism can inhibit the growth of other microorganisms, occur in the early twentieth century. Years later, the researchers discovered that it is due to the toxic metabolites that produced by microorganisms. In fact, these compounds are biosynthesis by bacteria, fungi and plants. Antimicrobial and antifungal compounds commonly used in the pharmaceutical and packaging industry. Natural compounds produced by fungi are as a new source of antimicrobial and antifungal compounds.

One way for control of disease and spoilage that caused by pathogenic fungi, is using of biological control methods. Unlike synthetic agents, substances that produced from varieties effective biotechnical, have low toxicity and allergies, also they are decomposed easily. They do not accumulate in food products, as well as they are cheap and appropriate for use on an industrial scale.

Antifungal agent is a diverse group of Antimicrobial agent that growth inhibition of fungal. The antifungals are classified into several groups based on their structure and mechanisms of action. These classes include the polyenes, imidazoles, triazoles, allylamines and echinocandins as well as miscellaneous agents.

9.1 Antifungal Azoles

Imidazoles and triazoles ('azoles') are the largest class of antifungal agents that use in clinical and industry (Fig. 9.5). The imidazole and triazole (azoles) groups of antifungal compounds inhibit the enzyme cytochrome P450 14α-demethylase, their activity done by inhibiting the biosynthesis of ergosterol, in fungal membranes. These compounds also block steroid synthesis in humans. The molecules target of Azole antifungal compounds, are cytochromes P450–Erg11p or Cyp51p. Figure 9.5 shows this protein contains an iron protoporphyrin that located at the active site and the antifungal azoles bind to the iron atom via a nitrogen atom in the imidazole or triazole ring.

Imidazoles are included: Miconazole, Bifonazole, Ketoconazole, Butoconazole, Clotrimazole, Econazole, Mebendazole, Fenticonazole, Isoconazole, Oxiconazole, Sertaconazole, Sulconazole, Thiabendazole and Tiaconazole.

The triazoles are newer, and are less toxic and more effective, they are included: Fluconazole, Itraconazole, Ravuconazole and Posaconazole.

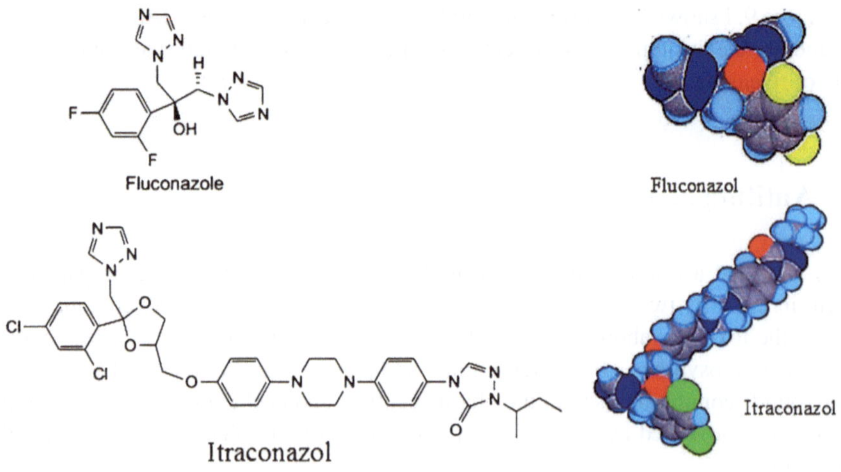

Fig. 9.5 Extensive chemical and globular structure of itraconazol and Fluconazol

Fig. 9.6 Extensive chemical and globular structure of flucytosine

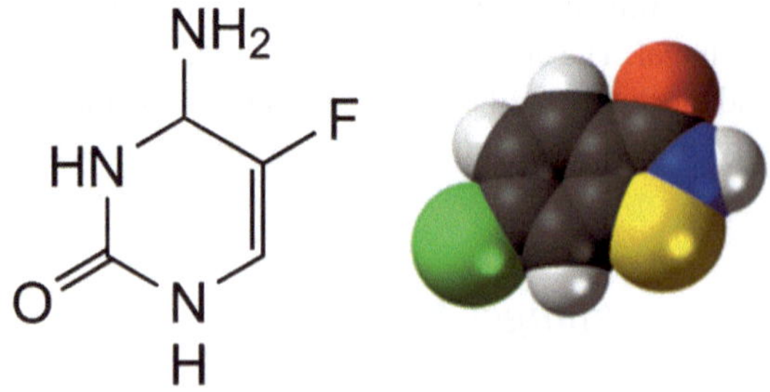

9.2 *Flucytosine*

Flucytosine (5-fluorocytosine; Fig. 9.6) is one antifungal agent that by conversion to 5-fluorouracil within target cells, growth inhibition of fungal. Fluorouracil by incorporating into RNA, causing premature chain termination, and it by effects on thymidylate synthase, inhibits DNA synthesis. In this mechanism, the target cells must Containing cytosine permease to internalize the flucytosine molecule, cytosine deaminase to convert it to 5-fluorouracil, and uracil phosphoribosyl transferase to convert 5-fluorouracil into a substrate for nucleic acid synthesis. Flucytosine is useful for pathogenic yeasts (Candida species and *C. neoformans*) but don't have any application for most filamentous fungi, because they lack these enzymes.

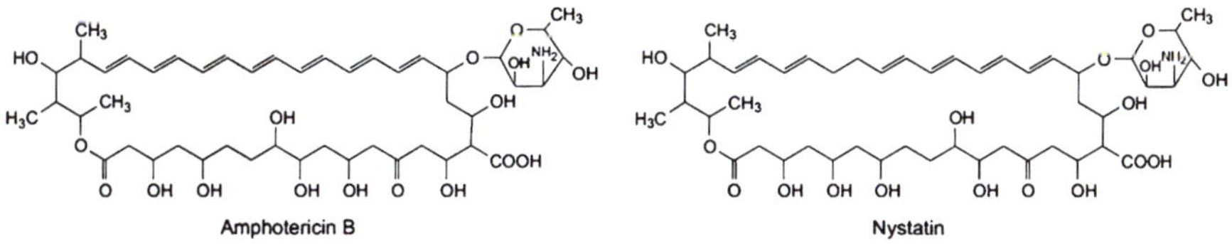

Fig. 9.7 Kinds of polyene

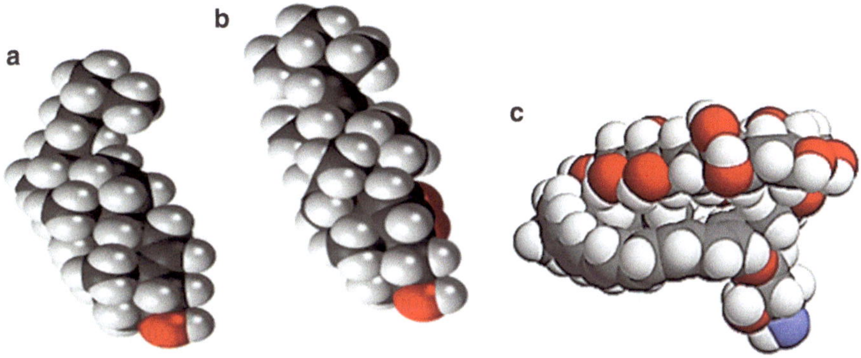

Fig. 9.8 The molecule of cholesterol, ergosterol and amphotericin B

9.3 *Polyene*

Amphotericin B (Fig. 9.7) is a Polyene that for many years was used to treat a visceral infection.

Figure 9.8 shows the ergosterol molecule of fungi and sterol in mammalian membranes, as can be seen, the ergosterol molecule has a cylindrical three-dimensional structure and cholesterol has a sigmoid shape. Reason is that amphotericin B for ergosterol over cholesterol binding affinity for this difference in structure. The polyene antibiotics bind with sterols in the fungal cell membrane, principally ergosterol. This causes the cell's contents to leak out and the cell dies. Animal cells contain cholesterol instead of ergosterol and so they are much less susceptible. They are including: Nystatin, Amphotericin B, Natamycin, Rimocidin, Filipin and Pimaricin.

9.4 *Allylamines and Morpholines*

The target pathway of allylamines and morpholines is the ergosterol biosynthetic pathway. Allylamines inhibit the enzyme squalling epoxidase, another enzyme required for ergosterol synthesis. These compounds are included: Terbinafine (marketed as Lamisil), Amorolfine, Naftifine and Butenafine (Odds et al. 2003).

9.5 *Echinocandins*

The echinocandins are fungal secondary metabolites. This compound contains a cyclic hexapeptide core with a lipid side chain responsible for antifungal activity. Compounds, anidulafungin, caspofungin and micafungin, in the late 1990s, entered clinical development. The three-dimensional configuration of all three molecules is similar. A central, common core bears a long, 'gun-barrel'-like side chain known to be a determinant of the spectrum of susceptible species (Fig. 9.9) and a hydroxylated side chain that appears opposite the 'gun barrel' in flat structural representations (Fig. 9.9) but is adjacent in energy-minimised 3-D structures (Fig. 9.9). Echinocandins inhibit the synthesis of glucan in the cell wall, probably via the enzyme 1, 3-β glucan synthase: Anidulafungin, Caspofungin and Micafungin.

9.6 *Chitosan*

Chitosan is biopolymer that has a good antimicrobial ability, because prevents of the growth of a wide range of fungi, yeasts and bacteria. In addition, after dissolving in acidic solution, forming a film alone (Bégin and Van Calsteren 1999). Chitosan is Cationic polysaccharide that has β (1–4) bonds, that is obtained from hard-shells or various fungi through acetylation of chitin. Commercial Chitosan products usually have a molecular weight ranging from 100,000 to 1,200,000 Dalton, but oligomer products with lower molecular weight can be prepared through thermal, enzymatic or chemical decomposition. Chitosan is readily soluble in acid various solvents, and has high antimicrobial activity against many pathogenic and spoilage microorganisms (TSAI et al. 2002). The antimicrobial activity of Chitosan depends on the molecular weight and degree of acetylation and the amount of chemical analysis of the.

9.6.1 The Effect of Chitosan Coating on Spoilage Microorganisms

Fruits and vegetables are damaged by the kinds of spoilage microorganisms postharvest that leads to the spoilage. Usually rot occurs in fruits and vegetables by fungi and bacteria. After coating with chitosan, the opportunity of microbes' decreases to contact with fruits and vegetables, so frees them from aggressive microbes. In addition, the amino group of Chitosan has bacteriostatic effect and can reduce the number of microbes. As well as other factors such as reducing of respiration rate, maintaining higher activity of protective enzymes and cell membrane integrity, reinforce the fruit's ability to protect against microbes. Even the coated fruit and vegetables with Chitosan show a downward trend (Youwei and Yinzhe 2013).

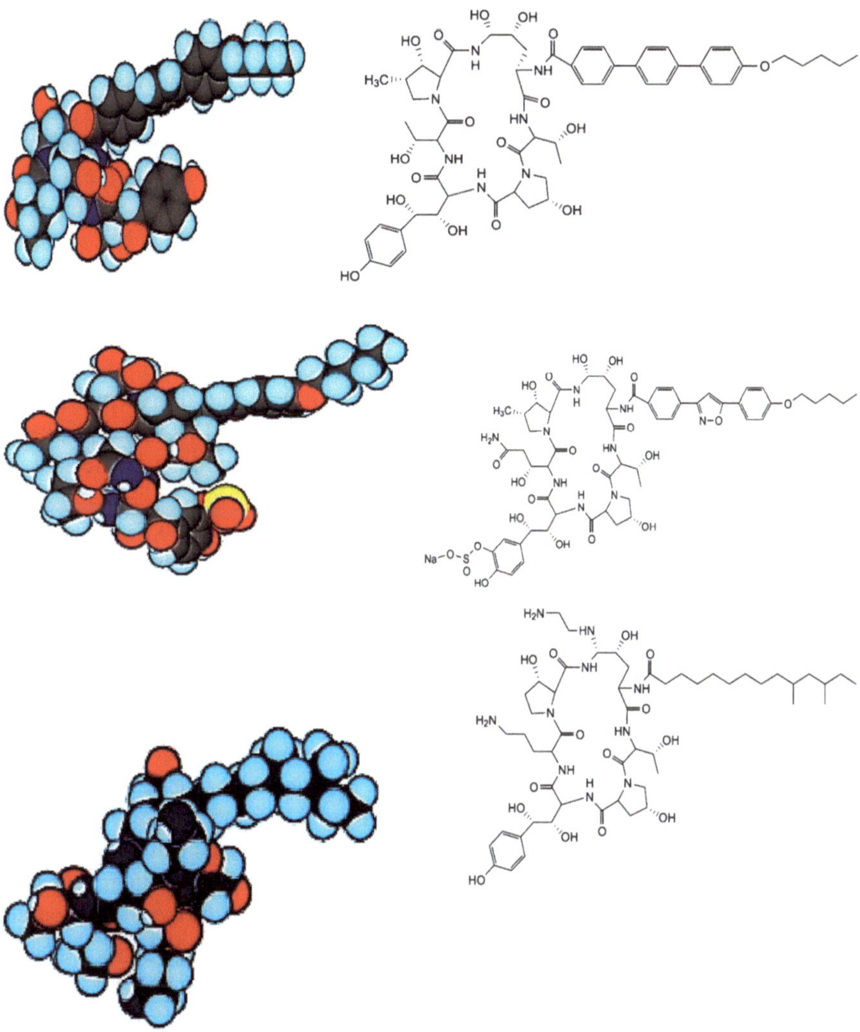

Fig. 9.9 Extensive chemical and globular structure of Anidulafungin, Caspofungin and Micafungin

9.6.2 Application of Chitosan Coating

Chitosan is a type of multi sugar that there is in exoskeleton of crustaceans and also many insects and edible mushrooms (Giatrakou et al. 2010; Vermeiren et al. 2002). The antimicrobial properties of chitosan are duo positively charge of amine groups. These groups are reacted whit cell membranes contain negatively charge that leads to precipitation out protein components and other intracellular components of microorganisms (Giatrakou et al. 2010). Nowadays, chitosan is used as a healthy additive and also as a film. Chitosan film produced in 1963 by Rigby by casting method at first time. This film is flexible, strong, transparent, and colorless and its tensile strength is 9000 pounds per square inch. Studies show that such films have proper mechanical properties and permeability to gases (Vermeiren et al. 2002).

Chitosan is non-toxic, biodegradable, functional and biocompatible. Chitosan has a strong antimicrobial and antifungal activity that can effectively control fruit rot. It can easily make the coating on fruits and vegetables and reduce the respiration rate of fruits and vegetables by adjusting permeability oxygen and carbon dioxide (De Reuck et al. 2009).

For the effective application of chitosan coating, chitosan is combined with other materials. In addition, chitosan coating often combined with only physical methods such as short-term heating, short-term gas sterilization, modified atmosphere packaging (MAP), etc. Modified atmosphere packaging is an important way to modify the atmosphere inside the package using of polymer films with or without pores that reduces loss of the quality and improve shelf life of the packaged fruits and vegetables by reducing of the moisture, microbial and metabolic activity (Vermeiren et al. 2002).

De Reuck et al. (2009) found that the combination of Chitosan (0.1 g L^{-1}) + MAP (control) was effective to prevent of the rot, browning and maintain crust color in McLean, so red Cultivar compared with MAP only. Chitosan (0.1 g L^{-1}) + MAP reduced polyphenol oxidase and peroxidase activity significantly, as well as maintained the integrity of the membrane and anthocyanin content and prevented the reduction of color of the crust during storage (Jianglian and Shaoying 2013).

Chitosan edible coating for increasing shelf life of pomegranate seeds for 12 days at 4 °C was used. Chitosan coating inhibited the growth of bacteria and fungi on the surface of pomegranates. Pomegranates water content that coated by chitosan at levels 0.5 and 1 %, was maintained during 12 days of storage. Chitosan decreased, increasing of pomegranates total soluble solids and titratable acidity during storage period. In contrast, the use of chitosan delayed reduction in total phenolic content, total anthocyanin and antioxidant capacity during storage period (Ghasemnezhad et al. 2013).

The use of chitosan coating of fresh and cut mushroom, delayed color change associated with reduction in the activities of polyphenol oxidase, peroxidase, catalase, phenylalanine ammonia lyase and laccase and also lower total phenol content, and reduced cellulase, total amylase, and α-amylase activity. Microbial development treated samples with Chitosan coating was inhibited compared to control samples (Tamer and Çopur 2009).

10 Conclusions

Antimicrobial packaging can prevent the growth of microorganisms, improve food safety and increase the shelf life of packaged foods. Many factors are involved in designing antimicrobial packaging, however, most factors are closely related to the properties of antimicrobial agents, packaged foods and target microorganisms.

References

Appendini P (1996) Immobilization of lysozyme on synthetic polymers for the application of food packages. Cornell University, Ithaca

Appendini P, Hotchkiss JH (1997) Immobilization of lysozyme on food contact polymers as potential antimicrobial films. Packag Technol Sci 10(5):271–279

Aziz N, Faraz M, Pandey R, Sakir M, Fatma T, Varma A, Barman I, Prasad R (2015) Facile algae-derived route to biogenic silver nanoparticles: synthesis, antibacterial and photocatalytic properties. Langmuir 31:11605–11612. doi:10.1021/acs.langmuir.5b03081

Bégin A, Van Calsteren M-R (1999) Antimicrobial films produced from chitosan. Int J Biol Macromol 26(1):63–67

Brody AL, Strupinsky E, Kline LR (2001) Active packaging for food applications. CRC, Boca Raton

Daeschel MA (1989) Antimicrobial substances from lactic acid bacteria for use as food preservatives. Food Technol 43:164–167

De Reuck K, Sivakumar D, Korsten L (2009) Effect of integrated application of chitosan coating and modified atmosphere packaging on overall quality retention in litchi cultivars. J Sci Food Agric 89:915–920. doi:10.1002/jsfa.3501

Floros J, Dock L, Han J (1997) Active packaging technologies and applications. Food Cosmet Drug Packag 20(1):10–17

Ghasemnezhad M, Zareh S, Rassa M, Sajedi RH (2013) Effect of chitosan coating on maintenance of aril quality, microbial population and PPO activity of pomegranate (*Punica granatum* L. cv. Tarom) at cold storage temperature. J Sci Food Agric 93(2):368–374

Giatrakou V, Ntzimani A, Savvaidis I (2010) Effect of chitosan and thyme oil on a ready to cook chicken product. Food Microbiol 27(1):132–136

Halek G, Garg A (1988) Fungal inhibition by a fungicide coupled to an ionomeric film. J Food Saf 9(3):215–222

Han JH (2003a) Antimicrobial food packaging. Food Technol 54:56–65

Han JH (2003b) Antimicrobial food packaging. Novel food packaging techniques. pp 50–70

Han JH, Floros JD (1997) Casting antimicrobial packaging films and measuring their physical properties and antimicrobial activity. J Plast Film Sheeting 13(4):287–298

Jianglian D, Shaoying Z (2013) Application of chitosan based coating in fruit and vegetable preservation: a review. J Food Process Technol 4:227

Krause TR, Sebranek J, Rust R, Honeyman M (2003) Use of carbon monoxide packaging for improving the shelf life of pork. J Food Sci 68(8):2596–2603

Lanciotti R, Belletti N, Patrignani F, Gianotti A, Gardini F, Guerzoni ME (2003) Application of hexanal,(E)-2-hexenal, and hexyl acetate to improve the safety of fresh-sliced apples. J Agric Food Chem 51(10):2958–2963

Miller W, Spalding D, Risse L, Chew V (1985) The effects of an imazalil-impregnated film with chlorine and imazalil to control decay of bell peppers. Proceedings of the annual meeting of the Florida State Horticulture Society, USA

Nadarajah D, Han J, Holley R (2002) Use of allyl isothiocyanate to reduce *Escherichia coli* O157: H7 in packaged ground beef patties. Book of abstracts, IFT annual meeting, Institute of Food Technologists, Chicago

Nam S, Han J, Scanlon M, Izydorczyk M (2002) Use of extruded pea starch containing lysozyme as an antimicrobial and biodegradable packaging. Book of Abstracts

Odds FC, Brown AJP, Gow NAR (2003) Antifungal agents: mechanisms of action. Trends Microbiol 11:272–279

Olsmats C, Wallteg B (2009) Packaging is the answer to world hunger. World Packaging Organisation (WPO) and International Packaging Press Organisation (IPPO). http://www.worldpackaging.org/i4a/pages/index.cfm

Prasad R (2014) Synthesis of silver nanoparticles in photosynthetic plants. J Nanopart 963961. doi:10.1155/2014/963961

Prasad R, Kumar V, Prasad KS (2014) Nanotechnology in sustainable agriculture: present concerns and future aspects. Afr J Biotechnol 13(6):705–713

Prasad R, Pandey R, Barman I (2015) Engineering tailored nanoparticles with microbes: quo vadis. WIREs Nanomed Nanobiotechnol. doi:10.1002/wnan.1363

Shin J, Selke EM (2014) Food packaging. Food processing: principles and applications. pp 249–273

Steven M, Hotchkiss J (2003) Non-migratory bioactive polymers (NMBP) in food packaging. Woodhead, Cambridge

Suppakul P, Miltz J, Sonneveld K, Bigger SW (2003) Antimicrobial properties of basil and its possible application in food packaging. J Agric Food Chem 51(11):3197–3207

Tamer CE, Çopur O (2009) Chitosan: an edible coating for fresh-cut fruits and vegetables. VI international postharvest symposium, pp 619–624

Tsai GJ, Su WH, Chen HC, Pan CL (2002) Antimicrobial activity of shrimp chitin and chitosan from different treatments and applications of fish preservation. Fish Sci 68(1):170–177

Vermeiren L, Devlieghere F, Debevere J (2002) Effectiveness of some recent antimicrobial packaging concepts. Food Addit Contam 19(S1):163–171

Youwei Y, Yinzhe R (2013) Effect of chitosan coating on preserving character of post-harvest fruit and vegetable: a review. J Food Process Technol 4:254. doi:10.4172/2157-7110.1000254

Chapter 10
Fungal Nanoparticles: An Emerging Tool in Medical Biology

Anurag Yadav, Amit Verma, and Kusum Yadav

1 Introduction

The word "nano" of nanotechnology is derived from the Greek word-meaning dwarf. The term "nanotechnology" (NT) evolved during 1980s and gained impetus from various fields of research. It has been noticed that bulk materials when fragmented into nanoparticles (NPs) change their properties. This vary change of nature can be exploited to alter cellular processes at nano-scale level. Various cellular processes which operate at miniscule levels in the body are governed by immune recognition. The NT deals with the synthesis of nano-scale ($\sim 10^{-9}$ m) materials, which can be put to different applications. Presently nanomaterials are synthesized through physical and chemical methods, which involve lithography, aerosolization, UV irradiation, laser ablation, ultrasonic fields, photochemical method etc. Nanomaterial production through these methods is costlier and involves hazardous chemicals which limits their commercial production (Makarov et al. 2014).

The recent advances in biological science are replacing the environment unfriendly conventional methods of nanomaterial development with newer technologies. The biological synthesis of nanomaterials, which is popularly known as "Nanobiotechnology", holds the future of nanotechnology due to its low production

A. Yadav (✉)
Department of Microbiology, College of Basic Science and Humanities, S.D. Agricultural University, S.K. Nagar, Banaskantha, Gujarat, India
e-mail: anuragyadav123@gmail.com

A. Verma
Department of Biochemistry, College of Basic Science and Humanities, S.D. Agricultural University, S.K. Nagar, Banaskantha, Gujarat, India

K. Yadav
Department of Biochemistry, University of Lucknow, Lucknow, Uttar Pradesh, India

R. Prasad (ed.), *Advances and Applications Through Fungal Nanobiotechnology*, Fungal Biology, DOI 10.1007/978-3-319-42990-8_10

Table 10.1 Comparison of different methods of nanoparticle synthesis

Parameters	Physical methods	Chemical methods	Biological methods
Protocol ease	Hectic protocols, with lot of energy inputs	Hectic protocols, with lot of energy and chemical inputs	Easy and simple protocols
Achievement of desired dimensions	Achievable	Achievable	Requires further research
Large scale production	Yes	Yes	Yes
Ecofriendly nature	No	No	Yes
Cost effectiveness	No	No	Yes
Examples	Lithography, attrition, pyrolysis etc.	Microwave method, chemical reduction, electrochemical methods etc.	Use of biological systems as "Nanofactories" viz. bacteria, fungi, algae, plants etc.

cost and less hazardous nature (Prasad 2014; Prasad et al. 2014, 2015). The comparison of various nanomaterial synthesis methods are shown in Table 10.1.

On global scale nanobiotechnology based methods are gaining momentum for harnessing the "metal-microorganism" mediated nanomaterial production. The biological synthesis of NPs has face-lifted the nanotechnology as a "*Green Technology*" with features of cost effective production, particle modification and application in various other fields. Although biological NPs can be synthesized through plants, fungi, algae and bacteria but the cost involved in downstream processing and lower production levels limit their commercial application (Prasad et al. 2015). The use of fungi in nanotechnology for commertial production seems preferable due to their better adaptability under metal stress. As a result, fungal nanotechnology has evolved as separate branch called "*Myconanotechnology*" (MNT). With the increase in MNT methods for nanomaterial production, their applications have spread in varied sectors ranging from general industries to life saving medical technologies as elaborated further in this chapter.

2 Fungal Synthesis of Nanoparticles

Human kind has learned to utilize fungi as a source of food, antibiotics, enzymes and biosurfactants. Using fungus for nanoparticle synthesis is yet another application in this link. Although nanoparticle production through various organism types has their pros and cons (Table 10.2), nevertheless the use of fungi in nanoparticle production is providing the greener and cheaper alternative for nanoparticle synthesis. The elucidation of exact biochemical mechanism of NPs synthesis can open the way for cheaper industrial scale production. Due to the filamentous nature, the fungi can produce large amounts of NPs. They also produce large quantities of non-homologus

Table 10.2 Comparison of different organism types for nanoparticle production

	Viruses	Bacteria	Fungi	Plants
Advantages	Due to their capability to form structural assembly they are preferred for nanomaterial synthesis like "nanowires"	Easy to handle and genetic manipulation possible as per requirement	High biomass production as well as presence of a range of different endogenous redox systems. Performs well in metal stress	One step synthesis protocol, presence of stabilizing and capping agents necessary for NP synthesis
Disadvantages	Requires extensive research	Downstream processing is very complex and requires specialized techniques	Genetic constitution and metabolic machinery requires study for efficient NP synthesis	Complex genetic build up and interrelated biochemical machinery requires hectic effort to modify for NP synthesis
Reference	Ki et al. (2006)	–	Faheem and Banu (2014), Prasad et al. (2015)	Baker et al. (2013), Prasad (2014)

proteins in the culture medium (Punt et al. 2011). Such vivid attributes of fungi make them efficient "biofactories" for industrial NPs synthesis. The key factor behind the efficienty of fungal nanofactories is the range of redox systems, which help in reducing metal ions to NPs. These redox systems depend mainly on reductases enzymes along with other stabilizing agents. Several workers have reported the synthesis of NPs by purified reductases from fungi (Kumar et al. 2007). Yeasts of the class ascomycetes of fungi are good candidates for NP synthesis (Fig. 10.1).

2.1 Silver Nanoparticle Production

The spherical silver NPs, having a diameter of 20–30 nm, significantly inhibit the growth of medically important pathogenic Gram-positive bacteria (e.g. *Bacillus subtilis* and *Enterococcus faecalis*), Gram-negative bacteria (e.g. *Escherichia coli* and *Salmonella typhimurium*) and fungus *Candida albicans*. Different genera of fungi such as *Fusarium*, *Aspergillus*, *Penicillium* and *Verticillium* are known for their ability to fabricate silver NPs (El-Rafie et al. 2010; Jaidev and Narasimha 2010; Verma et al. 2010; Prasad and Swamy 2013). Moreover, in the last few years, fabrication of Ag NPs has increased extensively owing to their important biomedical applications (Jose Ruben et al. 2005). The fungus *Aspergillus fumigatus* is the organism of choice for extracellular biosynthesis of silver due to its rapid rate of NP synthesis and size range of 5–15 nm (Ahmad et al. 2003).

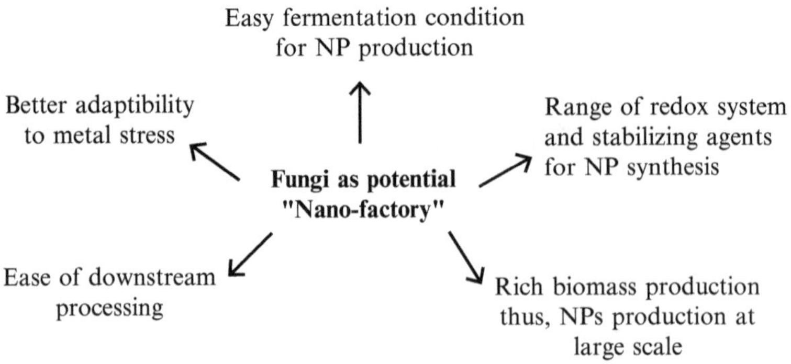

Fig. 10.1 Fungal based nanoparticle synthesis

2.2 *Gold Nanoparticle Production*

Gold NPs (GNPs) due to their optical and electronic properties are utilized in many disease diagnostics and treatments. Faheem and Banu (2014) extensively reviewed the use of GNPs in cancer diagnosis, cancer treatment and nanoparticle synthesis. Gold nanoparticles find their affinity towards DNA, RNA and other biological materials and are preferred choice over other multistep processes for delivery in cells (DeLong et al. 2010). GNPs synthesized by *Penicillium rugulosum* have been tested for their affinity with the genomes of *Escherichia coli* and *Staphylococcus aureus* (Mishra et al. 2012). Currently, GNPs are synthesized by several methods, nevertheless the synthesis through MNT is gaining popularity. GNPs are reported to be synthesized from various fungal sources viz. *Aspergillus* sp. *Fusarium* sp., *Helminthosporum* sp., *Sclerotium* sp., *Thermonospora* sp., *Tricothecium* sp. and *Verticillium* sp. (Mukherjee et al. 2002; Binupriya et al. 2010; Narayanan and Sakthivel 2011a, b; Faheem and Banu 2014). They are used in disease diagnostic due to their high-energy absorption and better contrast capability in comparison to standard agents. In addition, they can be effectively applied in medical therapy owing to their high binding capability and targeted drug delivery. Further research is required in pilot scale production of GNPs for faster biosynthesis.

2.3 *Miscellaneous Nanoparticle Production*

Apart from silver and gold nanoparticle production, nanoparticles of copper, selenium can also be produced. Selenium and copper nanoparticles are already in use as antimicrobial agent for dermal infections (Lee et al. 2011). Such particles have broader spectrum of antimicrobial activity against several animals and human pathogens but due to the high cost involved, they are infrequently used.

3 Nanoparticle Based Devices

3.1 *Nano Electro Mechanical Sensors (NEMS)*

NEMS are the "nanoscale" cantilevers which are used to detect very low concentrations of bacterial infection. They are also helpful in the tracking immune response against bacterial infection and for finding the drug efficacy against bacterial infections. NEMS are flexible boards of nanoscale dimensions which are synthesized using semiconductor lithography for microarray preparation. They have a coat of substrate specific molecules for high throughput detection of DNA or proteins (Fig. 10.2).

3.2 *Dendrimers*

These are the new class of nano-entities which are utilized in biological as well as material sciences. They are highly branched nanoscale macromolecules made up of the central core with interior branched structure and exterior surface with functional groups for target delivery. These structural arrangements are now-a-days utilized in target drug delivery for enhanced drug transport across membrane barriers. In the core region dendrimers entrap high molecular structures to attach drug (Fig. 10.3).

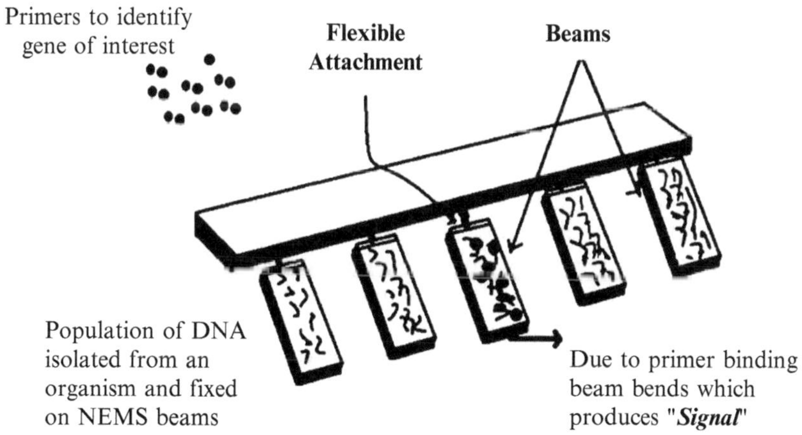

Fig. 10.2 General scheme of nanoelectromechanical sensors

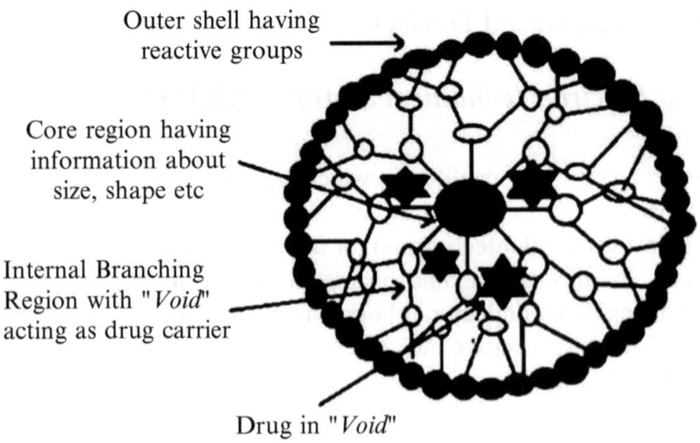

Fig. 10.3 A typical dendrimer

3.3 Liquid Crystals

Liquid crystals are made from organic crystals and are synthesized to interact with biomolecules like proteins or lipids. They are considered safe for targeted drug delivery to the specific regions of body tissues and are capable of detecting tumors. Liquid crystals find their use in display panels, as sensor and drug delivery agents (Hegmann et al. 2007) (Fig. 10.4).

3.4 Nanobots and Nanostars

Nanobots are robotic machines of nanometer dimention. They are also known by alternative names like nanites, nanoids, nanomachines or nanomites. Due to the very small size their large number is required for microscopic or macroscopic tasks. The research in this field, although in infancy stage, is known by the name "Nanobotics". It involves the combination of devices in the 0.1–10 μm range which are composed of nanoscale molecular components. These nanobots can revolutionize the medical diagnostics due to their potential in identification and determination of cancerous cells (Douglas et al. 2012). The use of such bots is under trial in medical nanotechnology. DNA based nanorobots are under developement to target cancerous cells (Fig. 10.5).

Fig. 10.4 Nanocrystal

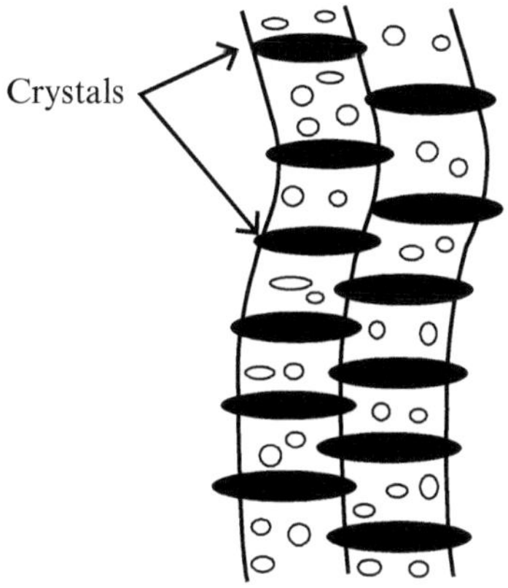

3.5 Nanoelmulsions

Nanoelmulsions are yet another set of nanotechnological tools for efficient drug delivery, which consist of fine oil in water dispersions with droplet dimensions of less than 100 nm. Nanoemulsions are paving a new hope in medical sciences as an efficient drug carrier for administrating pharmaceutically active entities. Due to their metastability (Sonneville-Aubrun et al. 2004) nanoemulsions are suitable for all delivery routes. They also find application in cosmetics and genetic engineering (Chime et al. 2014).

3.6 Nanoshells

Nanoshells are nano-locating agents which are now-a-days utilized for localizing necrotic centers for cancer treatment. They possess dielectric core covered with thin metallic shells, usually of gold. Nanoshells work by forming quasiparticles, causing electron oscillation with respect to their ions. In conjunction with monocytes they are carried over to tumor site where monocytes differentiate into macrophages releasing naoshells at the main necrotic center. Once ample amount of nanoshells deposit on the necrotic region the near infrared light is used to destroy the macrophages attached with tumors.

Nanoshells possess dielectric core covered with thin metallic shell (gold). Such kind of nanoparticles cause the formation of quasiparticles resulting in electron oscillation with respect to all the ions. Tumor cells have size equivalent to macrophages and once monocytes are brought into the tumor cells, it differentiates into macrophages, which would also be needed to maintain the cargo nanoparticles. Once nanoshells reach near the necrotic center, near-infrared radiation destroys the tumor-associated macrophages (Fig. 10.6).

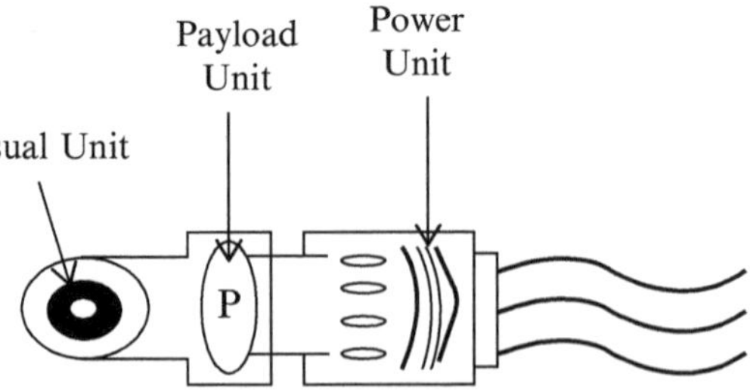

Fig. 10.5 Nanobots: general components

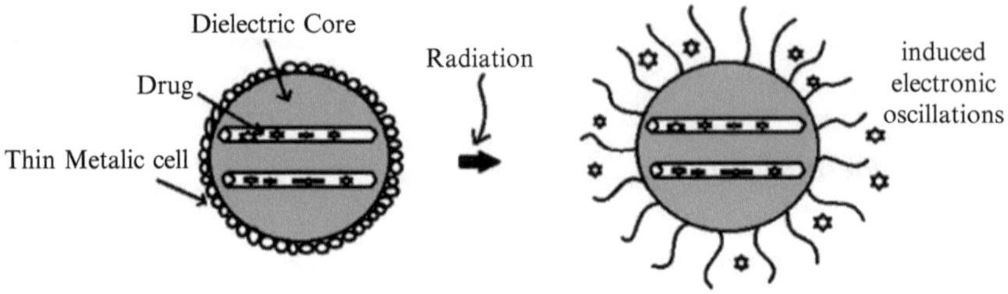

Fig. 10.6 Nanoshells: arrangement and mechanism of action

3.7 Nanotubes

Nanotubes are used in health monitoring for targeted drug delivery and biosensing. The unique chemical composition, very small size, optical, electrical and structural properties of carbon nanotubes make them a unique agent in drug delivery and biosensing. Carbon nanotubes (CNTs) can be designed to get structures with high drug loading capacity and cell penetration. Due to the large inner volume of nanotubes they can be used as an effective drug delivery agent. CNTs possess the property of electrical resistance, which is harnessed to develop electrical resistance based biosensors. Many overlapping nanotubes form an electrically conducting network, which can be electrically measured. CNTs have been used in developing glucose detection biosensors (Muguruma et al. 2007), DNA detection biosensors (Clendenin et al. 2007) and modified electrode biosensors (Timur et al. 2007) (Fig. 10.7).

3.8 Quantum Dots

Quantum dots are minuscule particles of nanometer range which are made up of hundreds or thousands of atoms. They have intermediate property between semiconductor and general molecules and are known to exhibit quantum mechanical properties. Quantum dots are preferred in medical applications for their superior photoemission and photostable effect (Zhang and Monteiro-Riviere 2009). Quantum

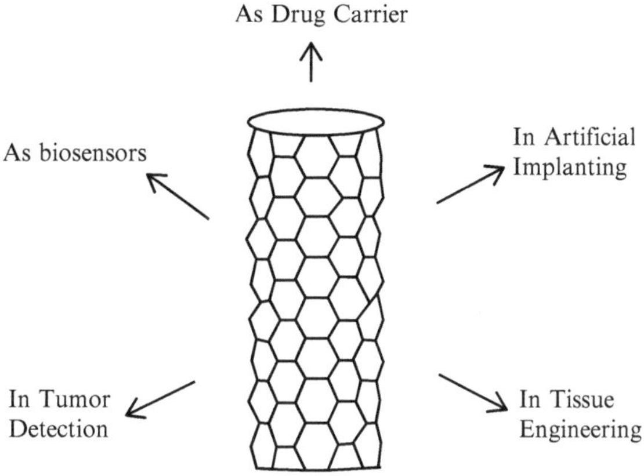

Fig. 10.7 Nanotubes: potential applications

dots are reported to be an efficient method for delivering small biomolecules like Si RNA in cells for gene silencing (Chen 2005). Their sensitive photoemission and photostable effect has made them a preferred medical nano-tool (Fig. 10.8).

4 Applications of Nanoparticles in Medical Sciences

MNT emerged as a momentous tool for nanomaterial synthesis that finds application in medical science. Nanomaterials are also applicable in the biological sciences due to their small size, which is same as active biomolecules found in organisms (Table 10.3). Due the same size, nanomaterials are capable of interacting with the biological structures like protein complexes, cell membranes, viruses etc. (Mahapatro and Singh 2011).

4.1 Antimicrobial Activity

NPs synthesized by biological means are vastly applicable in biomedicine. For example, mycologically synthesized silver NPs are now-a-days used as antibacterial agents. The high aspect ratio of silver NPs (AgNPs) facilitate them to interact with other particles and increase their antimicrobial efficacy by multifold. Silver NPs have been reported with very high antibacterial efficacy; even as little as 1 g of NPs imparts antibacterial properties to hundreds of square meters of substrate material (Hamouda et al. 2001; Gibbins and Warner 2005; Kim et al. 2007; Banu and Kumar 2009; El-Rafie et al. 2010; Huang et al. 2010; Jaidev and Narasimha 2010; Martinez-Gutierrez et al. 2010; Verma et al. 2010; Afreen and Ranganath 2011; Duncan 2011; Raheman et al. 2011; Savithramma et al. 2011; Elzatahry

Fig. 10.8 General components of a typical quantum dot

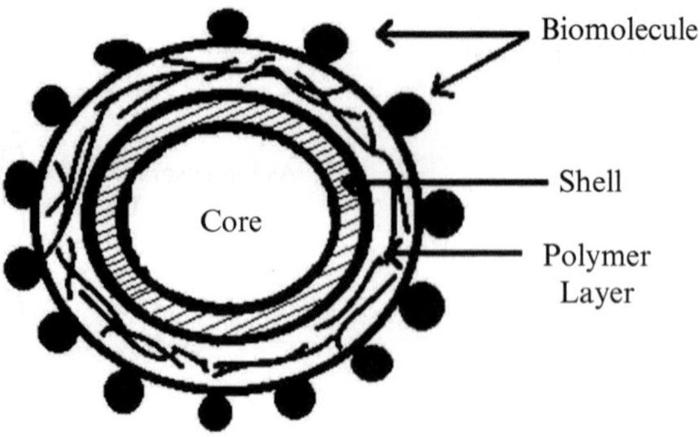

et al. 2012; Dar et al. 2013; Pasupuleti et al. 2013; Steelandt et al. 2014; Fatima et al. 2015; Aziz et al. 2015). The high antibacterial activity of Ag based NPs is due to their better anchoring and penetration in the bacterial cell wall, causing alternation in cellular signaling pathways by dephosphorylating putative key peptide substrates on tyrosine residues (Shrivastava et al. 2007). Ag ions interfere with thiol-group enzymes of microorganisms to inactivate them, causing inhibition of enzymatic activity (Klasen 2000). AgNPs possess the quality to retain their antibacterial efficacy in colloidal system. Antimicrobial effect of AgNPs has been well defined by several other workers (Gibbins and Warner 2005; Kim et al. 2007; El-Rafie et al. 2010; Dar et al. 2013; Pasupuleti et al. 2013; Prasad et al. 2015). Extracellularly produced silver or gold NPs using *Fusarium oxysporum* can be used to develop special kind of clothings. The cloths embedded with *F. oxysporum* produced silver NPs possess antibacterial properties and used in hospitals to prevent/minimize infection with pathogenic bacteria such as *Staphylococcus aureus* (Durán et al. 2007). The recent development of technologies for the impregnation of AgNPs seems to inhibit multiple drug resistant (MDR) strains of pathogenic microorganisms. Microbes are unlikely to develop resistance against silver as they commonly do against conventional narrow spectrum antibiotics. AgNPs are used in medical science for building Ag coated medical devices such as nanolotions and nanogels (Rai et al. 2009). Owing to their important biomedicine properties fungal based silver NPs can be produced intra- or extracellularly in the fermentation medium. In addition to AgNPs, other metallic NPs viz. copper (Cu), gold (Au), selenium (Se) etc. are also syntesized. Gold NPs have a broad spectrum of antimicrobial activity against several animal and human pathogens but due to higher cost involved in the production they are minimally used in antimicrobial study. Se and Cu NPs are antimicrobial agents utilized for purposes ranging from dermal infection cure to open wound injury treatment (Lee et al. 2011) (Fig. 10.9).

Table 10.3 Fungal NPs in medical science: Their source, dimensions, production parameters and applications

Fungal source	Nanoparticles	Salt concentration used	Dimensions (nm)	Application	Reference
Alternaria alternate	Gold	1 mM $HAuCl_4$	12		Sarkar et al. (2012)
Aspergillus clavatus	Silver	1 mM $AgNO_3$	10–25	Antimicrobial effect	Verma et al. (2010)
Aspergillus flavus	Gold	100, 200, 300, 400 and 500 μM $HAuCl_4$	22–26	Medical applications	Gupta and Bector (2013)
Aspergillus fumigatus	Gold	100, 200, 300, 400 and 500 μM $HAuCl_4$	22–26	Medical applications	Gupta and Bector (2013)
Aspergillus niger	Silver	1 mM $AgNO_3$	3–30	Antimicrobial activity	Jaidev and Narasimha (2010)
Aspergillus niger	Gold	1 mM $HAuCl_4$	10–30	Vector control for mosquito-borne disease	Soni and Prakash (2012)
Aspergillus oryzae var. *viridis*	Gold	1 mM $HAuCl_4$	10–60	–	Binupriya et al. (2010)
Aspergillus terreus	Silver	–	1–20	Antimicrobial	Li et al. (2012)
Bipolaris tetramera	Silver	1 mM $AgNO_3$	54.8–73.5	Antimicrobial and immunomodulatory efficacy	Fatima et al. (2015)
Bipolaris tetramera	Gold	1 mM $HAuCl_4$	58.4–261.7	Antimicrobial and immunomodulatory efficacy	Fatima et al. (2015)
Candida albicans	Gold	1 mM $HAuCl_4$	20–40	–	Chauhan et al. (2011)
Colletotrichum sp.	Gold	1 mM $HAuCl_4$	8–40	–	Shankar et al. (2003)
Cryphonectria sp.	Silver	1 mM $AgNO_3$	30–70	Antimicrobial effect	Dar et al. (2013)
Cylindrocladium floridanum	Gold	1 mM $HAuCl_4$	5–35	–	Narayanan and Sakthivel (2011b)
Fusarium oxysporum	Silver	–	5–15	–	Ahmad et al. (2003)
Fusarium oxysporum	Gold	1 mM $HAuCl_4$	8–40	–	Mukherjee et al. (2002)
Fusarium solani	Silver	0.5 mM $AgNO_3$	3–8	Antimicrobial effect	El-Rafie et al. (2010)

(continued)

Table 10.3 (continued)

Fungal source	Nanoparticles	Salt concentration used	Dimensions (nm)	Application	Reference
Ganoderma neo-japonicum	Silver	1 mM $AgNO_3$	–	Cytotoxic agent against breast cancer cells	Gurunathan et al. (2013)
Magnetospirillum gryphiswaldense	Gold	1 mM $HAuCl_4$	10–40	–	Cai et al. (2011)
Penicillium rugulosum	Gold	1 mM $HAuCl_4$	10–50	Binding affinity to isolated bacterial genomic DNA	Mishra et al. (2012)
Pestalotia sp.	Silver	1 mM $AgNO_3$	–	Antimicrobial effect	Raheman et al. (2011)
Phoma glomerata	Silver	1 mM $AgNO_3$	–	Bactericidal agents	Birla et al. (2009)
Rhizopus stolonifer	Silver	1 mM $AgNO_3$	5–50	Antibacterial effect	Afreen and Ranganath (2011)
Sclerotium rolfsii	Gold	1 mM $HAuCl_4$	25.2	–	Narayanan and Sakthivel (2011a)
Volvariella volvacea	Gold	1 mM $HAuCl_4$	20–150	–	Philip (2009)

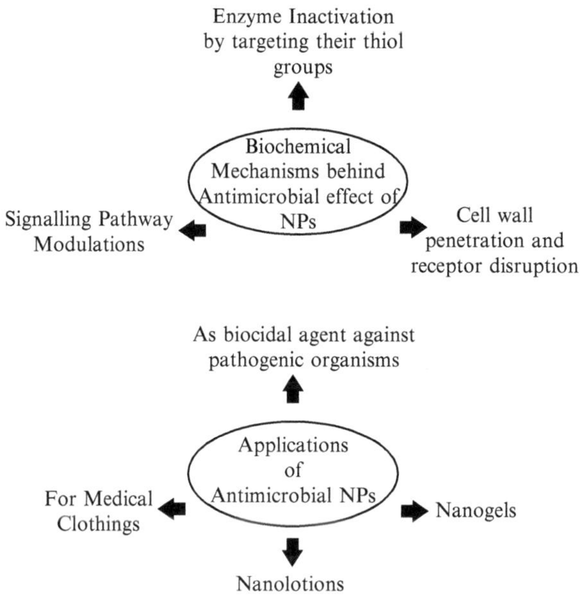

Fig. 10.9 Process of antimicrobial activity

4.2 *Disease Diagnosis and Detection*

4.2.1 Detection of Viral Diseases

According to World Health Organization (WHO) report, 34 million people have died so far from the human immunodeficiency virus (HIV) based diseases since the first cases were reported in 1981. The problem is severe in developing countries where HIV is the major cause of mortality.

Viruses are commonly identified and quantified by plaque assay, immunoassays, transmission or scanning electron microscopy, polymerase chain reaction (PCR) - based testing of viral DNA or RNA, tissue culture, lateral flow immunoassay, enzyme-linked immuno sorbent assay (ELISA) and PCR. The molecular HIV detection methods like PCR and real time PCR are not cost effective, prone to contamination and have analysis time of few hours. Moreover, these methods are not reliable to provide effective and rapid HIV detection and require large amounts of viral sample for testing. More accessible nanoscale technologies, capable of performing a CD4$^+$ T cell count and HIV viral load measurement are gaining their pace in the market (Damhorst et al. 2013).

4.2.2 Detection of Bacterial Diseases

Tuberculosis (TB) is one of the fatal infectious diseases of HIV-infected patients which increases death rate and drug resistance of immune suppressed patients. The multi drug resistant strains of mycobacteria are increasing at an alarming rate and require a thorough review. Conventional methods for TB diagnosis like sputum–smear microscopy are relatively less sensitive and poorly reliable. Nanotechnology based drug system is providing new hope for faster and reliable detection of mycobacterial strains to effectively eradicate mycobacterial infections. Reports show that TB can be successfully diagnosed by employing quantum dots, protein chips, sparse cell detection and NMR with microfluidic system (Cheepsattayakorn and Cheepsattayakorn 2013).

4.2.3 Detection of Protozoan Diseases

Rapid malaria detection in infected persons is crucial for effective treatment and prevention. In addition, malarial misdiagnosis is the leading cause of wrong treatment leading to unexpected side effects in patients (Amexo et al. 2004). Effective malaria treatment requires accurate and reliable diagnostic devices to reduce unnecessary use of anti-malarial drugs or antibiotics. To sustain this approach, gold nanoparticle-based fluorescence immunoassay have been designed for malarial antigen detection (Guirgis et al. 2012).

4.3 Disease Detection Nanotechnologies

4.3.1 Immunohistochemical Analysis

A method has been described for preparing electrically charged Cd–Se/ZnS core/shell nanocrystals (NCs) (Wang et al. 2004). NC–antibody conjugates can be used for specific detection of antigens from cancerous tissues using immunostaining. Epifluorescent microscopy can detect single NCs, ensuring a detection limit of single molecule attached with NC. NC–Ab conjugates may serve as ultrasensitive, photostable labels for immunohistochemical detection.

4.3.2 Nano-DNA Technology for Disease Diagnosis

Targeting and identifying microbial diseases have been made possible by detecting nucleic acid sequences unique to specific bacteria and viruses or by detecting abnormal concentration of certain proteins that signal the presence of various cancers and diseases (Rosi and Mirkin 2005). Nanomaterial-based assays are currently evaluated as sensitive protein detection methods. Sequences of nucleic acids are identified using PCR coupled with molecular fluorophore assays. PCR based methods of diagnosis, although highly sensitive, suffer from drawbacks of complexity, sensitivity to contamination, higher cost and lack of portability (Rosi and Mirkin 2005).

With the advancement of nanoscience technology various DNA-protein conjugates have been synthesized which in combination with PCR techniques have increased the sensitivity of immunoassays as high as 1000 folds compared to traditional ELISA. Niemeyer et al. (1999) reported the application of self-assembled DNA-streptavidin nanostructures as the effective reagent in immune-PCR assay. Niemeyer et al. (2001) reported the use of DNA-streptavidin nanocircles as reagent for ultrasensitive analysis via competitive immuno-PCR assay. Apartly, these self-assembled DNA-Protein conjugates are applicable in scanning probe microscopy (Gao et al. 2001; Pignataro et al. 2002). Such conjugates have potential to be utilized in nanomedicine and biomolecular engineering (Niemeyer 2002; Jain 2012).

4.3.3 Single Nucleotide Polymorphism (SNP) Analysis

Single nucleotide polymorphisms (SNPs) are bulk of informations related to human genome sequences which can be utilized in biotechnological, pharmaceutical and academic research. The SNP studies can help in revealing the relation between genetic variation and disease occurrences on one hand and novel disease diagnosis and therapy development in another. Galvin (2002) reported and compiled the use of nanotechnology and SNP information that can be applied in different sectors of disease diagnosis.

4.3.4 Nanoproteomic Analysis

Proteins are the effector molecules of metabolism and their detection and quantification uniquely elaborates the physiological conditions of the organism. ELISA is one of the reliable methods for protein detection and quantification. The ELISA implication in the medical field is generally limited by the concentration of proteins required for detection and usually results in identification at an advanced stage of disease. The MNT based methods are paving way for developing sensitive methods for protein detection.

NPs are extremely sensitive in protein detection. Specifically, magnetized nanoparticles coated with antibodies are used to bind specifically to the specific target, thus unraveling a potential to detect disease-causing agents at the nanotitre level (Rosi and Mirkin 2005). A sensory array with six non-covalent gold nanoparticles-fluorescent polymer conjugate is used to quantify proteins. Gold nanoparticles are used to quench fluorescence of a polymer. The presence of protein in sample interferes with nanoparticle-polymer interaction, causing distinctive fluorescent patterns. Such patterns are highly reproducible and are the characteristics of respective protein at nanomolar concentrations and quantification (You et al. 2007).

A molecular-imprint nanosensor method for accurate protein detection has also been reported (Cai et al. 2010). Adoption of MNT based methods can accurately detect proteins and quantify samples for medical diagnosis.

4.3.5 Nanobiosensors

Nanoparticles also have role in enhancing biosensor functionality (Hushiarian et al. 2014). Magnetic nanoparticles are becoming an integral tool in DNA bioassay where they are used in optimization of DNA hybridization and separation of target DNA (Clendenin et al. 2007).

Nanomaterials are state of the art chemical and biological biosensors due to their extreme sensitivity to other elements. A variety of nanobiosensors like nanowire biosensors, ion channel switch biosensors, electronic nanobiosensors viral nanosensor, pebble nanosensors, optical biosensors, surface plasmon resonance biosensors, laser nanosensors and nanoshell biosensors are used in medical and life sciences (Jain 2005b; Aziz et al. 2015). Nanobiosensors are useful in detecting unknown microorganism from samples, monitoring of metabolites in body secretions and detection of tumor cells.

4.4 Nanotechnology in Disease Treatment

4.4.1 Alzheimer's Disease

Alzheimer's disease (AD) is a neurodegenerative disorder which causes progressive loss of functions like impairment of intellectual capacity, deficit in attention, memory and thinking, inability to perform "motor functions" and alternation of personality. New technologies derived from nanotechnology-based approaches are helping in early diagnosis of AD. Early diagnosis could provide opportunities to treat patients at risk of AD development, thereby preventing the onset of irreversible neuronal damages (Brambilla et al. 2011; Nazem and Mansoori 2008).

4.4.2 Cancer

Early detection of cancer is crucial to provide effective treatment as timely detection increases the probability of curing diseases and reducing the mortality rate (Cuenca et al. 2006; Grodzinski et al. 2006; Romero-Morelos et al. 2011). The imaging methods in use these days can only detect visible change to a tissue; by that time several thousand cells may proliferate to initiate cancer. Nanotechnology, through the state of the art technologies is paving the way for early detection and visualization of cancerous cells. Antibodies that specifically bind to overexpressed cancerous cells by coating with metal oxide based nanoparticles, produce high contrast signals on Magnetic Resonance Images (MRI) or Computed Tomography (CT) scans. After binding to cancerous cells such antibodies can be easily visualized under the scanner. The fusion of tissue engineering and cancer science is leading the way for developing 3D models of tumors for cancer diagnosis (da Rocha et al. 2014; Prasad et al. 2015).

Since all the drugs entering the bloodstreams are also pumped into the blood vascular system; anticancer drugs which are made to kill cancerous cells, also kill healthy body cells. Attachment of drugs to nanoparticles can make the targeted delivery of drugs to tumor sites (Gaguski 2008). Nanotechnology finds its application for the treatment of prostate cancer (Cherian et al. 2014), breast cancer (Bruce 2006; Johnson et al. 2013), head and neck cancer (El-Sayed 2010), gastric cancer (Elingarami et al. 2014), cervical cancer, oral cancer and pancreatic cancer (McCarroll et al. 2014). Currently, cancer is treated by thermal therapy by photothermal ablation using silica nanoshells, photo-thermal ablation using carbon nanotubes, magnetic field-induced ablation using magnetic nanoparticles; photodynamic therapy using quantum dots, ceramic-based nanoparticles; chemotherapy using nano-structured polymer capsules, dendrimers, nanocells and radiotherapy using dendrimers, carbon nanotubes and gold nanoparticles (Jain 2005a; Roszek et al. 2005; Bamrungsap et al. 2012; Johal et al. 2015; Prasad et al. 2016).

4.4.3 HIV

Nanoparticles seem to possess anti-HIV effect. Capsid of HIV can be targeted for viral replication. The *in sillico* and wet lab results have identified the compounds that can be used to inhibit HIV capsid assembly. HIV adherence to host cells, toxicity, drug resistance and viral reservoirs make the treatment difficult. Studies show that the HIV capsid can be a targeted to design structure-based drugs to inhibit viral replication (Lisziewicz et al. 1999; Couvreur and Vauthier 2006; Kingsley et al. 2006; Yih and Al-Fandi 2006; Ganser-Pornillos et al. 2008; Riddler et al. 2008; Amiji et al. 2009; Pornillos et al. 2009; das Neves et al. 2010; Mamo et al. 2010; Sharma and Garg 2010; Fakruddin et al. 2012; Parboosing et al. 2012). The in vitro study shows that silver nanoparticles inhibit HIV from binding to host cells by preferentially binding to the gp120 glycoprotein knobs (Elechiguerra et al. 2005).

4.4.4 Diabetes

The diabetes is a metabolic disease of worldwide concern with more than 300 million affected individuals. Diabetes mellitus (DM) is a common chronic disease which risks the health of human beings (Couvreur and Vauthier 2006; Subramani, 2006; Pickup et al. 2008; Harsoliya 2012; Subramani et al. 2012). Type 1 diabetes mellitus is treated by controlling blood sugar level through modified dietary sugar intake, physical exercise, insulin therapy and oral medications. Medical nanotechnology research from the past few years is focusing on the treatment of type 1 diabetes through effective drug delivery.

4.4.5 Others

The core anti-filarial drugs of human lymphatic filariasis are hydrophobic and associated with a number of pharmacokinetics and pharmacodynamics issues leading to inadequate anti-filarial chemotherapy, demanding higher dose schedules with reduced efficacy and high side effects. The use of nanomedical technology improves drug efficacy, reduce dose frequency and diminish toxicity; thus holding many possibilities to cure this morbid disease (Makhsin et al. 2012; Santhoshkumar et al. 2011; Ali et al. 2014).

NPs are recently gaining application in development of stain resistant clothings, cosmetics and sunscreens due to better interaction at the molecular level. Nanoemulsions are also used in cosmetic and lotion formulations (Sonneville-Aubrun et al. 2004). Pure silver has been recently engineered into nanometer-sized particles (diameter <100 nm) for the treatment of wounds. For nanoparticle to be effective in skin treatment, it has to breach the barrier of the stratum corneum though receptor mediated process (Zhang and Monteiro-Riviere 2009). The skin therapeutic applications focus in the area of skin cancer imaging, immunomodulation and vaccine delivery, antimicrobials and wound healing. Metal oxide based nanoparticles are in use for manufacturing ultraviolet-B radiation protective sunscreens and topical cosmetics (Burnett and Wang 2011).

4.5 Gene Transfection

Surface-functionalized NPs are highly permeable to cell membranes compared to non-functionalized NPs (Lewin et al. 2000). Such NPs can be used as active carriers for delivering genetic material into living cell through transfection. Silica based nanoparticles labelled with positively charged ammonium groups can bind to polyanionic DNA through electrostatic interactions (Kneuer et al. 2000). These nanoparticles ensure quicker and effective delivery of DNA into cells.

4.6 Tissue Engineering

Nanotechnology is also applicable in peripheral nerve regeneration (Cunha et al. 2011; Ma et al. 2005; Shi et al. 2010), bone tissue engineering (Zhang et al. 2008) and heart valve engineering (Ma et al. 2005). Nerve cells and other organs can also be regenerated using nanofiber based scaffolds. Peptide nanofiber scaffolds have been reported to be used in the regeneration of axonal tissues of hamster (Ellis-Behnke et al. 2006).

4.7 Vaccine Development

DNA/polynucleotide vaccines function by inserting specific genes encoding for disease organism specific antigen into the host cell. They are expressed in the vicinity of antigen presenting cells providing immune response. Such vaccines have vast potential in medical science because they are cheaply produced, stored and handled better in varied environmental fluctuations than many of the protein based vaccines. Due to these qualities the polynucleotide vaccines are going to replace many conventional vaccines for immunotherapy in the near future. However, like any other technology, such vaccines have flaw of improper polynucleotide delivery into cells (Mohanraj and Chen 2007). As a modification, plasmid DNA coated with nanoparticles can be used as effective delivery agents. This system may have faster escape of DNA from degradative enzymes in endo-lysosomal compartment (Panyam et al. 2002). Also, nanoparticles cause release of plasmid DNA at moderate rate causing continuous gene expression (Hedley et al. 1998). This method of gene delivery finds application in many fields of life sciences.

Pathogens try to overpower the immune system by hiding intracellularly to cease off cellular intolerance by shedding surface markers and by serially switching dominant antigenic epitopes (El-Sayed et al. 2005). Development of an effective vaccine is crucial for human health which requires a vast amount of input in terms of time and money. With the advancement of technology newer tools are becoming available for developing next generation vaccines. Nanotechnology is providing the way for effective vaccine development with easier applicability, stability and site specific targeting.

4.8 Bio-Separation

Due to the ever-increasing human population the demand for quality therapeutic proteins is increasing. Such therapies are still very costly due to the time and energy involved in downstream processing of therapeutic proteins. Infact, separation of any biological component is important but difficult task in fermentation industry during downstream processing. Nanotechnology can help in bioseparation through employment of nanotubes. Nanotube membranes are the channels causing selective transport of ions and solutions on both sides of the membrane (Jirage et al. 1997). Nanotubes with inner diameter of less than 1 nm are suitable for separating small molecules, while nanotubes with inner larger inside diameters (20–60 nm) may be suitable to separate proteins (Martin and Kohli 2003). Several size specific nanotubes possess molecular recognition abilities and therefore can be used as nanophase extractors to remove specific molecules from solution (Martin and Kohli 2003).

The use of magnetic nanoparticles (MNPs) hold the future of cost effective separation of therapeutic proteins. Industrially the affinity chromatography is the preferred method for protein separation. However, the method suffers from the flaws of column operation and clogging of packed bed adsorbents. Such problems can be sorted by employing MNPs in affinity chromatography. MNPs can impart magnet-based separation to cause least diffusion in columns by permitting their rapid and convenient removal from heterogeneous reaction mixtures, without the use of filtration and centrifugation (Santana 2011).

4.9 MRI

Recent developments in MRI are due to the involvement of complex NPs as potential contrast agents for various studies. NPs based MRI has resulted in better understanding of tumor tissues, apoptosis mechanism (Strijkers et al. 2010), stem cell tracing etc. Presently, nanotechnology has developed biofunctionalized NPs, which are efficient contrast agent for MRI diagnosis at cellular and molecular levels (Strijkers et al. 2010). Also, NPs application in MRI has the additional advantage of low toxicity, high relaxivity and cell internalization capability (Matson and Wilson 2010).

4.10 Phagokinetic Studies

The ability of taking colloidal semiconductor nanocrystals by eukaryotic cells is related to cell motility. This property of eukaryotic cells has been exploited to develop motility-based nanosensors for targeting primary cancer cells. Nanocrystals are used as vital tool to study metastasis in cancer. Phagokinetics has found effective in comparison to other methods viz. the use of organic dyes, chemical agents etc. due to its photochemically robust nature (Parak et al. 2002).

4.11 Drug Delivery

NPs are important component for enhancing drug efficacy at lower doses, thus reducing side effects on body cells. The application of latest nanotechnology-based drug delivery systems like cyclodextrin complexes, gold nanoparticles, hydrogels, liquid crystals, nanostructured lipid carriers, polymeric nanoparticles and solid lipid nanoparticles are ensuring effective and safe delivery to biological systems (Calixto et al. 2014). In general, nanocarriers may enhance drug absorption through facilitated diffusion by improving intracellular drug penetration and distribution, by modifying pharmacokinetic and drug tissue distribution profile and by protecting drug degradation.

4.12 Nano-Remediation for Healthcare

NPs are becoming effective oxidizing agents to tackle the environmental contaminants. They are capable of high penetration and reactivity against redox contaminants, hydrophobic organic pollutants, and chlorinated hydrocarbons. Use of NPs is helping in environmental cleanup to sustain healthy human life. Research reports show the utilization of NPs for bioremediation (Nurmi et al. 2005; Quan et al. 2005).

Being a newer tool nanotechnology still finds limited application in environmental sciences and is facing an important challenge to cleanup anthropologically contaminated water drinking sites in the developing countries. The literature supports the role of zero-valent iron (nZVI) in clearing ground water contamination (Shirazi et al. 2013). The nanoremediation methods employ reactive nano-materials for catalysis and mitigation of pollutants. Nanoremediation is preferred in environmental remediation because the technology can omit need for water to be pumped out and soil transportation for treatment (Otto et al. 2008). The concept of "*Nanoremediation*" is gaining momentum and seems efficient in decontamination of contaminated surroundings cost-effectively.

5 Conclusions and Future Prospects

Owing to the recent reported utilities of NPs in disease diagnosis, for the treatment and control of various physiological conditions and diseases, the medical myconanotechnology requires an effort for searching novel species of microorganism to be used as "*Nanofactories*". Some selective fungi can be efficient in biosynthesis, due to easiness of handling, easier downstream processing and production of large amounts of biomass. With the increasing applications of fungi derived NPs in various fields, their use has received impetus, creating a branch of nanotechnology, "*Fungal Nanotechnology or Myconanotechnology* ". The advent of MNT methods of NP production has eliminated the need of physico-chemical processes which utilize environmentally hazardous chemicals. Use of fungi as nanofactory is paving way for "*Green Synthesis*" of NPs at cost effective levels. Fungi have been reported in the productions of Ag, Au, and Fe NPs along with other NPs. Research requires an emphasizing effort for elucidating biochemical mechanisms involved in the reduction of metals to metal NPs. This understanding will boost large-scale NPs synthesis in pilot plants for industrial production, which is presently confined to the flask level in research laboratories. Further biochemical understanding will enable the use of genetic engineering methods to modify fungal nanoparticles into desired dimensions.

References

Afreen RV, Ranganath E (2011) Synthesis of monodispersed silver nanoparticles by *Rhizopus stolonifer* and its antibacterial activity against MDR strains of *Pseudomonas aeruginosa* from burnt patients. Int J Environ Sci 1(7):1582–1592

Ahmad A, Mukherjee P, Senapati S, Mandal D, Khan MI, Kumar R, Sastry M (2003) Extracellular biosynthesis of silver nanoparticles using the fungus *Fusarium oxysporum*. Colloids Surf B Biointerfaces 28(4):313–318

Ali M, Afzal M, Kaushik U, Bhattacharya SM, Ahmad FJ, Dinda AK (2014) Perceptive solutions to anti-filarial chemotherapy of lymphatic filariasis from the plethora of nanomedical sciences. J Drug Target 22(1):1–13

Amexo M, Tolhurst R, Barnish G, Bates I (2004) Malaria misdiagnosis: effects on the poor and vulnerable. Lancet 364(9448):1896–1898

Amiji MM, Vyas TK, Shah LK (2009) Role of nanotechnology in HIV/AIDS treatment: potential to overcome the viral reservoir challenge. Discov Med 6(34):157–162

Aziz N, Faraz M, Pandey R, Sakir M, Fatma T, Varma A, Barman I, Prasad R (2015) Facile algae-derived route to biogenic silver nanoparticles: synthesis, antibacterial and photocatalytic properties. Langmuir 31:11605–11612. doi:10.1021/acs.langmuir.5b03081

Baker S, Harini BP, Rakshith D, Satish S (2013) Marine microbes: invisible nanofactories. J Pharm Res 6:383–388

Bamrungsap S, Zhao Z, Chen T, Wang L, Li C, Fu T, Tan W (2012) Nanotechnology in therapeutics: a focus on nanoparticles as a drug delivery system. Nanomedicine (Lond) 7(8):1253–1271

Banu GS, Kumar G (2009) Preliminary screening of endophytic fungi from medicinal plants in India for antimicrobial and antitumor activity. Int J of Pharm Sci Nanotechnol 2(2):566–571

Binupriya AR, Sathishkumar M, Vijayaraghavan K, Yun SI (2010) Bioreduction of trivalent aurum to nanocrystalline gold particles by active and inactive cells and cell free extract of *Aspergillus oryzae* var. viridis. J Hazard Mater 177:539–545

Birla SS, Tiwari VV, Gade AK, Ingle AP, Yadav AP, Rai MK (2009) Fabrication of silver nanoparticles by *Phoma glomerata* and its combined effect against Escherichia coli, *Pseudomonas aeruginosa* and *Staphylococcus aureus*. Lett Appl Microbiol 48:173–179

Brambilla D, Le Droumaguet B, Nicolas J, Hashemi SH, Wu L-P, Moghimi SM, Couvreur P, Andrieux K (2011) Nanotechnologies for Alzheimer's disease: diagnosis, therapy, and safety issues. Nanomed Nanotechnol Biol Med 7(5):521–540

Bruce SD (2006) Nanotechnology: metastatic breast cancer and beyond. ONS News 21(Suppl 8):5–6

Burnett ME, Wang SQ (2011) Current sunscreen controversies: a critical review. Photodermatol Photoimmunol Photomed 27(2):58–67

Cai D, Ren L, Zhao H, Xu C, Zhang L, Yu Y et al (2010) A molecular-imprint nanosensor for ultrasensitive detection of proteins. Nat Nanotechnol 5(8):597–601

Cai F, Li J, Sun J, Ji Y (2011) Biosynthesis of gold nanoparticles by biosorption using *Magnetospirillum gryphiswaldense* MSR-1. Chem Eng J 175:70–75

Calixto G, Bernegossi J, Fonseca-Santos B, Chorilli M (2014) Nanotechnology-based drug delivery systems for treatment of oral cancer: a review. Int J Nanomedicine 9:3719–3735

Chauhan A, Zubair S, Tufail S, Sherwani A, Sajid M, Raman SC et al (2011) Fungus-mediated biological synthesis of gold nanoparticles: potential in detection of liver cancer. Int J Nanomedicine 6:2305–2319

Cheepsattayakorn A, Cheepsattayakorn R (2013) Roles of nanotechnology in diagnosis and treatment of tuberculosis. J Nanotechnol Diagnos Treat 1(1):19–25

Chen AA (2005) Quantum dots to monitor RNAi delivery and improve gene silencing. Nucleic Acids Res 33(22):e190

Cherian AM, Nair SV, Lakshmanan VK (2014) The role of nanotechnology in prostate cancer theranostic applications. J Nanosci Nanotechnol 14(1):841–852

Chime SA, Kenechukwu FC, Attama AA (2014) Nanoemulsions—Advances in formulation, characterization and applications in drug delivery. In: Sezer AD (ed) Application of nanotechnology in drug delivery. InTech, Rijeka
Clendenin J, Kim J-W, Tung S (2007) An aligned carbon nanotube biosensor for DNA detection. 2nd IEEE international conference on Nano/Micro Engineered and Molecular Systems (NEMS'07), IEEE, pp 1028–1033
Couvreur P, Vauthier C (2006) Nanotechnology: intelligent design to treat complex disease. Pharm Res 23(7):1417–1450
Cuenca AG, Jiang H, Hochwald SN, Delano M, Cance WG, Grobmyer SR (2006) Emerging implications of nanotechnology on cancer diagnostics and therapeutics. Cancer 107(3):459–466
Cunha C, Panseri S, Antonini S (2011) Emerging nanotechnology approaches in tissue engineering for peripheral nerve regeneration. Nanomed: Nanotechnol Biol Med 7(1):50–59
da Rocha EL, Porto LM, Rambo CR (2014) Nanotechnology meets 3D in vitro models: tissue engineered tumors and cancer therapies. Mater Sci Eng C Mater Biol Appl 34:270–279
Damhorst GL, Watkins NN, Bashir R (2013) Micro- and nanotechnology for HIV/AIDS diagnostics in resource-limited settings. IEEE Trans Biomed Eng 60(3):715–726
Dar MA, Ingle A, Rai M (2013) Enhanced antimicrobial activity of silver nanoparticles synthesized by *Cryphonectria* sp. evaluated singly and in combination with antibiotics. Nanomed: Nanotechnol Biol Med 9(1):105–110
das Neves J, Amiji MM, Bahia MF, Sarmento B (2010) Nanotechnology-based systems for the treatment and prevention of HIV/AIDS. Adv Drug Deli Rev 62(4):458–477
DeLong RK, Reynolds CM, Malcolm Y, Schaeffer A, Severs T, Wanekaya A (2010) Functionalized gold nanoparticles for the binding, stabilization, and delivery of therapeutic DNA, RNA, and other biological macromolecules. Nanotechnol Sci Appl 3:53–63
Douglas SM, Bachelet I, Church GM (2012) A logic-gated nanorobot for targeted transport of molecular payloads. Science 335(6070):831–834
Duncan TV (2011) Applications of nanotechnology in food packaging and food safety: barrier materials, antimicrobials and sensors. J Colloid Interface Sci 363(1):1–24
Durán N, Marcato PD, De Souza GI, Alves OL, Esposito E (2007) Antibacterial effect of silver nanoparticles produced by fungal process on textile fabrics and their effluent treatment. J Biomed Nanotechnol 3(2):203–208
Elechiguerra JL, Burt JL, Morones JR, Camacho-Bragado A, Gao X, Lara HH, Yacaman MJ (2005) Interaction of silver nanoparticles with HIV-1. J Nanobiotechnol 3(6):1–10
Elingarami S, Liu M, Fan J, He N (2014) Applications of nanotechnology in gastric cancer: detection and prevention by nutrition. J Nanosci Nanotechnol 14(1):932–945
Ellis-Behnke RG, Liang Y-X, You S-W, Tay DK, Zhang S, So K-F, Schneider GE (2006) Nano neuro knitting: peptide nanofiber scaffold for brain repair and axon regeneration with functional return of vision. Proc Natl Acad Sci U S A 103(13):5054–5059
El-Rafie MH, Mohamed AA, Shaheen TI, Hebeish A (2010) Antimicrobial effect of silver nanoparticles produced by fungal process on cotton fabrics. Carbohydr Polym 80(3):779–782
El-Sayed IH (2010) Nanotechnology in head and neck cancer: the race is on. Curr Oncol Rep 12(2):121–128
El-Sayed NM, Mylor PJ, Blandin G, Berriman M, Crabtree J, Aggarwal G, Caler E, Renauld H, Worthey EA, Hertz-Fowler C, Ghedin E, Peacock C, Bartholomeu DC, Haas BJ, Tran A-N, Wortman JR, Alsmark UCM, Angiuoli S, Anupama A, Badger J, Bringaud F, Cadag E, Carlton JM, Cerqueira GC, Creasy T, Delcher AL, Djikeng A, Embley TM, Hauser C, Ivens AC, Kummerfeld SK, Pereira-Leal JB, Nilsson D, Peterson J, Salzberg SL, Shallom J, Silva JC, Sundaram J, Westenberger S, White O, Melville SE, Donelson JE, Andersson B, Stuart KD, Hall N (2005) Comparative genomics of Trypanosomatid parasitic protozoa. Science 309(5733):404–409
Elzatahry AA, Al-Enizi AM, Elsayed EA, Butorac RR, Al-Deyab SS, Wadaan MA, Cowley AH (2012) Nanofiber composites containing N-heterocyclic carbene complexes with antimicrobial activity. Int J Nanomedicine 7:2829–2832

Faheem SM, Banu H (2014) Gold nanoparticles in cancer diagnosis and treatment: a review. Austin J Biotechnol Bioeng 1(6):5

Fakruddin M, Hossain Z, Afroz H (2012) Prospects and applications of nanobiotechnology: a medical perspective. J Nanobiotechnol 10(1):31

Fatima F, Bajpai P, Pathak N, Singh S, Priya S, Verma SR (2015) Antimicrobial and immunomodulatory efficacy of extracellularly synthesized silver and gold nanoparticles by a novel phosphate solubilizing fungus *Bipolaris tetramera*. BMC Microbiol 15:52

Gaguski ME (2008) Nanotechnology treats cancer at the atomic level. ONS Connect 23(7):23

Galvin P (2002) A nanobiotechnology roadmap for high-throughput single nucleotide polymorphism analysis. Psychiatr Genet 12(2):75–82

Ganser-Pornillos BK, Yeager M, Sundquist WI (2008) The structural biology of HIV assembly. Curr Opin Struct Biol 18(2):203–217

Gao S, Chi L, Lenhert S, Anczykowski B, Niemeyer C, Adler M, Fuchs H (2001) High-quality mapping of DNA–protein complexes by dynamic scanning force microscopy. ChemPhysChem 2(6):384–388

Gibbins B, Warner L (2005) The role of antimicrobial silver nanotechnology. Med Device Diagn Ind Mag 1:1–2

Grodzinski P, Silver M, Molnar LK (2006) Nanotechnology for cancer diagnostics: promises and challenges. Expert Rev Mol Diagn 6(3):307–318

Guirgis BS, e Cunha CS, Gomes I, Cavadas M, Silva I, Doria G, Blatch GL, Baptista PV, Pereira E, Azzazy HM (2012) Gold nanoparticle-based fluorescence immunoassay for malaria antigen detection. Anal Bioanal Chem 402(3):1019–1027

Gupta S, Bector S (2013) Biosynthesis of extracellular and intracellular gold nanoparticles by *Aspergillus fumigatus* and *A. flavus*. Antonie Van Leeuwenhoek 103:1113–1123

Gurunathan S, Han JW, Eppakayala V, Jeyaraj M, Kim JH (2013) Cytotoxicity of biologically synthesized silver nanoparticles in MDA-MB-231 human breast cancer cells. Biomed Res 2013:1–10

Hamouda T, Myc A, Donovan B, Shih AY, Reuter JD, Baker JR (2001) A novel surfactant nanoemulsion with a unique non-irritant topical antimicrobial activity against bacteria, enveloped viruses and fungi. Microbiol Res 156(1):1–7

Harsoliya M (2012) Recent advances and applications of nanotechnology in diabetes. Int J Pharm Biol Arch 3(2):255–261

Hedley ML, Curley J, Urban R (1998) Microspheres containing plasmid-encoded antigens elicit cytotoxic T-cell responses. Nat Med 4(3):365–368

Hegmann T, Qi H, Marx V (2007) Nanoparticles in liquid crystals: synthesis, self-assembly, defect formation and potential applications. J Inorg Organomet Polym Mater 17(3):483–508

Huang L, Terakawa M, Zhiyentayev T, Huang Y-Y, Sawayama Y, Jahnke A, Tegos GP, Wharton T, Hamblin MR (2010) Innovative cationic fullerenes as broad-spectrum light-activated antimicrobials. Nanomed: Nanotechnol Biol Med 6(3):442–452

Hushiarian R, Yusof NA, Abdullah AH, Ahmad SA, Dutse SW (2014) A novel DNA nanosensor based on CdSe/ZnS quantum dots and synthesized Fe_3O_4 magnetic nanoparticles. Molecules 19(4):4355–4368

Jaidev LR, Narasimha G (2010) Fungal mediated biosynthesis of silver nanoparticles, characterization and antimicrobial activity. Colloids Surf B Biointerfaces 81(2):430–433

Jain KK (2005a) Nanotechnology-based drug delivery for cancer. Technol Cancer Res Treat 4(4):407–416

Jain KK (2005b) Nanotechnology in clinical laboratory diagnostics. Clin Chim Acta 358(1–2):37–54

Jain KK (2012) The handbook of nanomedicine. Springer, New York

Jirage KB, Hulteen JC, Martin CR (1997) Nanotubule-based molecular-filtration membranes. Science 278(5338):655–658

Johal HS, Garg T, Rath G, Goyal AK (2015) Advanced topical drug delivery system for the management of vaginal candidiasis. Drug Deliv 21:1–14

Johnson R, Sabnis N, McConathy WJ, Lacko AG (2013) The potential role of nanotechnology in therapeutic approaches for triple negative breast cancer. Pharmaceutics 5(2):353–370

Jose Ruben M, Jose Luis E, Alejandra C, Katherine H, Juan BK, Jose Tapia R, Miguel Jose Y (2005) The bactericidal effect of silver nanoparticles. Nanotechnology 16(10):2346

Ki TN, Dong-Wan K, Yoo PJ, Chung-Yi C, Nonglak M, Hammond PT et al (2006) Virus-enabled synthesis and assembly of nanowires for lithium ion battery electrodes. Science 312(5775):885–888

Kim JS, Kuk E, Yu KN, Kim J-H, Park SJ, Lee HJ, Kim SH, Park YK, Park YH, Hwang C-Y et al (2007) Antimicrobial effects of silver nanoparticles. Nanomed: Nanotechnol Biol Med 3(1):95–101

Kingsley JD, Dou H, Morehead J, Rabinow B, Gendelman HE, Destache CJ (2006) Nanotechnology: a focus on nanoparticles as a drug delivery system. J Neuroimmune Pharmacol 1(3):340–350

Klasen HJ (2000) Historical review of the use of silver in the treatment of burns. I. Early uses. Burns 26(2):117–130

Kneuer C, Sameti M, Bakowsky U, Schiestel T, Schirra H, Schmidt H, Lehr C-M (2000) A nonviral DNA delivery system based on surface modified silica-nanoparticles can efficiently transfect cells in vitro. Bioconjug Chem 11(6):926–932

Kumar SA, Abyaneh MK, Gosavi SW, Kulkarni SK, Pasricha R, Ahmad A, Khan MI (2007) Nitrate reductase-mediated synthesis of silver nanoparticles from $AgNO_3$. Biotechnol Lett 29:439–445

Lee SH, Heng D, Ng WK, Chan HK, Tan RB (2011) Nano spray drying: a novel method for preparing protein nanoparticles for protein therapy. Int J Pharm 403:192–200

Lewin M, Carlesso N, Tung CH, Tang XW, Cory D, Scadden DT, Weissleder R (2000) Tat peptide-derivatized magnetic nanoparticles allow in vivo tracking and recovery of progenitor cells. Nat Biotechnol 18(4):410–414

Li G, He D, Qian Y, Guan B, Gao S, Cui Y, Koji Y, Li W (2012) Fungus-mediated green synthesis of silver nanoparticles using *Aspergillus terreus*. Int J Mol Sci 13(1):466–476

Lisziewicz J, Rosenberg E, Lieberman J, Jessen H, Lopalco L, Siliciano R, Walker B, Lori F (1999) Control of HIV despite the discontinuation of antiretroviral therapy. N Engl J Med 340(21):1683–1684

Ma Z, Kotaki M, Inai R, Ramakrishna S (2005) Potential of nanofiber matrix as tissue-engineering scaffolds. Tissue Eng 11(1–2):101–109

Mahapatro A, Singh DK (2011) Biodegradable nanoparticles are excellent vehicle for site directed in-vivo delivery of drugs and vaccines. J Nanobiotechnol 9:55

Makarov VV, Love AJ, Sinitsyna OV et al (2014) Green nanotechnologies: synthesis of metal nanoparticles using plants. Acta Nat 6(1):35–44

Makhsin SR, Razak KA, Noordin R, Zakaria ND, Chun TS (2012) The effects of size and synthesis methods of gold nanoparticle-conjugated MαHIgG4 for use in an immunochromatographic strip test to detect brugian filariasis. Nanotechnology 23(49):495719

Mamo T, Moseman EA, Kolishetti N, Salvador-Morales C, Shi J, Kuritzkes DR, Langer R, Uv A, Farokhzad OC (2010) Emerging nanotechnology approaches for HIV/AIDS treatment and prevention. Nanomedicine (Lond) 5(2):269–285

Martin CR, Kohli P (2003) The emerging field of nanotube biotechnology. Nat Rev Drug Discov 2(1):29–37

Martinez-Gutierrez F, Olive PL, Banuelos A, Orrantia E, Nino N, Sanchez EM, Ruiz F, Bach H, Av-Gay Y (2010) Synthesis, characterization, and evaluation of antimicrobial and cytotoxic effect of silver and titanium nanoparticles. Nanomed: Nanotechnol Biol Med 6(5):681–688

Matson ML, Wilson LJ (2010) Nanotechnology and MRI contrast enhancement. Future Med Chem 2(3):491–502

McCarroll J, Teo J, Boyer C, Goldstein D, Kavallaris M, Phillips PA (2014) Potential applications of nanotechnology for the diagnosis and treatment of pancreatic cancer. Front Physiol 5:2

Mishra A, Tripathy SK, Yun S-I (2012) Fungus mediated synthesis of gold nanoparticles and their conjugation with genomic DNA isolated from *Escherichia coli* and *Staphylococcus aureus*. Process Biochem 47(5):701–711

Mohanraj V, Chen Y (2007) Nanoparticles—a review. Trop J Pharm Res 5(1):561–573
Muguruma H, Matsui Y, Shibayama Y (2007) Carbon nanotube? Plasma polymer-based amperometric biosensors: enzyme-friendly platform for ultrasensitive glucose detection. Jpn J Appl Phys 46(9R):60–78
Mukherjee P, Senapati S, Mandal D, Ahmad A, Khan MI, Kumar R, Sastry M (2002) Extracellular synthesis of gold nanoparticles by the fungus *Fusarium oxysporum*. Chembiochem 3:461–463
Narayanan KB, Sakthivel N (2011a) Facile green synthesis of gold nanostructures by NADPH—dependent enzyme from the extract of *Sclerotium rolfsii*. Colloids Surf A: Biointerfaces 380:156–161
Narayanan KB, Sakthivel N (2011b) Synthesis and characterization of nano-gold composite using *Cylindrocladium floridanum* and its heterogeneous catalysis in the degradation of 4-nitrophenol. J Hazard Mater 189:519–525
Nazem A, Mansoori GA (2008) Nanotechnology solutions for Alzheimer's disease: advances in research tools, diagnostic methods and therapeutic agents. J Alzheimers Dis 13(2):199–224
Niemeyer CM (2002) Tools for the biomolecular engineer. Science 297(5578):62–63
Niemeyer CM, Adler M, Pignataro B, Lenhert S, Gao S, Lifeng C, Fuchs H, Dietmar B (1999) Self-assembly of DNA-streptavidin nanostructures and their use as reagents in immuno-PCR. Nucleic Acids Res 27(23):4553–4561
Niemeyer CM, Wacker R, Adler M (2001) Hapten-functionalized DNA-streptavidin nanocircles as supramolecular reagents in a competitive immuno-PCR assay. Angew Chemi Int Ed 40(17):3169–3172
Nurmi JT, Tratnyek PG, Sarathy V, Baer DR, Amonette JE, Pecher K, Wang C, Linehan JC, Matson DW, Penn RL, Driessen MD (2005) Characterization and properties of metallic iron nanoparticles: spectroscopy, electrochemistry, and kinetics. Environ Sci Technol 39(5):1221–1230
Otto M, Floyd M, Bajpai S (2008) Nanotechnology for site remediation. Remediat J 19(1): 99–108
Panyam J, Zhou W-Z, Prabha S, Sahoo SK, Labhasetwar V (2002) Rapid endo-lysosomal escape of poly (DL-lactide-co-glycolide) nanoparticles: implications for drug and gene delivery. FASEB J 16(10):1217–1226
Parak WJ, Boudreau R, Le Gros M, Gerion D, Zanchet D, Cm M, Sc W, Ap A, Larabell C (2002) Cell motility and metastatic potential studies based on quantum dot imaging of phagokinetic tracks. Adv Mater 14(12):882–885
Parboosing R, Maguire GEM, Govender P, Kruger HG (2012) Nanotechnology and the treatment of HIV infection. Viruses 4(4):488–520
Pasupuleti VR, Prasad, Shiekh RA, Balam SK, Narasimhulu G, Reddy CS, Ab Rahman I, Gan SH (2013) Biogenic silver nanoparticles using *Rhinacanthus nasutus* leaf extract: synthesis, spectral analysis, and antimicrobial studies. Int J Nanomed 8:3355–3364
Philip D (2009) Biosynthesis of Au, Ag and Au-Ag nanoparticles using edible mushroom extract. Spectrochim Acta A Mol Biomol Spectrosc 73:374–381
Pickup JC, Zhi ZL, Khan F, Saxl T, Birch DJ (2008) Nanomedicine and its potential in diabetes research and practice. Diabetes Metab Res Rev 24(8):604–610
Pignataro B, Chi L, Gao S, Anczykowski B, Niemeyer C, Adler M, Fuchs H (2002) Dynamic scanning force microscopy study of self-assembled DNA-protein nanostructures. Appl Phys A 74(3):447–452
Pornillos O, Ganser-Pornillos BK, Kelly BN, Hua Y, Whitby FG, Stout CD, Sundquist WI, Hill CP, Yeager M (2009) X-ray structures of the hexameric building block of the HIV capsid. Cell 137(7):1282–1292
Prasad R (2014) Synthesis of silver nanoparticles in photosynthetic plants. J Nanopart 963961. doi:10.1155/2014/963961
Prasad R, Swamy VS (2013) Antibacterial activity of silver nanoparticles synthesized by bark extract of *Syzygium cumini*. J Nanopart. doi:10.1155/2013/431218
Prasad R, Kumar V, Prasad KS (2014) Nanotechnology in sustainable agriculture: present concerns and future aspects. Afr J Biotechnol 13(6):705–713

Prasad R, Pandey R, Barman I (2015) Engineering tailored nanoparticles with microbes: quo vadis. WIREs Nanomed Nanobiotechnol. doi:10.1002/wnan.1363

Prasad R, Pandey R, Varma A, Barman I (2016) Polymer based nanoparticles for drug delivery systems and cancer therapeutics. In: Kharkwal H, Janaswamy S (eds) Natural polymers for drug delivery. CABI, Oxfordshire, pp 53–70

Punt PJ, Levasseur A, Visser H, Wery J, Record E (2011) Fungal protein production: design and production of chimeric proteins. Annu Rev Microbiol 65:57–69

Quan X, Yang S, Ruan X, Zhao H (2005) Preparation of titania nanotubes and their environmental applications as electrode. Environ Sci Technol 39(10):3770–3775

Raheman F, Deshmukh S, Ingle A, Gade A, Rai M (2011) Silver nanoparticles: novel antimicrobial agent synthesized from an endophytic fungus *Pestalotia* sp. isolated from leaves of *Syzygium cumini* (L). Nano Biomed Eng 3(3):174–178

Rai M, Yadav A, Gade A (2009) Silver nanoparticles as a new generation of antimicrobials. Biotechnol Adv 27(1):76–83

Riddler SA, Haubrich R, DiRienzo AG, Peeples L, Powderly WG, Klingman KL, Garren KW, George T, Rooney JF, Brizz B et al (2008) Class-sparing regimens for initial treatment of HIV-1 infection. N Engl J Med 358(20):2095–2106

Romero-Morelos P, Peralta-Rodriguez R, Mendoza-Rodriguez M, Valdivia-Flores A, Marrero-Rodriguez D, Paniagua-Garcia L, Rodriguez-Cabrales J, Parra-Melquiadez M, Salcedo-Vargas M (2011) The nanotechnology as a support for diagnosis and prognosis in cancer research. Rev Med Inst Mex Seguro Soc 49(6):621–630

Rosi NL, Mirkin CA (2005) Nanostructures in biodiagnostics. Chem Rev 105(4):1547–1562

Roszek B, de Jong WH, Geertsma RE (2005) Nanotechnology in medical applications: state-of-the-art in materials and devices. RIVM report 265001001, RIVM, National Institute for Public Health and the Environment, Bilthoven

Santana SDF (2011) Magnetic nanoparticles for biocatalysis and bioseparation. ACS Appl Mater Interfaces 4:5907–5914

Santhoshkumar T, Rahuman AA, Rajakumar G, Marimuthu S, Bagavan A, Jayaseelan C, Zahir AA, Elango G, Kamaraj C (2011) Synthesis of silver nanoparticles using *Nelumbo nucifera* leaf extract and its larvicidal activity against malaria and filariasis vectors. Parasitol Res 108(3):693–702

Sarkar J, Ray S, Chattopadhyay D, Laskar A, Acharya K (2012) Mycogenesis of gold nanoparticles using a phytopathogen *Alternaria alternata*. Bioprocess Biosyst Eng 35:637–643

Savithramma N, Rao ML, Rukmini K, Devi PS (2011) Antimicrobial activity of silver nanoparticles synthesized by using medicinal plants. Int J ChemTech Res 3(3):1394–1402

Shankar SS, Ahmad A, Pasricha R, Sastry M (2003) Bioreduction of chloroaurate ions by geranium leaves and its endophytic fungus yields gold nanoparticles of different shapes. J Mater Chem 13:1822–1826

Sharma P, Garg S (2010) Pure drug and polymer based nanotechnologies for the improved solubility, stability, bioavailability and targeting of anti-HIV drugs. Adv Drug Deliv Rev 62(4):491–502

Shi J, Votruba AR, Farokhzad OC, Langer R (2010) Nanotechnology in drug delivery and tissue engineering: from discovery to applications. Nano Lett 10(9):3223–3230

Shirazi E, Torabian A, Nabi-Bidhendi G (2013) Carbamazepine removal from groundwater: effectiveness of the TiO_2/UV, nanoparticulate Zero-Valent Iron, and Fenton (NZVI/H_2O_2) processes. Clean: Soil, Air, Water 41(11):1062–1072

Shrivastava S, Bera T, Roy A, Singh G, Ramachandrarao P, Dash D (2007) Characterization of enhanced antibacterial effects of novel silver nanoparticles. Nanotechnology 18(22):225103

Soni N, Prakash S (2012) Efficacy of fungus mediated silver and gold nanoparticles against *Aedes aegypti* larvae. Parasitol Res 110(1):175–184

Sonneville-Aubrun O, Simonnet JT, L'Alloret F (2004) Nanoemulsions: a new vehicle for skincare products. Adv Colloid Interface Sci 108:145–149

Steelandt J, Salmon D, Gilbert E, Almouazen E, Renaud FN, Roussel L, Haftek M, Pirot F (2014) Antimicrobial nanocapsules: from new solvent-free process to in vitro efficiency. Int J Nanomed 9:4467–4474

Strijkers GJ, van Tilborg GA, Geelen T, Reutelingsperger CP, Nicolay K (2010) Current applications of nanotechnology for magnetic resonance imaging of apoptosis. Methods Mol Biol 624:325–342

Subramani K (2006) Applications of nanotechnology in drug delivery systems for the treatment of cancer and diabetes. Int J Nanotechnol 3(4):557–580

Subramani K, Pathak S, Hosseinkhani H (2012) Recent trends in diabetes treatment using nanotechnology. Dig J Nanomat Biostruct 7(1):85–95

Timur S, Anik U, Odaci D, Gorton L (2007) Development of a microbial biosensor based on carbon nanotube (CNT) modified electrodes. Electrochem Commun 9(7):1810–1815

Verma VC, Kharwar RN, Gange AC (2010) Biosynthesis of antimicrobial silver nanoparticles by the endophytic fungus *Aspergillus clavatus*. Nanomedicine (Lond) 5(1):33–40

Wang H, Wang J, Deng X, Sun H, Shi Z, Gu Z, Liu Y, Zhaoc Y (2004) Biodistribution of carbon single-wall carbon nanotubes in mice. J Nanosci Nanotechnol 4(8):1019–1024

Yih TC, Al-Fandi M (2006) Engineered nanoparticles as precise drug delivery systems. J Cell Biochem 97(6):1184–1190

You CC, Miranda OR, Gider B, Ghosh PS, Kim IB, Erdogan B, Krovi SA, Bunz UH, Rotello VM (2007) Detection and identification of proteins using nanoparticle-fluorescent polymer ‘chemical nose’ sensors. Nat Nanotechnol 2(5):318–323

Zhang LW, Monteiro-Riviere NA (2009) Mechanism of quantum dot nanoparticle cellular uptake. Toxicol Sci 110(1):138–155

Zhang Y, Venugopal JR, El-Turki A, Ramakrishna S, Su B, Lim CT (2008) Electrospun biomimetic nanocomposite nanofibers of hydroxyapatite/chitosan for bone tissue engineering. Biomaterials 29(32):4314–4322

Chapter 11
Intervention of Fungi in Nano-Particle Technology and Applications

Ramesh Raliya, Vinod Saharan, Ramprakash Saran, Kailash Choudhary, Jagadish C. Tarafdar, and Pratim Biswas

Abstract Biosynthesis of nanomaterial is of particular attention for material scientists due to its environmentally benign perspective and durability in a natural medium. Nanoparticles synthesized by using the whole cell, either inside the biological entity (intracellular) or extract/lysate/peptide-template (extracellular) believed to have a wide range of biological application. The chapter focuses primarily on the mechanistic investigation of metal and metal oxide nanoparticle synthesis and their potential applications in the agricultural and biomedical sector. So far fungus is explored more for silver nanoparticle synthesis among all other nanoparticles and their use as an antimicrobial agent either bare nanoparticles or as a synergetic agent with existing counterparts. In addition, fungus-nanotechnology explored for the synthesis of agriculturally important nutrient for native phosphorus mobilization and enhancement in photosynthetic activity.

1 Introduction

Fungi, belongs to the group of eukaryotic organism, have been extensively used to produce industrial chemical and enzymes for various purposes, notably from food to medicine (Carlile et al. 2001; Prasad et al. 2015). With the advent of modern nanotechnology, researchers have practiced harnessing fungal strains to provide an

R. Raliya (✉) • P. Biswas
Washington University, St. Louis, MO 63130, USA
e-mail: rameshraliya@wustl.edu

V. Saharan
Maharana Pratap University of Agriculture and Technology, Udaipur, Rajasthan 313001, India

R. Saran
J. N. Vyas University, Jodhpur, Rajasthan 342005, India

K. Choudhary
Lachoo Memorial College of Science and Technology, Jodhpur, Rajasthan 342001, India

J.C. Tarafdar
Central Arid Zone Research Institute, Jodhpur, Rajasthan 342001, India

R. Prasad (ed.), *Advances and Applications Through Fungal Nanobiotechnology*, Fungal Biology, DOI 10.1007/978-3-319-42990-8_11

alternative greener approach of nanoparticle production. A nanoparticle or nanostructure is an intentionally created, engineered particle of 1–100 nm at least in one dimension. Due to extremely small size, nanoparticles have high surface area to mass ratio that brings extraordinary physicochemical properties than its bulk counterpart (Suman et al. 2010; Prasad 2014; Aziz et al. 2015). Nanoparticle research is currently an area of intense scientific interest of the community due to a wide variety of potential applications in agriculture, biomedical, energy, environmental, electronic and optical fields (Prasad et al. 2014; NSF 2015).

Producing nanomaterial's sustainably in part means using less harmful chemicals in nanoparticle production (Murphy 2008). One of the ways nanoparticles can be created without using harmful chemicals is by exploiting natural biological processes, and fungus is one of the organisms used in nanoparticle technology to create nanoscale objectives. To the best of authors' literature survey in the scientific domain, historically, Dameron et al. (1989) biosynthesized nanometer scale (2 nm) semiconductor quantum crystallites of cadmium sulfide (CdS) using yeasts *Candida glabrata* and *Schizosaccharomyces pombe*, cultured in the presence of cadmium salts. The biosynthesized quantum CdS crystallites were more monodisperse than CdS nanoparticles synthesized chemically. Later, in 2001, a team of National Chemical Laboratory, India, used fungus *Verticillium* to reduce aqueous silver nitrate solution into silver nanoparticle of 25 nm (Mukherjee et al. 2001a). The reduction of the metal ions occurs on the surface of the mycelia (extracellular) leading to the formation of silver nanoparticles of fairly well-defined dimensions and tolerable monodispersity. In-spite of different proposed synthesis mechanism (intracellular vs. extra-cellular), both of these early landmark studies suggested that protein/peptide play a key role in the reduction of metal salt.

The advantage of biosynthesis reaction is obvious from the sustainability point of view (e.g., recycling or utilization of bio-based objects for chemical production) (Murphy 2008). The fungal mediated synthesis of nanoparticles has many advantages over other biosynthesis approaches (using whole cell(s), secretion, extract or lysate of bacteria, plant or animal source) such as ease of scale-up process, minimum downstream processing, economically viable, easy to maintain the culture and handling of biomass (Prasad et al. 2015). Compared to bacteria, fungi are saprophytic in nature, the large surface area covered by mycelia (help in intracellular synthesis) and also known to secret much higher amounts of protein (help in extracellular synthesis), thereby fungi has significant attention to using as nanoparticle production.

2 Fungal Mediated Nanoparticle Synthesis and Mechanistic Aspects

Fungi are drawing the attention of nanotechnology researchers for the bottom up and top down synthesis of nanoparticles due to metal tolerance, bioaccumulation, and saprophytic ability (Sastry et al. 2003; Thakkar et al. 2010; Raliya et al. 2013). Over the several nanoparticle biosynthesis sources such as bacteria, plant, and

Table 11.1 Synthesis of metal and metal oxide nanoparticles using fungi

Fungi species/name	Type of nanoparticles	Size (nm)	References
Verticillium sp.	Ag, Au	25 ± 12	Mukherjee et al. (2001a, b) and Sastry et al. (2003)
Fusarium oxysporum	Ag, Au, CdS, Zirconia, $BaTiO_3$, Ag-Au bi-metallic	4–40	Ahmad et al. (2002, 2003), Mukherjee et al. (2002), Senapati et al. (2005) and Bansal et al. (2006)
Aspergillus oryzae	Fe	10–24.6	Raliya and Tarafdar (2013a)
Aspergillus fumigatus	Ag	5–25	Bhainsa and D'Souza (2006)
Aspergillus terreus	Ag	2.5	Raliya and Tarafdar (2012)
Aspergillus flavus	Ag	8.92 ± 1.61	Vigneshwaran et al. (2007)
Aspergillus tubingensis	TiO_2, Phosphorous	1.5–30	Tarafdar et al. (2012, 2013)
Rhizoctonia bataticola	Au	6.2	Raliya and Tarafdar (2013a, b, c, 2014)
Pochonia chlamydosporium, Aspergillus fumigatus, Curvularia lunata, Chaetomium globosum, Aspergillus wentii, Aspergillus tubingensis, Aspergillus flavus, Aspergillus terreus	Zn, Ti, Mg and Fe	12–95	Kaul et al. (2012), Raliya and Tarafdar (2014) and Raliya et al. (2014a, b)

higher kingdom animal tissues, fungi are a potential candidate to scale-up nanoparticle synthesis process due to economic viability (Prasad et al. 2015). Table 11.1 summarizes the metal and metal oxide nanoparticles synthesize using fungi. Scientific literature survey suggests that fungi have been used primarily for metal and metal oxide nanoparticles, in particular, silver (Ag), gold (Au), zinc (Zn), zinc oxide (ZnO) and titanium di oxide (TiO_2) nanoparticles synthesis (Table 11.1).

The fungus can synthesize nanoparticles both inside the cell (intracellular) and outside the cell (extracellular). One of the early report on the use of fungus for intracellular nanoparticle synthesis by using *Verticillium sp.* was reported by Mukherjee et al. (2001a). Author exposed fungal biomass with aqueous silver nitrate suspension. As a result of the intracellular reduction of the metal ions, 20–30 nm silver nanoparticles were observed below the cell wall surface (Fig. 11.1). It was also observed in the same study that silver nanoparticle didn't exert any toxic effect rather increased biomass after biosynthesis of silver nanoparticles. Similarly, the same *Verticillium sp.* used for the bioreduction of gold chloride and found similar to the size of the silver nanoparticle as reported above (Mukherjee et al. 2001b). The authors explained that internalization of a precursor salt of gold ions on the surface of fungal

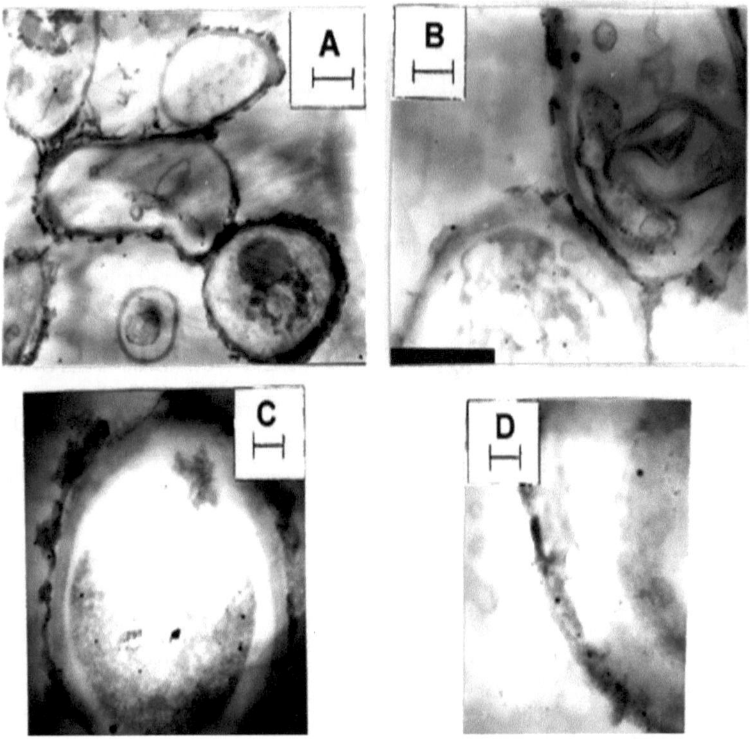

Fig. 11.1 TEM micrograph *Verticillium* cells after reaction with silver nitrate at different magnifications (**a–d**). The scale bars in (**a–d**) correspond to 1 μm, 500, 200, and 100 nm respectively (Mukherjee et al. 2001a)

cells perhaps occurs by electrostatic interactions with positively charged groups such as lysine residues in enzymes that existed in the mycelial cell wall. In addition to *Verticillium, Fusarium oxysporum* has also been used for the synthesis of CdS (Ahmad et al. 2002), $BaTiO_3$ (Bansal et al. 2006), Ag (Ahmad et al. 2003), Au (Mukherjee et al. 2002) and Ag-Au bi-metallic (Senapati et al. 2005). Diversion to the extracellular synthesis of nanoparticles using fungi is believed to simplify the purification steps for synthesized nanoparticles.

Mechanistically, it is proved that molecular machinery and various cellular proteins are involved in the bioreduction of the metal salt. Barwal et al. (2011) explained that ATPase, sedoheptulose-1, 7-bisphosphatase, carbonic anhydrase, ferredoxin NADP+ reductase, superoxide dismutase, oxygen evolving enhancer protein ribulose bisphosphate carboxylase and nuclear histone are involved in the bioreduction of silver nitrate by the unicellular algae *Chlamydomonas reinhardtii.* Raliya and Tarafdar (2012) explained the mechanism of extracellular synthesis of silver nanoparticles using the fungus *Aspergillus terreus* (Fig. 11.2). Authors also explained that protein capping enabled nanoparticle stability, monodispersity and

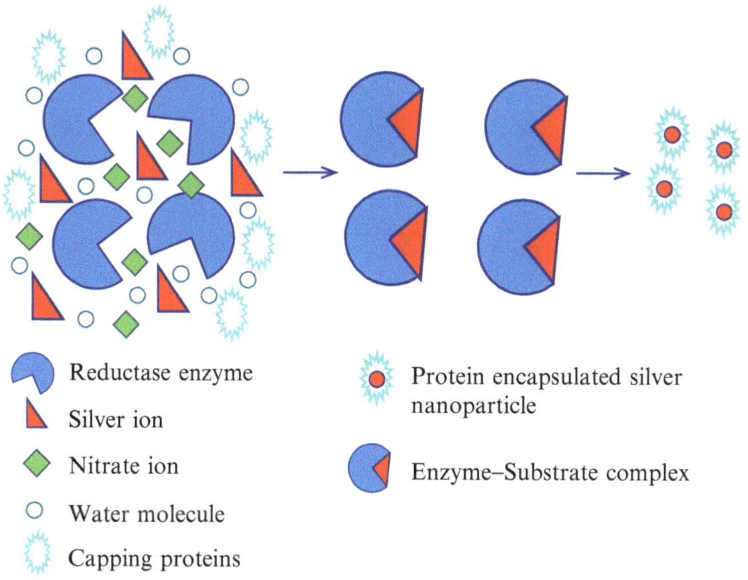

Fig. 11.2 Mechanism for biosynthesis of silver nanoparticles using *A. terreus* (Raliya and Tarafdar 2012)

environmentally benign (Raliya and Biswas 2015). In the recent development of nanoparticle biosynthesis, peptides (isolated from bio-source or in vitro synthesized) are being used for the precursor metal ion reduction. A patent granted to Belcher et al. (2015), showed a method for producing magnetic nanocrystals by using a biological molecule that has been modified to possess an amino acid oligomer that is capable of specific binding to a magnetic material. Unal Gulsuner et al. (2015) described multidomain (modular) peptides, which direct a cascade reaction is coupling the synthesis and surface functionalization of gold nanoparticles in a single step (Fig. 11.3). Synthesized gold particles have improved colloidal stability on the counter approach of nanoparticle synthesis. Design and construct of biological macromolecules control the assembly of inorganic material (metal and metal oxide). Peptide-mediated synthesis approach opened the door for the scale up of engineered nanoparticles.

3 Applications of Nanoparticles Synthesized by Fungus

The nanoparticle is being used for various applications in biomedical engineering, medicine, environment, manufacturing and material, energy and electronics, pesticides and fertilizers. Owing to high inputs of energy and use of harmful chemicals; physical and chemical methods are less preferred. More emphasis has been given

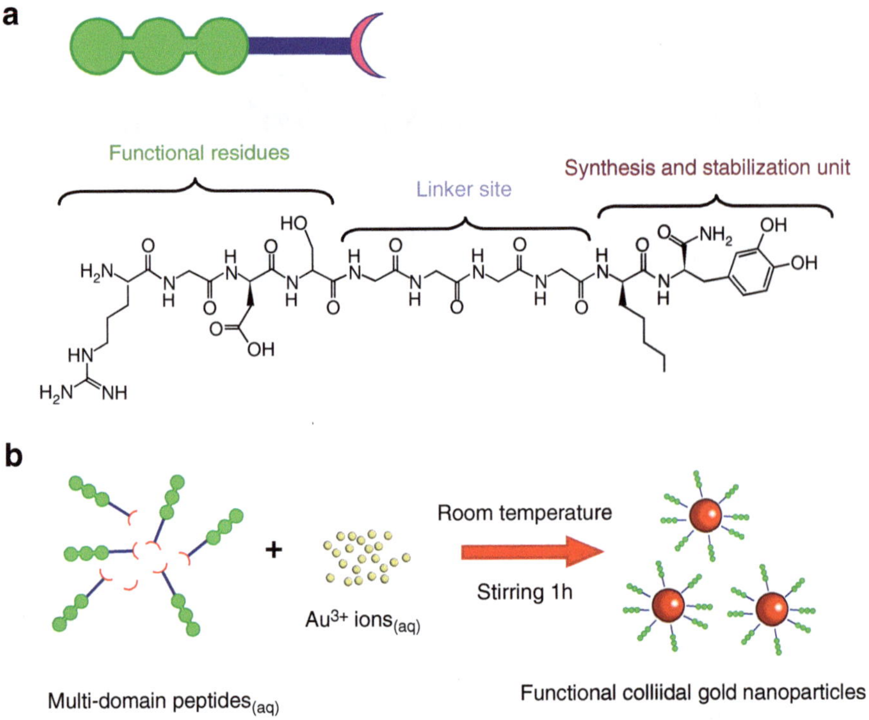

Fig. 11.3 Schematic illustration of the MDP design for one-step synthesis and surface modification of AuNPs. (**a**) Chemical sketch of a proof-of-concept MDP, RGDSGGGGKDopa-Am, where Dopa and Lys serve as synthesis and stabilization units, Gly4 functions as a steric linker to Arg-Gly-Asp-Ser, an integrin-binding peptide sequence. (**b**) Proposed synthesis and surface modification (capping) of AuNPs with the peptides (Unal Gulsuner et al. 2015)

currently for the synthesis of nanoparticle in a sustainable approach like using of fungi bacteria and plants. With advantages of fewer inputs of energy and devoid of harmful chemicals, use of fungi for nanoparticles synthesis becomes a prime choice. Another possible reason is perhaps due to recent advancement in the use of fungus for the scale of nanoparticle synthesis. There are very few reports on the application of fungus mediated nanoparticle synthesis. So far, metal, in particular, Ag and Au, and metal oxide (ZnO, MgO, and TiO_2) are dominant nanoparticles synthesized and explored by researchers (Table 11.2).

Fungus mediated Ag nanoparticles of various size and shape have been extensively harnessed for its antimicrobial properties either nascent particles or in combination with existing antimicrobial agents (Rai et al. 2009). Silver nanoparticles of 3–30 nm, synthesized by *Aspergillus niger* have antibacterial activity against *Bacillus sp.* and *E. coli* (Jaidev and Narasimha 2010). Mechanistically, silver nanoparticles cause dissipation of proton motive force, and the pitting of the bacterial cell membrane leads to cellular death. Fayaz et al. (2010) studied biogenic synthesis of silver nanoparticles and their synergetic effect with antibiotics against

Table 11.2 Application of biosynthesized nanoparticles

Nanoparticles	Fungus used to synthesize	Size (nm)	Application	Reference
Ag	*Alternaria alternata*	20–60	Activity against pathogenic fungi	Gajbhiye et al. (2009)
Ag	*Trichoderma viride*	5–40	Synergistic effect with antibiotics	Fayaz et al. (2010)
Zn	*Rhizoctonia bataticola*	15–25	Nanofertilizer for pearl millet	Tarafdar et al. (2014)
MgO	*Aspergillus flavus*	5.8	Nanonutrient for *Cyamopsis tetragonoloba*	Raliya et al. (2014a, b)
ZnO	*Aspergillus fumigatus*	1.2–6.8	Enhance native phosphorous mobilization in rhizosphere	Raliya and Tarafdar (2013c) Raliya et al. 2016
TiO_2	*Aspergillus flavus*	18	Physiological improvement in mung bean	Raliya et al. (2015a)

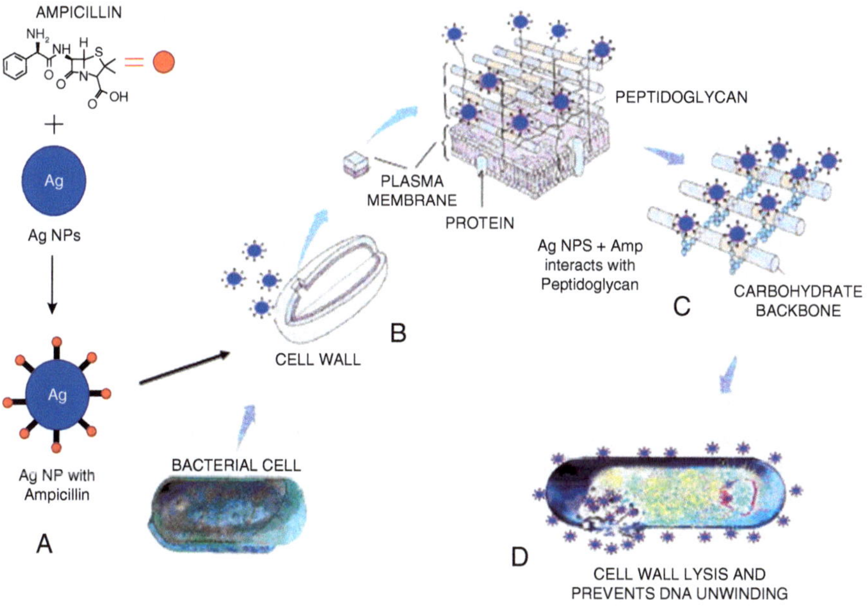

Fig. 11.4 Synergistic activity of AgNPs with ampicillin (Amp) against bacteria. (**a**) Formation of core silver nanoparticles with ampicillin. (**b**) Interaction of silver nanoparticles-Amp complex over the cell wall of bacteria. (**c**) Silver nanoparticles-Amp complex inhibits the formation of cross-links in the peptidoglycan layer (which provides rigidity to the cell wall), leading to cell wall lysis. (**d**) Silver nanoparticles-Amp complex prevents the DNA unwinding (Fayaz et al. 2010)

gram-positive and gram-negative bacteria. Author synthesized silver nanoparticles of 5–40 nm using the fungus *Trichoderma viride*. It was observed that silver nanoparticles exert synergistic antibacterial effect with antibiotics such as ampicillin, kanamycin, erythromycin, and chloramphenicol. Figure 11.4 shows a schematic of synergetic activity of silver nanoparticle with ampicillin.

In contrast to antibiotics, fungus originated silver nanoparticle also enhances the activity of antifungal agents. Gajbhiye et al. (2009) synthesized silver nanoparticles of 20–60 nm using the fungus *Alternaria alternata* and evaluated antifungal activity along with commercial counterpart fluconazole. Disk diffusion method was used to evaluate in vitro antifungal activity of fluconazole against pathogenic fungi *Phoma glomerata, Phoma herbarum, Fusarium semitectum, Trichoderma sp.*, and *Candida albicans.* To determine the synergitic antifungal effect, each standard paper disk was saturated with 20 μL of the freshly prepared silver nanoparticles. The antifungal activity of fluconazole increased significantly in the presence of silver nanoparticles. Antimicrobial effect of nanoparticles depends on particle size, concentration, and surface zeta potential that causes reactive oxygen species formation, cellular leakage as a result of membrane pore and electrostatic interaction involved in the binding of nanoparticles on the surface of a microbial agent (Rai et al. 2009).

Recently, fungus mediated nanoparticles in particular metal oxides are being used as nano nutrient fertilizer, delivered either by soil or root application. It is believed that due to smaller size nanoparticles based nutrient uptake rate is quite higher than conventional fertilizer applied through the soil (Wang et al. 2013; Raliya et al. 2015a, b). Enhanced uptake of nutrients by plants may help to avoid eutrophication in the aquatic body, maintain soil health and economically viable too. A group of Indian Council of Agricultural Research isolated a fungus *Aspergillus fumigatus* to synthesize zinc oxide nanoparticles using a precursor salt zinc nitrate (Raliya and Tarafdar 2013c). Zinc act as a cofactor of various phosphorous mobilizing enzymes such as phytase, alkaline, and acid phosphatase, have the potential to mobilize native phosphorous in the rhizospheric soil. It is important to mention that maximum proportion of conventional phosphorous fertilizer applied in soil getting fixed as a stable inorganic complex with calcium, iron or aluminum. Such complex is unavailable to plants for uptake and runoff with water that ultimately causes eutrophication by increasing phosphorous availability in water-body. The zinc oxide nanoparticle (1.2–6.8 nm) synthesized by *A. fumigatus,* significantly improve plant biomass (27.1 %), chlorophyll content (276.2 %), total soluble leaf protein (27.1 %), rhizospheric microbial population (11–14 %), acid phosphatase (73.5 %), alkaline phosphatase (48.7 %), and phytase (72.4 %) activity in clusterbean rhizosphere (Raliya and Tarafdar 2013c). Similar effect were also found in pearl millet (*Pennisetum americanum*) as a result of zinc nanofertilizer applied through foliar spray (Tarafdar et al. 2014).

To enhance solar light absorption by plant leaves to boost plant photosynthesis, fungus originated titanium dioxide nanoparticles and magnesium oxide nanoparticles were used because of their photocatalytic activity and essential part of pigment (chlorophyll) structure, respectively. *Aspergillus flavus* mediated titanium di oxide nanoparticles of 12–15 nm enhances chlorophyll content in the mung bean plant leaves by 46.4 % (Raliya et al. 2015b). Similarly, magnesium oxide nanoparticles (5.8 nm) synthesized by *A. flavus* increase in chlorophyll content by 76.1 % by the application of biologically synthesized MgO nanoparticle at 15 Mg L^{-1} concentration on 2 week old *Cyamopsis tetragonoloba* plants (Raliya et al. 2014b).

4 Conclusions

The fungus is a preferential source for nanoparticle synthesis over other biological sources such as bacterial, animal tissue lysate or plant cell due to easy and scalable mass culture, saprophytic nature, low downstream processing, environmentally benign and economically viable. Fungi used more for metal and metal oxide nanoparticle synthesis. Among the entire synthesized particle, fungus explored more for silver nanoparticle synthesis. Nanoparticle synthesis reaction is mediated by oxidation-reduction reaction mechanism carried out by fungus enzymatic protein and application used as an antimicrobial agent and also exert synergetic effect when to combine with antibiotics or antifungal agents. Rhizospheric fungus harnessed for the synthesis of agriculturally important nanoparticles help to mobilize native nutrient mobilization by boosting plant physiological and metabolic activities.

References

Ahmad A, Mukherjee P, Mandal D, Senapati S, Khan MI, Kumar R, Sastry M (2002) Enzyme mediated extracellular synthesis of CdS nanoparticles by the fungus, *Fusarium oxysporum*. J Am Chem Soc 124:12108–12109

Ahmad A, Mukherjee P, Senapati S, Mandal D, Khan MI, Kumar R, Sastry M (2003) Extracellular biosynthesis of silver nanoparticles using the fungus *Fusarium oxysporum*. Colloids Surf B Biointerfaces 28:313–318

Aziz N, Faraz M, Pandey R, Sakir M, Fatma T, Varma A, Barman I, Prasad R (2015) Facile algae-derived route to biogenic silver nanoparticles: synthesis, antibacterial and photocatalytic properties. Langmuir 31:11605–11612. doi:10.1021/acs.langmuir.5b03081

Bansal V, Poddar P, Ahmad A, Sastry M (2006) Room-temperature biosynthesis of ferroelectric barium titanate nanoparticles. J Am Chem Soc 128:11958–11963

Barwal I, Ranjan P, Kateriya S, Yadav SC (2011) Cellular oxido-reductive proteins of *Chlamydomonas reinhardtii* control the biosynthesis of silver nanoparticles. J Nanobiotechnology 9:56

Belcher AM, Reiss BD, Mao C, Solis DJ, inventors; Board of Regents, assignee. Peptide mediated synthesis of metallic and magnetic materials. United States patent US 8,969,252. 2015 Mar 3.

Bhainsa KC, D'Souza SF (2006) Extracellular biosynthesis of silver nanoparticles using the fungus *Aspergillus fumigatus*. Colloids Surf B Biointerfaces 47:160–164

Carlile MJ, Watkinson SC, Gooday GW (2001) The fungi, 2nd edn. Academic, London

Dameron CT, Reese RN, Mehra RK, Kortan AR, Carroll PJ, Steigerwald ML, Brus LE, Winge DR (1989) Biosynthesis of cadmium sulphide quantum semiconductor crystallites. Nature 338:596–597

Fayaz AM, Balaji K, Girilal M, Yadav R, Kalaichelvan PT, Venketesan R (2010) Biogenic synthesis of silver nanoparticles and their synergistic effect with antibiotics: a study against gram-positive and gram-negative bacteria. Nanomed: Nanotechnol, Biol Med 6:103–109

Gajbhiye M, Kesharwani J, Ingle A, Gade A, Rai M (2009) Fungus-mediated synthesis of silver nanoparticles and their activity against pathogenic fungi in combination with fluconazole. Nanomed: Nanotechnol, Biol Med 5:382–386

Jaidev LR, Narasimha G (2010) Fungal mediated biosynthesis of silver nanoparticles, characterization and antimicrobial activity. Colloids Surf B Biointerfaces 81:430–433

Kaul RK, Kumar P, Burman U, Joshi P, Agrawal A, Raliya R, Tarafdar JC (2012) Magnesium and iron nanoparticles production using microorganisms and various salts. Mater Sci Pol 30:254–258

Mukherjee P, Ahmad A, Mandal D, Senapati S, Sainkar SR, Khan MI, Ramani R, Parischa R, Ajayakumar P, Alam M (2001a) Bioreduction of $AuCl4^{-}$ ions by the fungus, Verticillium sp. and surface trapping of the gold nanoparticles formed. Angew Chem Int Ed 40:3585–3588

Mukherjee P, Ahmad A, Mandal D, Senapati S, Sainkar SR, Khan MI, Parishcha R, Ajaykumar PV, Alam M, Kumar R, Sastry M (2001b) Fungus-mediated synthesis of silver nanoparticles and their immobilization in the mycelial matrix: a novel biological approach to nanoparticle synthesis. Nano Lett 1:515–519

Mukherjee P, Senapati S, Mandal D, Ahmad A, Khan MI, Kumar R, Sastry M (2002) Extracellular synthesis of gold nanoparticles by the fungus *Fusarium oxysporum*. ChemBioChem 3:461–463

Murphy CJ (2008) Sustainability as an emerging design criterion in nanoparticle synthesis and applications. J Mater Chem 18:2173–2176

NSF (2015) NSF National Nanotechnology Initiative. http://www.nsf.gov/crssprgm/nano/. Accessed 12 Dec 2015

Prasad R (2014) Synthesis of silver nanoparticles in photosynthetic plants. J Nanoparticles, Article ID 963961, doi:10.1155/2014/963961

Prasad R, Kumar V, Prasad KS (2014) Nanotechnology in sustainable agriculture: present concerns and future aspects. Afr J Biotechnol 13(6):705–713

Prasad R, Pandey R, Barman I (2015) Engineering tailored nanoparticles with microbes: quo vadis. WIREs Nanomed Nanobiotechnol. doi:10.1002/wnan.1363

Rai M, Yadav A, Gade A (2009) Silver nanoparticles as a new generation of antimicrobials. Biotechnol Adv 27:76–83

Raliya R, Biswas P (2015) Environmentally benign bio-inspired synthesis of Au nanoparticles, their self-assembly and agglomeration. RSC Adv 5:42081–42087

Raliya R, Tarafdar JC (2012) Novel approach for silver nanoparticle synthesis using *Aspergillus terreus* CZR-1: mechanism perspective. J Bionanosci 6:12–16

Raliya R, Tarafdar JC (2013a) Rapid, low-cost, and ecofriendly approach for iron nanoparticle synthesis using *Aspergillus oryzae* TFR9. J Nanoparticles 2013:1–4

Raliya R, Tarafdar J (2013b) Biosynthesis of gold nanoparticles using *rhizoctonia bataticola* TFR-6. Adv Sci Eng Med 5:1073–1076

Raliya R, Tarafdar J (2013c) ZnO nanoparticle biosynthesis and its effect on phosphorous-mobilizing enzyme secretion and gum contents in Clusterbean (*Cyamopsis tetragonoloba* L.). Agric Res 2:48–57

Raliya R, Tarafdar J (2014) Biosynthesis and characterization of zinc, magnesium and titanium nanoparticles: an eco-friendly approach. Int Nano Lett 4:1–10

Raliya R, Rathore I, Tarafdar J (2013) Development of microbial nanofactory for zinc, magnesium, and titanium nanoparticles production using soil fungi. J Bionanosci 7:590–596

Raliya R, Tarafdar J, Choudhary K, Mal P, Raturi A, Gautam R, Singh S (2014a) Synthesis of MgO nanoparticles using *Aspergillus tubingensis* TFR-3. J Bionanosci 8:34–38

Raliya R, Tarafdar J, Singh S, Gautam R, Choudhary K, Maurino VG, Saharan V (2014b) MgO nanoparticles biosynthesis and its effect on chlorophyll contents in the leaves of Clusterbean (*Cyamopsis tetragonoloba* L.). Advanced Science. Eng Med 6:538–545

Raliya R, Biswas P, Tarafdar JC (2015a) TiO2 nanoparticle biosynthesis and its physiological effect on mung bean (Vigna radiata L.). Biotechnol Rep 5:22–26

Raliya R, Nair R, Chavalmane S, Wang W-N, Biswas P (2015b) Mechanistic evaluation of translocation and physiological impact of titanium dioxide and zinc oxide nanoparticles on the tomato (*Solanum lycopersicum* L.) plant. Metallomics 7:1584–1594

Raliya R, Tarafdar JC, Biswas P (2016) Enhancing the Mobilization of Native Phosphorus in the Mung Bean Rhizosphere Using ZnO Nanoparticles Synthesized by Soil Fungi. J Agri Food Chem 64:3111–3118

Sastry M, Ahmad A, Islam Khan M, Kumar R (2003) Biosynthesis of metal nanoparticles using fungi and actinomycete. Curr Sci 85:162–170

Senapati S, Ahmad A, Khan MI, Sastry M, Kumar R (2005) Extracellular biosynthesis of bimetallic Au-Ag alloy nanoparticles. Small 1:517–520

Suman V, Prasad R, Jain VK, Varma A (2010) Role of nanomaterials in symbiotic fungus growth enhancement. Curr Sci 99:1189–91

Tarafdar J, Raliya R, Rathore I (2012) Microbial synthesis of phosphorous nanoparticle from tricalcium phosphate using *Aspergillus tubingensis* TFR-5. J Bionanosci 6:84–89

Tarafdar A, Raliya R, Wang W-N, Biswas P, Tarafdar J (2013) Green synthesis of TiO2 nanoparticle using *Aspergillus tubingensis*. Adv Sci Eng Med 5:943–949

Tarafdar JC, Raliya R, Mahawar H, Rathore I (2014) Development of Zinc nanofertilizer to enhance crop production in Pearl Millet (*Pennisetum americanum*). Agric Res 3:257–262

Thakkar KN, Mhatre SS, Parikh RY (2010) Biological synthesis of metallic nanoparticles. Nanomed: Nanotechnol, Biol Med 6:257–262

Unal Gulsuner H, Ceylan H, Guler MO, Tekinay AB (2015) Multi-domain short peptide molecules for in situ synthesis and biofunctionalization of gold nanoparticles for integrin-targeted cell uptake. ACS Appl Mater Interfaces 7:10677–10683

Vigneshwaran N, Ashtaputre N, Varadarajan P, Nachane R, Paralikar K, Balasubramanya R (2007) Biological synthesis of silver nanoparticles using the fungus *Aspergillus flavus*. Mat Lett 61:1413–1418

Wang W-N, Tarafdar JC, Biswas P (2013) Nanoparticle synthesis and delivery by an aerosol route for watermelon plant foliar uptake. J Nanopart Res 15:1–13

Chapter 12
Microbial Laccases and Nanobiotechnology: Environmental Perspective

Sunita J. Varjani

Abstract In recent years advances in nanotechnology have open up a new era in industrial technology. Majority of nanoparticles are incorporated into products. Laccases (EC 1.10.3.2) are oxidoreductases belonging to multi-copper oxidases, which have been subject of intensive research in last decade due to their ability to oxidize both phenolic and nonphenolic lignin related compounds. They are widely distributed in bacteria, fungi, plants and insects. Among all these sources fungi is reported as the best laccase producer. They have been described in different genera of ascomycetes and deuteromycetes, however mainly laccases are found in basidiomycetes. Laccases have number of industrial and environmental applications including bioremediation, pulp and paper industry, textile industry, food technology, nanobiotechnology, medicine and cosmetology. More recently laccase is also used in design of biosensors, biofuel cells and as bioremediation agent to clean up herbicides, pesticides and certain explosives in soil. Especially they are used in the field of biodegradation and treatment of xenobiotics. Though it has been studied from last century it remains a topic of research till today due to its enormous hidden potential. This paper reviews the occurrence, general properties and applications of laccases. This review is also emphasized on role of nanobiotechnology in environmental and various industrial sectors.

1 Introduction

The modern technology accepts that the concept of interdisciplinary research in engineering and sciences leads to creation of environmentally benign "green processes" (Varjani et al. 2015) with special concern to nanoscience and nanotechnology (Kannan and Subbalaxmi 2011). Recently enzymes are widely used in diverse industries. Many such potential enzymes are present in nature; laccases are one among them (Madhavi and Lele 2009). Laccases are the oldest and most studied enzymatic systems (Pannu and Kapoor 2014). Study of nanotechnology with

---

S.J. Varjani (✉)
School of Biological Sciences and Biotechnology, University and Institute of Advanced Research, Gandhinagar 382007, Gujarat, India
e-mail: drsvs18@gmail.com

R. Prasad (ed.), *Advances and Applications Through Fungal Nanobiotechnology*, Fungal Biology, DOI 10.1007/978-3-319-42990-8_12

biotechnology is referred to as nanobiotechnology. Nanobiotechnology offers new ways to improve methods used to solve environmental and industrial problems. It can contribute to reduce energy consumption and waste production as well as to achieve sustainable industrial and societal development. Nanobiotechnology has emerged up as integration between nanotechnology and biotechnology for developing bioactive, biosynthetic and ecofriendly technology for synthesis of nanomaterials (Kannan and Subbalaxmi 2011; Prasad and Swamy 2013). Norio Taniguchi, Tokyo Science University was the scientist who first defined the term Nanotechnology in 1974. Nanotechnology deals with structures in the size range between 1 and 100 nm and involves developing materials or devices within that size (Suman et al. 2010; Arivalagan et al. 2011). Methods employed for synthesis of nanomaterial are physical, mechanical and chemical. However, these methods are very expensive and some of them which involve hazardous chemicals. Therefore, there is emergent need to develop environmentally benign and sustainable methods for nanomaterial synthesis (Shraddha et al. 2011; Prasad 2014). Green chemistry processes led to environmental friendly method of synthesis and safe process as compared to other methods (Aziz et al. 2015; Varjani et al. 2015). Biological sources such as fungi, bacteria, viruses, actinomycetes, algae and plant materials, etc. can catalyze specific reactions to produce ecofriendly nanomaterial (Shraddha et al. 2011; Aziz et al. 2015; Prasad et al. 2015). These can be used as alternate source for physical, chemical methods, ultraviolet irradiation, lithography, laser ablation, ultrasonic fields, and photochemical reduction techniques. Current perspective of biological synthesis process should focus towards use of highly structured physical and biosynthetic activities of microbial cells to achieve controlled manipulation of the size and shape of nanomaterial (Kannan and Subbalaxmi 2011).

Laccase (p-diphenol:O_2 oxidoreductase, EC 1.10.3.2) is a glycoprotein made up of 500–600 amino acids and the carbohydrate fraction constitutes 10–40 % of molecular weight (Gavnholt et al. 2002). Laccases exhibit activity on wide substrate range like diphenols, polyphenols, aromatic amines, hydroxyindols, benzenethiols and even some inorganic compounds such as iodine (Viswanath et al. 2008). However most other enzymes are generally substrate specific. This is the major reason for attractiveness of laccases for several biotechnological applications (Nyanhongo et al. 2002; Riva 2006; Viswanath et al. 2008). Due to energy saving and biodegradable nature laccase based biocatalysts are preferred for development of highly efficient, sustainable and eco-friendly industrial processes (Xu 2005; Riva 2006). Laccases are found to have several biotechnological applications in various industries, synthetic chemistry, cosmetics, bioremediation i.e. removal of toxic or recalcitrant pollutants like herbicides, pesticides, dyes (Novotny et al. 2004). Laccases are encoded by multigene family. Laccases have been characterized mostly from fungi till now while limited research has been performed in plants, bacteria and insects (Wang et al. 2015). Immobilization techniques such as layer-by-layer, micro patterning and self-assembled monolayer technique can be used to preserve the enzymatic activity of laccase (Pannu and Kapoor 2014). Recently laccases have been efficiently applied to nanobiotechnology. Laccases catalyze electron transfer reactions without additional cofactor (Shraddha et al. 2011).

There is a emergent need to assemble information available on microbial laccases from different sources in order to throw light their importance in environmental and industrial sectors which can yield much more potential laccase source and broaden their applications which are ecofriendly and economic. The present review will encompass the information about nanobiotechnology, laccases as enzyme, microbial sources of laccases and their potential application in bioremediation & selected industrial sectors. It also explores the untapped potential of laccases.

2 Nanobiotechnology and Nanomaterial

Nanotechnology is emerging as a rapidly growing field due to its versatile applications in science and technology for manufacturing of new materials at nanoscale level (Albrecht et al. 2006). Nanotechnology was originally a field within the physical sciences, the intersection of nanotechnology and biotechnology was realized from the potential applications of nanomaterials to environmental as well as biotechnological problems (Viswanath et al. 2008; Agrawal and Rathore 2014). Nanobiotechnology is an interdisciplinary area of research that operates at interface of biotechnology, environmental science, chemistry, biology, materials science, engineering and medicine (Kannan and Subbalaxmi 2011; Swamy and Prasad 2012). Figure 12.1 illustrates application of nanobiotechnology in scientific research.

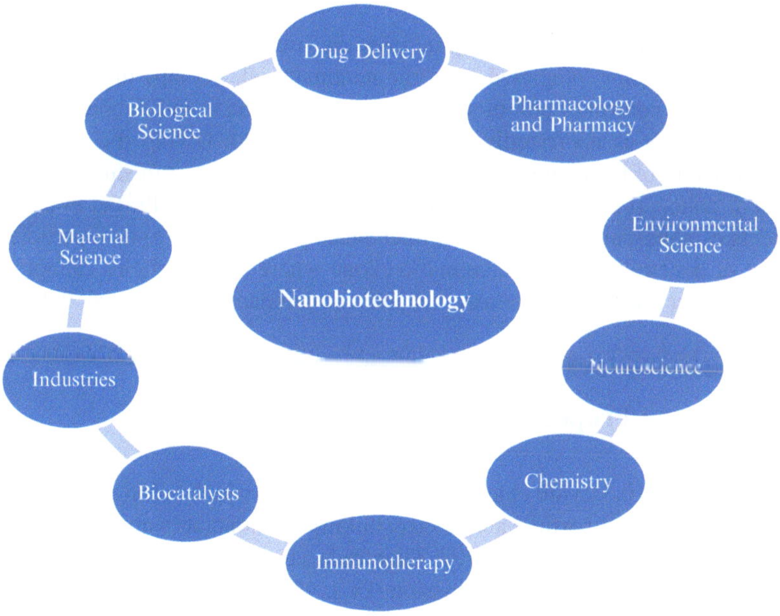

Fig. 12.1 Application of nanobiotechnology in scientific research (*Source*: Kannan and Subbalaxmi 2011)

Nanobiotechnology is defined as an area that applies the nanoscale principles and techniques to understand and transform biosystems (living or non-living). Various terms have been used to describe nanomaterials according to their chemical composition; carbon related nanomaterials, dendrimers, semi-conductor nanocrystals, zero-valent metal nanoparticles and metal containing nanoparticles (Klaine et al. 2008). Nanomaterials can be nanoparticles, nanofibres or nanoplates respectively, having three, two or one dimensions within nano-scale. Nanorods and nanotubes examples of nanofibres. Nanowires are electrically conductive nanofibres. Nanoprism have been used to describe various polyhedral nanoparticles (Agrawal and Rathore 2014). A nanosphere is a general term used to describe a quasi-spherical nanoparticle but in practice perfect spheres are not achieved instead many near-spherical nanoparticles have a multi-faceted topology (Pal et al. 2007; Agrawal and Rathore 2014). Nanostructured materials such as nanoparticles and nanofibers have been used extensively as carrying materials for preparing biosensor and biofuel cell. The high potential impacts of nanotechnology almost cover all fields of human activity viz. environmental, economy, industrial, clinical and health-related etc. (Albrecht et al. 2006; Agrawal and Rathore 2014; Prasad et al. 2014). In biotechnological analysis nanomaterial can be used as different materials such as bacteriostatic, antibacterial, antistatic, cryogenic superconducting, and biosensor etc. (Cao et al. 2002; Prasad 2014; Palza 2015).

2.1 Advantages of Nanomaterials

In nanobiotechnology biological principles and materials are used to create new devices and systems integrated from the nanomaterial. Perhaps the most striking example is the control of pollution (D'Souza 2001; Kannan and Subbalaxmi 2011). This technology is ecofriendly (Albrecht et al. 2006). Owing the environment friendly nature of biosynthetic process, there is a great interest in studying the synthesis of nanoparticle by biological route (Klaine et al. 2008; Prasad et al. 2015). Benefits of nanoscale materials over microscale particles are (1) nanoparticles can diffuse or penetrate into a contamination zone where microparticles cannot reach, and (2) higher reactivity to redox-amenable contaminants (Nurmi et al. 2005).

3 Laccases: Properties and Occurrence

Laccases (EC 1.10.3.2) are chief ligninolytic enzyme, which belongs to blue multi-copper oxidases and participates in crosslinking of monomers and degradation of xenobiotics (Call and Miicke 1997; Gianfreda et al. 1999). They are defined as oxidoreductases in Enzyme Commission (EC), which oxidizes diphenol and allied substances & use molecular oxygen as an electron acceptor. There are diverse sources of laccase producing organisms like bacteria, fungi, plants and insects

(Pannu and Kapoor 2014). But they have been widely characterized in fungi than in higher plants. Due to broad range of application there is a growing need for isolation and identification of new laccase producing organisms to be used in industries (Madhavi and Lele 2009).

3.1 *Properties of Laccases*

Laccase catalyses reduction of O_2 to water by oxidising phenolic substrates. Laccase molecule binds four copper atoms responsible for electron transfer (Viswanath et al. 2008). Copper-binding regions of the enzyme are highly conserved throughout evolution (Lewis et al. 1999). It was initially thought that laccases would only be able to oxidize phenolic substrates as they have a lower redox potential (150–800 mV) than those of lignolytic peroxidases (>1 V) (Kersten et al. 1990). But later on laccases have received attention of researchers due to their ability to oxidize both phenolic and nonphenolic lignin-related compounds as well as highly recalcitrant environmental pollutants (Pannu and Kapoor 2014).

The enzymatic catalysis by laccases in different industrial applications such as textile dye bleaching, pulp bleaching and bioremediation could serve as a more environmentally benign alternative than currently used chemical processes (Octavio et al. 2006). Apart from its benefits there are some problems to commercialize use of laccases. One of them is lack of sufficient enzyme stock. Thus, efforts have to be made in order to achieve cheap overproduction of laccase in heterologous hosts (Manzanares et al. 1995; Shraddha et al. 2011) Laccases can also be modified by chemical means or protein engineering to obtain more robust and active enzyme (Shraddha et al. 2011). Another additional problem is cost and toxicity of redox mediators. Further investigation should be focused on less polluting mediators (Pannu and Kapoor 2014).

3.2 *Sources of Laccases*

Laccases are ubiquitous in nature. The presence of laccases has been found in plants, fungi, bacteria and insects (Pannu and Kapoor 2014). Laccases are generally distributed in a wide range of higher plants and fungi. In fungi specifically white rot basidiomycete laccases appear more than the higher plants (Shraddha et al. 2011). In general plant and insect laccases have lower redox potential. Microbial laccases are divided in three groups according to redox potential of laccases (a) Bacterial laccases possess low-redox-potential, (b) basidiomycete white-rot fungi laccases have high redox-potential and (c) majority of ascomycete and some basidiomycete fungi laccases show middle-redox-potential (Mate and Alcalde 2014; Pardo and Camarero 2015).

3.2.1 Fungi

Fungal laccases have been comprehensively identified from ascomycetes, deuteromycetes, and basidiomycetes. Fungi seem to hold more laccase producers when compared to other sources (Table 12.1) (Pannu and Kapoor 2014). Research on laccase is mainly focused on basidiomycetes fungi because among all fungi the white-rot basidiomycetes are the best laccase producers (Madhavi and Lele 2009). These are involved in lignin degradation, morphogenesis, pathogenesis and stress defense (Baldrian 2006).

Fungal laccases are produced in cells, secreted and accumulated outside the hyphal filaments i.e. fungal laccases are of secretory type (Shraddha et al. 2011; Subramanian et al. 2014). Commonly laccases are produced by wood rotting fungi such as *Phanerochaete chrysosporium, Theiophora terrestris, Trametes versicolor, Trametes hirsuta, Trametes ochracea, Trametes villosa, Trametes gallica, Cerena maxima, Coriolposis polyzona, Lentinus tigrinus, Pleurotus eryngii* (Assavanig et al. 1992; Viswanath et al. 2008; Pannu and Kapoor 2014). Pannu and Kapoor (2014) have reported production of laccases from *Phlebia radiate, Neurospora crassa* and *Pleurotus ostreatus*. First laccase was characterized from ascomycete, *Monocillium indicum* which showed peroxidase activity (Thakker et al. 1992).

3.2.2 Bacteria and Archea

Laccase enzymes are widely distributed in plants and fungi, laccase activity has also been reported in few bacteria viz. *Azospirillum lipoferum, Marinomonas mediterranea, Streptomyces griseus,* and *Bacillus subtilis* (Diamantidis et al. 2000; Hullo et al. 2001; Octavio et al. 2006). Hullo et al. (2001) reported CotA in *Bacillus subtilis,* which participates in pigment biosynthesis and protection against UV light.

Laccase have also been identified in archaea (Sharma and Kuhad 2009). Thermostable and salt tolerant laccase LccA was studied from halophilic archaeon *Haloferax volcanii* (Uthandi et al. 2010).

Table 12.1 Production of laccases by different sources

S. No.	Source	Name of organisms
1.	Fungi	*Phanerochaete chrysosporium, Theiophora terrestris, Lenzites, betulina, Phlebia radiate, Pleurotus ostreatus, Trametes* sp., *Neurospora crassa, Cerena maxima, Coriolposis polyzona, Lentinus tigrinus, Pleurotus eryngii, Monocillium indicum*
2.	Plant	*Arabidopsis thaliana, Brassica napus, Oryza sativa, Zea mays, Liridendron tulipifera, Nicotiana tabacum, Lolium perenne, Saccharum officinarum*
3.	Bacteria and archea	*Azospirillum lipoferum, Marinomonas mediterranea, Streptomyces griseus, Bacillus subtilis, Haloferax volcanii*
4.	Insect	*Monochamus alternatus, Bombyx mori, Apis mellifera*

3.2.3 Insects

Laccases are mainly associated with cuticle sclerotization of insects. Various insects such as *Monochamus alternatus, Bombyx mori* and *Apis mellifera* have been found to have laccases (Yatsu and Asano 2009; Elias-Neto et al. 2010; Shraddha et al. 2011).

3.3 Laccases Application

Ecologically oxidoreductases play an essential role in mobilization of carbon into the ecosystem. Laccase is an oxidative enzyme that has the ability to oxidize lignin using molecular oxygen, which is reduced to water (Madhavi and Lele 2009). The oxidative ability of laccase is employed in a number of industrial and environmental applications including bioremediation, nanotechnology, pulp and paper industry, textile industry, food technology and pharmaceutical (medical and cosmetology) industry (Octavio et al. 2006; Kannan and Subbalaxmi 2011; Pannu and Kapoor 2014). Application of laccase enzyme is shown in Fig. 12.2.

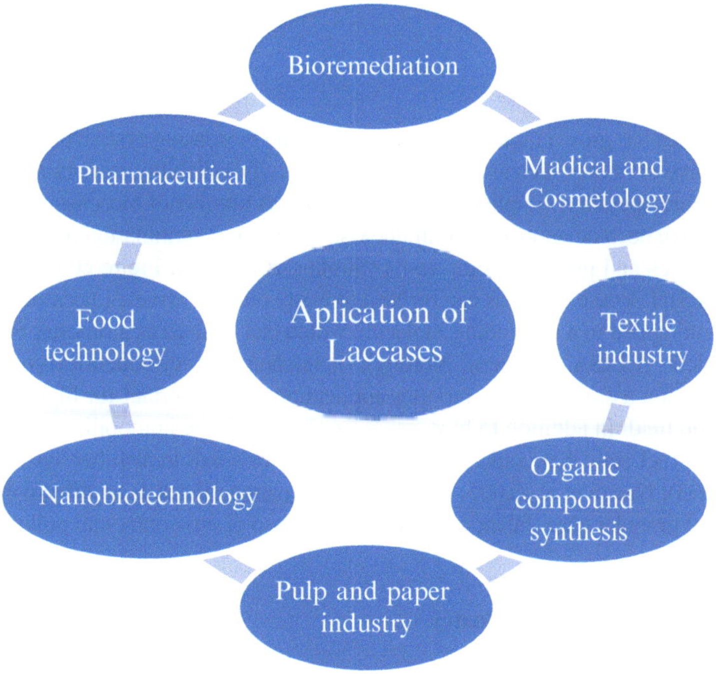

Fig. 12.2 Application of laccase enzyme

3.3.1 Bioremediation

Laccases have many possible applications in bioremediation (Madhavi and Lele 2009). Laccases can be used to decolorize dye effluents that are hardly decolorized by conventional sewage treatment plants (Novotny et al. 2004). In addition to dye house effluents, laccases can decolorize wastewater from olive oil mills and pulp mills (Manzanares et al. 1995; D'Annibale et al. 2000) by removing colored phenolic compounds. With the application of nanoparticles microbial cells can be immobilized that can degrade or biorecover specific chemicals. These immobilized microbial cells can also be used as biocatalysts for reductive dechlorination (Nurmi et al. 2005).

Another potential environmental application for laccases is bioremediation of contaminated soils, as laccase and laccase mediator system (LMS) are able to oxidize toxic recalcitrant organic pollutants such as various xenobiotics, poly aromatic hydrocarbons (PAHs), chlorophenols and other contaminants (Alcalde et al. 2006; Madhavi and Lele 2009; Pannu and Kapoor 2014). Phenolic compounds are present in wastes from several industrial processes, coal conversion, petroleum refining, production of organic chemicals and olive oil production (Aggelis et al. 2003). Immobilized laccase was found to be useful for removal of phenolic and chlorinated pollutants (Hublik and Schinner 2000; Octavio et al. 2006; Pannu and Kapoor 2014). Laccase can be used to reduce concentration of synthetic heterocyclic compound such as halogenated organic pesticides in soil. LMS is being included in several enzymatic bioremediation programs (Alcalde et al. 2006; Pannu and Kapoor 2014).

3.3.2 Nanobiotechnology

Nanoscience has grown rapidly in last decade. Good biosensing system possesses various positive attributes such as its specificity, sensitivity, reliability, portability, real time analysis and operation simplicity (D'Souza 2001). A number of biosensors containing laccase have been developed for immunoassays and for determination of glucose, aromatic amines and phenolic compounds (Simkus et al. 1996; Freire et al. 2002; Park et al. 2003). Laccase catalysis can be used to assay other enzymes. Laccases are covalently conjugated to a biobinding molecule which can be used as a reporter for immunochemical (ELISA, Western blotting), histochemical, cytochemical or nucleic acid-detection assays. The bio reporter applications are important for high-sensitivity diagnostic field. In addition to biosensors laccases could be immobilized on the cathode of biofuel cells that could provide power eg. for small transmitter systems (Park et al. 2003). Nurmi et al. (2005) used dimetallic nanoparticles as an effective oxidant instead of granular zero-valency metal in the cleanup of environmental pollutants.

3.3.3 Pulp and Paper Industry

In industrial preparation of paper, separation and degradation of lignin in wood pulp are conventionally obtained using chlorine- or oxygen-based chemical oxidants. Non-chlorine bleaching of pulp with laccase was first patented in 1994 using an enzyme treatment to obtain a brighter pulp with low lignin content (Luisa et al. 1996). Environment polluting chlorine-based methods have been replaced by oxygen delignification process in last decade. In spite of this new method, the pretreatments of wood pulp with laccase can provide milder and cleaner strategies of delignification that also respect the integrity of cellulose (Barreca et al. 2003; Gamelas et al. 2005). Laccases can delignify pulp when they are used together with mediators. The mediator is oxidized by laccase and oxidized mediator molecule further oxidizes subunits of lignin that otherwise would not be laccase substrates (Pannu and Kapoor 2014). Potential of laccase for cross-linking and functionalizing ligninaceous compounds was discovered by Guebitz and Paulo (2003). Laccases can also be used for binding fiber-, particle- and paper boards (Bajpai 1999).

3.3.4 Textile Industry

Laccase is used in textile industries to improve whiteness in conventional bleaching of cotton and biostoning process. Advantages of laccase in textile industry include chemicals, energy and water saving. In 1996 Novozyme (Novo Nordisk, Denmark) launched a new industrial application of laccase enzyme in denim finishing. DeniLite® is the first industrial laccase and the first bleaching enzyme acting with help of a mediator. Laccase can be used in situ to convert dye precursors for better and more efficient fabric dyeing (Pannu and Kapoor 2014). Laccases find potential applications for cleansing viz. as cloth washing and dish washing. Laccase may be included in a cleansing formulation to eliminate odor on fabrics, including cloth, sofa surface and curtain or in a detergent to eliminate the odor generated during cloth washing (Kirk et al. 2002). Lantto et al. (2004) found that wool fibers can be activated with LMS. Therefore the use of laccase for anti-shrink treatment of wool seems very attractive.

4 Conclusion

Nanobiotechnology is an intersection between nanotechnology and biotechnology. In this branch of science novel applications for very small materials are being realised at an alarming rate. Owing to its versatility laccases are continuously under investigation for new applications. They are multicopper enzymes. They are capable of degrading lignin and are present abundantly in white rot basidiomycetes. They decolorize and detoxify the industrial effluents and help in wastewater treatment. They act on both phenolic and nonphenolic lignin-related compounds as well as highly recalcitrant environmental pollutants, which help researchers to put them in various biotechnological applications. They can be actively used in paper and pulp industry, textile industry, xenobiotic degradation and bioremediation. Laccases in biotechnological applications act as nanobiosensor. Laccase has been applied to nanobiotechnology, which is an increasing research field and catalyzes electron transfer reactions without additional cofactors. Laccases are promising enzymes to replace conventional chemical processes to control environmental pollution. The benefits and environmental shortcomings of nanobiotechnology will only be realized after extensive investigation.

Acknowledgement I express my sincere gratitude to academic and administrative staff of University and Institute of Advanced Research for their kind support.

References

Aggelis G, Iconomou D, Christou M, Bokas D, Kotzailias S, Christou G, Tsagou V, Papanikolaou S (2003) Phenolic removal in a model olive oil mill wastewater using *Pleurotus ostreatus* in bioreactor cultures and biological evaluation of the process. Water Res 37:3897–3904

Agrawal S, Rathore P (2014) Nanotechnology pros and cons to agriculture: a review. Int J Curr Microbiol App Sci 3(3):43–55

Albrecht MA, Evans CW, Raston CL (2006) Green chemistry and the health implications of nanoparticles. Green Chem 8(5):417–432

Alcalde M, Ferrer M, Plou FJ, Ballesteros A (2006) Environmental biocatalysis from remediation with enzymes to novel green processes. Trends Biotechnol 24:281–287

Arivalagan K, Ravichandran S, Rangasamy K (2011) Nanomaterials and its potential applications. Int J ChemTech Res 3:534–538

Assavanig A, Amornkitticharoen B, Ekpaisal N, Meevootisom V, Flegel TW (1992) Isolation, characterization and function of laccase from *Trichoderma*. Appl Microbiol Biotechnol 38:198–202

Aziz N, Faraz M, Pandey R, Sakir M, Fatma T, Varma A, Barman I, Prasad R (2015) Facile algae-derived route to biogenic silver nanoparticles: synthesis, antibacterial and photocatalytic properties. Langmuir. doi:10.1021/acs.langmuir.5b03081

Bajpai P (1999) Application of enzymes in the pulp and paper industry. Biotechnol Prog 15:147–157

Baldrian P (2006) Fungal laccases-occurrence and properties. FEMS Microbiol Rev 30(2):215–242

Barreca AM, Fabbrini M, Galli C, Gentili P, Ljunggren S (2003) Laccase/mediated oxidation of a lignin model for improved delignification procedures. J Mol Catal B Enzym 26:105–110

Call HP, Miicke I (1997) History, overview and applications of mediated ligninolytic systems, especially laccase-mediator-systems (Lignozyme®-process). J Biotechnol 53:163–172

Cao YC, Jin R, Mirkin CA (2002) Nanoparticles with Raman spectroscopic fingerprints for DNA and RNA detection. Science 297:1536–1540

D'Annibale A, Stazi SR, Vinciguerra V, Giovannozzi Sermanni G (2000) Oxirane immobilized *Lentinula edodes* laccases: stability and phenolics removal efficiency in olive mill wastewater. J Biotechnol 77:265–273

D'Souza SF (2001) Microbial biosensors. Biosens Bioelectron 16:337–353

Diamantidis G, Effosse A, Potier P, Bally R (2000) Purification and characterization of the first bacterial laccase in rhizospheric bacterium *Azospirillum lipoferum*. Soil Biol Biochem 32:919–927

Elias-Neto M, Soares MPM, Simoes ZLP, Hartfelder K, Bitondi MMG (2010) Developmental characterization, function and regulation of a Laccase 2 encoding gene in the honey bee, *Apis mellifera* (Hymenoptera, Apinae). Insect Biochem Mol Biol 40(3):24–51

Freire RS, Duran N, Wang J, Kubota LT (2002) Laccase-based screen printed electrode for amperometric detection of phenolic compounds. Anal Lett 35:29–38

Gamelas JAF, Tavares APM, Evtuguin DV, Xavier AMB (2005) Oxygen bleaching of kraft pulp with polyoxometalates and laccase applying a novel multi-stage process. J Mol Catal B Enzym 33:57–64

Gavnholt B, Larsen K, Rasmussen SK (2002) Isolation and characterisation of laccase cDNAs from meristematic and stem tissues of ryegrass (*Lolium perenne*). Plant Sci 162(6):873–885

Gianfreda L, Xu F, Bollag JM (1999) Laccases: a useful group of oxidoreductive enzymes. Bioremed J 3:1–25

Guebitz GM, Paulo CA (2003) New substrates for reliable enzymes: enzymatic modification of polymers. Curr Opin Biotechnol 14:577–582

Hublik G, Schinner F (2000) Characterization and immobilization of the laccase from *Pleurotus ostreatus* and its use for the continuous elimination of phenolic pollutants. Enzyme Microb Technol 7:330–336

Hullo MF, Moszer I, Danchin A, Martin-Verstraete I (2001) CotA of *Bacillus subtilis* is a copper-dependent laccase. J Bacteriol 183(18):5426–5430

Kannan N, Subbalaxmi S (2011) Biogenesis of nanoparticles—a current perspective. Rev Adv Mater Sci 27:99–114

Kersten PJ, Kalyanaraman B, Hammel KE, Reinhamrnar B, Kirk YK (1990) Comparison of lignin peroxidase, horseradish peroxidase and laccase in the oxidation of methoxybenzenes. Biochem J 23:475–480

Kirk O, Borchert TV, Fuglsang CC (2002) Industrial enzyme applications. Curr Opin Biotechnol 13:345–351

Klaine SJ, Alvarez PJ, Batley GE, Fernandes TF, Handy RD, Lyon DY, Mahendra S, McLaughlin MJ, Lead JR (2008) Nanomaterials in the environment: behavior, fate, bioavailability, and effects. Environ Toxicol Chem 27(9):1825–1851

Lantto R, Schonberg C, Buchert J, Heine E (2004) Effects of laccase-mediator combinations on wool. Text Res J 74:713–717

Lewis NG, Davin LB, Sarkanen S (1999) The nature and function of lignins. In: Barton DHR, Nakasaki K, Methcohn O (eds) Comprehensive natural products chemistry, vol 3. Pergamon Press, Oxford, pp 617–745

Luisa M, Goncalves FC, Steiner W (1996) Purification and characterization of laccase from a newly isolated wood-decaying fungus. In: Jeffries TW, Viikari IL (eds) Enzymes for pulp and paper processing. American Chemical Society, Washington, DC, pp 258–266

Madhavi V, Lele SS (2009) Laccase: properties and applications. Bioresources 4:1694–1717

Manzanares P, Fajardo S, Martin C (1995) Production of ligninolytic activities when treating paper pulp effluents by *Trametes versicolor*. J Biotechnol 43:125–132

Mate DM, Alcalde M (2014) Laccase engineering: from rational design to directed evolution. Biotechnol Adv 33(1):25–40

Novotny C, Svobodova K, Kasinath A, Erbanova P (2004) Biodegradation of synthetic dyes by Irpex lacteus under various growth conditions. Int Biodeterior Biodegrad 54:21523

Nurmi JT, Tratnyek PG, Sarathy V, Baer DR, Amonette JE, Pecher K, Wang C, Linehan WC, Matson DW, Penn RL, Driessen MD (2005) Characterization and propertied of metallic iron nanoparticles: spectroscopy, electrochemistry, and kinetics. Environ Sci Technol 39:1221–1230

Nyanhongo GS, Gomes J, Gubitz GM, Zvauya R, Read JS, Steiner W (2002) Decolorization of textile dyes by laccases from a newly isolated strain of *Trametes modesta*. Water Res 36:1449–1456

Octavio LC, Ma PP, Irma C, Rodrı́guez B, Ricardo J, Francisco VO (2006) Laccases. In: Guevara-González RG, Irineo Torres-Pacheco I (eds) Advances in agricultural and food biotechnology. Research Signpost, Trivandrum, pp 323–340

Pal S, Tak YK, Song JM (2007) Does the antibacterial activity of silver nanoparticles depend on the shape of the nanoparticle? A study of the gram-negative bacterium *Escherichia coli*. Appl Environ Microbiol 73(6):1712–1720

Palza H (2015) Antimicrobial polymers with metal nanoparticles. Int J Mol Sci 16:2099–2116

Pannu JS, Kapoor RK (2014) Microbial laccases: a mini-review on their production, purification and applications. Int J Pharm Arch 3(12):528–536

Pardo I, Camarero S (2015) Laccase engineering by rational and evolutionary design. Cell Mol Life Sci 72(5):897–910

Park DH, Vieille C, Zeikus JG (2003) Bioelectrocatalysts: engineered oxidoreductase system for utilization of fumarate reductase in chemical synthesis, detection, and fuel cells. Appl Biochem Biotechnol 111:41–53

Prasad R (2014) Synthesis of silver nanoparticles in photosynthetic plants. J Nanopart, Article ID 963961, doi:10.1155/2014/963961
Prasad R, Swamy VS (2013) Antibacterial activity of silver nanoparticles synthesized by bark extract of *Syzygium cumini*. J Nanopart, Article ID 431218, doi:10.1155/2013/431218
Prasad R, Kumar V, Prasad KS (2014) Nanotechnology in sustainable agriculture: present concerns and future aspects. Afr J Biotechnol 13(6):705–713
Prasad R, Pandey R, Barman I (2015) Engineering tailored nanoparticles with microbes: quo vadis. WIREs Nanomed Nanobiotechnol. doi:10.1002/wnan.1363
Riva S (2006) Laccases: blue enzyme for green chemistry. Trends Biotechnol 24:219–226
Sharma KK, Kuhad RC (2009) An evidence of laccases in archaea. Indian J Microbiol 49:142–150
Shraddha, Shekher R, Sehgal S, Kamthania M, Kumar A (2011) Laccase: microbial sources, production, purification, and potential biotechnological applications. Enzym Res 2011:0–11 (Article ID 217861). doi:10.4061/2011/217861
Simkus RA, Laurinavicius V, Boguslavsky L, Skotheim T, Tanenbaum S, Nakas JP, Slomczynski DJ (1996) Laccase containing sol-gel based optical biosensors. Anal Lett 29:1907–1919
Subramanian J, Ramesh T, Kalaiselvam M (2014) Fungal laccases—properties and applications: a review. Int J Pharm Biol Arch 5(2):8–16
Suman V, Prasad R, Jain VK, Varma A (2010) Role of nanomaterials in symbiotic fungus growth enhancement. Curr Sci 99:1189–1191
Swamy VS, Prasad R (2012) Green synthesis of silver nanoparticles from the leaf extract of *Santalum album* and its antimicrobial activity. J Optoelectron Biomed Mater 4(3):53–59
Thakker GD, Evans CS, Koteswara RK (1992) Purification and characterization of laccase from *Monocillium indicum* Saxena. App Microbiol Biotechnol 37:321–323
Uthandi S, Saad B, Humbard MA, Maupin-Furlow JA (2010) LccA, an archaeal laccase secreted as a highly stable glycoprotein into the extracellular medium by *Haloferax volcanii*. Appl Environ Microbiol 6(3):733–743
Varjani SJ, Rana DP, Jain AK, Bateja S, Upasani VN (2015) Synergistic ex-situ biodegradation of crude oil by halotolerant bacterial consortium of indigenous strains isolated from on shore sites of Gujarat, India. Int Biodeterior Biodegrad 103:116–124
Viswanath B, Chandra MS, Pallavi H, Rajasekhar-Reddy B (2008) Screening and assessment of laccase producing fungi isolated from different environmental samples. Afr J Biotechnol 7:1129–1133
Wang J, Feng J, Jia W, Chang S, Li S, Li Y (2015) Lignin engineering through laccase modification: a promising field for energy plant improvement. Biotechnol Biofuels 8:145–155. doi:10.1186/s13068-015-0331-y
Xu F (2005) Applications of oxidoreductases: recent progress. Industrial Biotechnol 1:38–50
Yatsu J, Asano T (2009) Cuticle laccase of the silkworm, *Bombyx mori*: purification, gene identification and presence of its inactive precursor in the cuticle. Insect Biochem Mol Biol 9(4):254–262

Chapter 13
Polymer Inorganic Nanocomposites: A Sustainable Antimicrobial Agents

Ayat F. Hashim, Mousa A. Alghuthaymi, Alexander Yu. Vasil'kov, and Kamel A. Abd-Elsalam

Abstract Certainly there is a vital necessitates to identify such more compounds to present more alternatives to some of the over-used antimicrobial compounds. Some of these new green and/or hybrid composites may reveal antimicrobial efficacy that differ mechanistically from other classical synthetic antimicrobials that being used. Additionally, using green nanotechnology to reduce probable ecological, plant and human health hazards linked with the drug and pesticides industries and use of nano-based agricultural products, and to find more eco-friendly bioactive materials. Biopolymers include plant-derived materials (starch, cellulose, other polysaccharides, proteins), animal products (proteins, polysaccharides), microbial products (polyhydroxybutyrate) and polymers synthesized chemically from naturally derived monomers (polylactic acid, PLA). Uses a combination of active ingredients from polymer inorganic nanocomposites may increase antimicrobial activity, reduce drug and pesticide dose. In the current article, synthesis and characterize a new green and/or hybird polymer inorganic nanocomposites will be reviewed to demonstrate, synthesis characterize, synergistic antimicrobial activity, toxicity and recyclable in soil and water environment, and understood toxicity dynamics of new nanocomposites. As a final point, we will discuss the applications and our future trends on how outlook research should be oriented to contribute in the replacement of synthetic materials with new polymer inorganic nanocomposites.

A.F. Hashim • K.A. Abd-Elsalam (✉)
Agricultural Research Center, Plant Pathology Research Institute, Giza, Egypt

Unit of Excellence in Nano-Molecular Plant Pathology Research (ARC), Giza, Egypt
e-mail: kamelabdelsalam@gmail.com

M.A. Alghuthaymi
Biology Department, Science and Humanities College, Shaqra University, Al-Quway'iyah, Saudi Arabia

A.Y. Vasil'kov
A.N. Nesmeyanov Institute of Organoelement Compounds (INEOS) of Russian Academy of Sciences, Moscow, Russia

R. Prasad (ed.), *Advances and Applications Through Fungal Nanobiotechnology*, Fungal Biology, DOI 10.1007/978-3-319-42990-8_13

1 Introduction

The biodegradable polymers created from bioorganic material including polysaccharides such as starch, cellulose and chitin/chitosan. Also, some polyester like poly lactic acid (PLA) is also naturally biodegradable polymers (Yu et al. 2006). Recently, the use of biocompatible and biodegradable polymers has risen because their environmentally-friendly nature and low dependence of non-renewable resources (Alix et al. 2013; Prasad et al. 2016). Noble metal nanoparticles (NPs) have been of great scientific interest due to their unique properties at the nanoscale and have a wide range of applicability for optical, catalytic, electrical, magnetic and antimicrobial activities (Kalita et al. 2009; McKenna 2009; Morones et al. 2005; Romanska and Mazur 2003; Schmid and Simon 2005). Metal NPs have attracted significant consideration as an antimicrobial agent due to their high surface area to volume ratio, which permits NPs to be effective in very small doses (Sundaresan et al. 2012). Nanoparticles covered with the biopolymer such as polyethylene glycol (PEG) that were coated with natural oil of garlic were screened for their biocidal activity against red flour beetle (*Tribolium castaneum*) in adult stage (Yang et al. 2009). Few researchers have investigated antimicrobial effects of the combined inorganic NPs with bioorganic pesticides for controlling plant pathogens in greenhouse and open field (Joselito and Soytong 2014; Xue et al. 2014). Uses a combination of active ingredients may increase antifungal activity, reduce pesticide dose and avoid development of fungal strain resistance (Soytong et al. 2013). An Ecofriendly hybrid nanofungicide enable minor doses of the active ingredients to be used efficiently over a given period of time and in that their nano engineered design allows them to resist the difficult environmental practices that work to eliminate conventionally applied pesticides in agro-ecological zone, i.e., leaching, evaporation and photolysis, chemical hydrolysis and biodegradation. Examples of polymer/metal composites designed to have antimicrobial activities, with a special focus on copper and silver metal nanoparticles and their mechanisms were described by Palza (Palza 2015). The main objective of current review article to discuss on the synthesis, characterize of metal-polymer nanoparticles, toxicity and its application in various fields especially as nano-antimicrobial (Fig. 13.1).

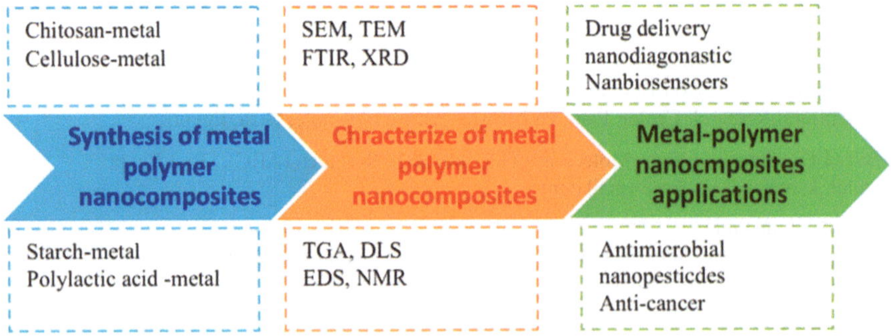

Fig. 13.1 Synthesis, characterize of metal-polymer nanoparticles, and its application in various fields especially as nanoantimicrobial

1.1 *Synthesis of Mono and Bimetallic Nanocomposites Using Metal-Vapor*

The metal-vapor synthesis (MVS) is an efficient route to produce biologically active metal nanoparticles and their based composites with biocompatible polymers for biomedical applications. The method is based on co-condensation of metal vapors with organic reactants in an evacuated reactor (10^{-2} Pa) whose walls are cooled with a low temperature cooling agent (e.g., liquid nitrogen). Metal vapor were produced by successive evaporation of metal from a tungsten rod upon resistive heating. The synthetic procedure involved metal-ligands co-condensation. As the synthesis completed, cooling was stopped, the co-condensate was heated until melting. Then the organosol obtained was used to impregnate in vacuum polymer kept in the Schlenk flask. The excess of the organosol was deleted, and the remaining product was dried in vacuum (Cárdenaz et al. 2009).

The structure of the resultant Au-containing composite materials was elucidated using X-ray and synchrotron techniques. The presence of gold nanoparticles within the composite materials is unambiguously supported by EXAFS/XANES, XRD and XPS (the Au 4f7/2 binding energy is 85.2 eV, which is only by 1.2 eV higher than that for bulk Au due to size effects). According to SAXS, the gold nanoparticles are essentially spherical and characterized by the predominant size of 6 nm, although the size distribution is asymmetric due to a fraction of larger particles with sizes of up to 30 nm (Belyakova et al. 2013).

The new method for formation of metal-polymeric nanocomposites based on biocompatible polymers such as ultrahigh molecular weight polyethylene, ultrafine polytetrafluoroethylene and cellulose, which combines modification of powder polymer samples with supercritical carbon dioxide (method of impulse modification) and introduction of metal nanoparticles with MVS has been developed (Nikitin et al. 2011). Modification of some polymers (polyethylene, poly tetrafluoroethylene, poly vinylpyrrolidone etc.) in supercritical carbon dioxide (SC CO_2) is complicated by the low values of swelling coefficients that makes it impossible to introduce the required amount of metal nanoparticles into polymer. The usage of MVS for formation of metal-polymeric nanocomposites requires rather porous materials. Therefore, an effective solution of this problem is to use ultrafine materials with developed surface, which is enough to stabilize significant amount of metal nanoparticles. The method of porous structure forming in polymers from SC CO_2 could be used to increase active surface of composites. It should be noted that the use of SC CO_2 is an effective way of modifying biocompatible polymers, polyvinyl pyrrolidone in particular, with drugs to create functional materials for dosage delivery of drugs.

A series of medical bandaging materials have been prepared via the modification of traditional cotton gauze with gold and silver nanoparticles synthesized by the MVS (Belyakova et al. 2013).

2 Characterization of Metal Polymer Nanocomposites

In order to understand the enhancement of the properties of the polymeric nanocomposites proper material characterization is mandatory. Many routinely used techniques for the analysis and characterization of nanomaterials (Abd-Elsalam and Khokhlov 2015). Dynamic light scattering (DLS) is also known as photon correlation spectroscopy (PCS). DLS is one of the most accepted light-scattering techniques used today because it is a quick procedure that allows recognition of particle size up to 1 nm in diameter. The invention of the atomic force microscope (AFM) which is also known as scanning probe microscope (SPM) has modernized the scientific revise of surfaces. Metallic or polymeric thin-film surfaces can be now imaged with nanometer high resolution. Electron microscopy (EM) has long been used for ultrastructural analysis of both biological and non-biological samples. There are basically two types of electron microscopy, transmission and scanning electron microscopy (TEM and SEM). TEM can picture internal subcellular structures from thin sliced cells, while SEM can yield more lifelike cell surface images. TEM is of high resolution, and it is one of the only few available apparatus capable of resolving the structural properties of nanoscale particles. The characteristic functional groups present in the molecules of synthesized nanoparticles were analyzed using Fourier Transform-Infra Red (FT-IR) spectroscopy. X-Ray Diffraction (XRD) technique was used to determine the crystalline structure of the particles. Thermogravimetric Analysis (TGA) technique is used to determine polymer degradation temperatures in polymer or composite materials (Alonso 2012).

3 Properties of Natural Polymers

Easy availability: In many countries, they are widely available in nature from different bio sources.

Economic: They are cheaper and their production cost is less than synthetic materials.

Biodegradable: Naturally occurring polymers are degradable by all living organisms. They will show no side effects on the environment or human being.

Biocompatible and non-toxic: chemically, nearly all of these plant materials are carbohydrates in nature and composed of repeating monosaccharide units. Hence they are non-toxic.

Safe and avoid side effects: They are widely available in natural sources and hence, safe having no side effects. In general natural polymers: having inherently biodegradable and extra ordinary properties, such as self-assembly, specific recognition of other molecules, and formation of reversible bonds (Petrak 1990).

4 Polymer/Metal Antimicrobial Activity

The main biomedical device that based on the polymer nanocomposite containing metal nanoparticles is the antimicrobial effects that composed of polymer and metal nanoparticles, which is a mostly silver nanoparticle (Shanmugam et al. 2006). Many reports investigated the deposition of metal nanoparticles to polymeric fabric substrates, i.e., metal/fabric nanohybrids, due to their strong antimicrobial activity (Geranio et al. 2009; Lee et al. 2003, 2007; Perelshtein et al. 2008; Smiechowicz et al. 2011). Polymer-metal complexes showed higher activity than the single metal. The incorporation of nanoparticles into polymer nanofiber attracts the interest of researchers who work in biomaterial, nano-agri products and drug delivery fields. Different types of nanomaterials like copper, zinc, titanium, magnesium, gold (Gu et al. 2003) alginate (Ahmad et al. 2006). Ag NPs have showed to be most effectual as they exhibit high potent antimicrobial efficacy against bacteria, viruses and other microorganisms. Ag NPs is used as a disinfectant drug (Gong et al. 2007).

4.1 *Metal Antimicrobial Activity*

Many literatures have investgated that both ionic and metallic silver exhibit antimicrobial activity (Chien et al. 2007). Different nano silver-coated barrier dressings were reported to exhibit antimicrobial activity and reduce infection in wounds (Supp et al. 2005). Silver nanoparticles have high therapeutic potential and exhibit good antimicrobial activity. It has a wide range of antimicrobial activities and exhibit high performance even at a very low concentration. It has been identified to possess good potential for the treatment of cancer (Sriram et al. 2010). Silver nanoparticles could inhibit the growth of the fungus *Raffaelea* spp. (Kim et al. 2009). It was demonstrated that the antimicrobial effect of silver nanoparticles strongly depends on their size. Nanoparticles showed greater antimicrobial activity than microparticles (Honary et al. 2011). One of the potential applications of silver is in management of plant diseases. Silver displays multiple modes of inhibitory action against plant pathogenic fungi (Park et al. 2006). Therefore, it may be used with relative safety for control of various plant pathogens, compared to synthetic fungicides (Min et al. 2009).

Some metal oxide nanoparticles including Zn, Cu and silver have antimicrobial effects (Raghupathi et al. 2011) For example, Panacek et al. (Panacek et al. 2009) reported that silver nanoparticle had better antifungal effect against candida by lesser concentration. The antibacterial and antifungal activity of ZnO microparticles has been reported by (Yamamoto 2001) and (Sawai and Yoshikawa 2004). Also it has been demonstrated ZnO nanoparticles possess significant antifungal properties against *Botrytis cinerea* and *Penicillium expansum* and the inhibitory effects increase by different concentrations (He et al. 2011). ZnO and CuO nanoparticles can be effective against *C. albicans*, and this subject can be studied more on different strains of this yeast, and other fungal species (Najafzadeh et al. 2015). Copper is an interesting candidate to be implemented in novel food safety strategies. Copper is an essential cofactor for metalloproteins and enzymes and, at high concentrations, is a wide-ranging antimicrobial effective against main food-borne pathogenic bacteria, such as *Salmonella enteric* and *Campylobacter jejuni* (Faundez et al. 2004). The antifungal activity of copper nanoparticles against plant pathogenic fungi was reported by Cioffi et al. (Cioffi et al. 2004). It is considered the most toxic element to microorganisms in the following sequence: Ag>Hg>Cu>Cd>Cr>Pb>Co>Au>Zn>Fe>Mn>Mo>Sn (Berger et al. 1976). Nanofungicides can be synthesized in an easy cost-effective manner are suitable for formulating new types of biohybrids nanocide materials would be used as a new environment friendly antimicrobial against different fungal pathogenic organisms (Abd-Elsalam and Alghuthaymi 2015).

4.2 Polymer Antimicrobial Activity

Chitosan reduces the fungal and bacterial growth some pathogens, showing broad spectra of antibacterial activity, high killing rate, and low toxicity toward mammalian cells (Fang et al. 1994; Kim et al. 1997; Liu et al. 2001; Rinaudo 2006). Chitosan, chitosan films have excellent antimicrobial properties, may used in food packaging (Tripathi et al. 2009). Several reports on antifungal activity of chitosan against plant pathogens have been carried out (Arul et al. 1992; Baba et al. 1996; Reddy et al. 1998) and reviewed (Rabea et al. 2003; Sautista-Banos et al. 2006). Chitosan/Ag/ZnO nanoparticle composite membranes were prepared via a sol-cast transformation method resulting in a good and homogeneous dispersion of ZnO and Ag nanoparticles within the chitosan matrix (Li et al. 2010). It was also found that chitosan/Ag/ZnO films had antimicrobial activities higher than those exhibited by Chitosan/Ag and Chitosan/ZnO films, a synergetic effect being then observed. Incorporation of titania in the obtainment of titania-chitosan nanocomposites increased the mechanical properties (Al-Sagheer and Merchant 2011; Kavitha et al. 2013) prepared different titania-chitosan nanocomposites varying the chitosan ratio by an in situ sol-gel method. The increase in chitosan content led to an enhancement in antibacterial activity against *S. aureus*.

Cellulose, which is a naturally occurring complex polysaccharide, is biodegradable and the most abundant renewable organic raw material at low costs in the world. Modification of cellulose by graft copolymerization and direct chemical modification techniques allows one to chemically change the cellulose chain by introducing functional groups, which leads to new cellulose products with new properties (Abu-Laiwi et al. 2003). Starch is one of the most abundant materials on earth due to its cheapness; it is widely-used in stabilizing and controlling size and shape of metal nanoparticles (Li et al. 2011). The possibility of Preparing nano-sized metals and metal oxides, mainly silver (Ag), titanium dioxide (TiO_2), zinc oxide (ZnO) and cooper II oxide (CuO), has brought about the development of novel ranges of biocides. Poly lactic acid (PLA) is widely used in various medical applications such as surgical implants, tissue culture, resorbable surgical sutures, wound closure, and controlled release systems (Jain 2000; Mikos et al. 1994; Park et al. 1992; Taylor et al. 1994). Other composites films with antimicrobial properties based on chitosan and poly(lactic acid) (Sebastien et al. 2006), cellulose (Shih et al. 2009) have been reported for food packaging applications, among others.

4.3 Chitosan Metal Antimicrobial Activity

Recently, hybrid materials based on chitosan (CS) have been synthesized, including conducting polymers, metal nanoparticles, and oxide agents, due to their excellent antimicrobial properties (Li et al. 2010). The bactericidal effects of chitosan tripolyphosphate nanoparticles loaded with diverse metal ions was tested, the results showed that antibacterial activity was improved by the loaded metal ions (Du et al. 2008). The monocomponent chitosan is antimicrobial agent is not already full filling the requirements of some conditions. For instance, the combination of chitosan with other inorganic agents such as Ag, Zn, SiO_2, and TiO_2 and among them chitosan-Ag nanoparticles composite had significantly have high antibacterial activity with only small presence of Ag-nanoparticles which exhibit potential antifungal properties (Li et al. 2010). The chitosan have a high chelating capability with diverse metal ions such as Ag, Cu, Zn, Mn and Fe in acidic conditions and this chitosan metal complex is stronger in their antimicrobial activity (Kong et al. 2010).

Nanochitosans are effective against different types of microorganisms (Ahn et al. 2009; Alt et al. 2004; Friedman and Juneja 2010; Poosti et al. 2013; Yang et al. 2012). Biosynthesis silver nanoparticles-chitosan composite was applied to develop an effective antibacterial (Di et al. 2012). A novel chitosan-Ag-nanoparticle composite was synthesized and investigated a significantly higher antimicrobial activity than its components at the relevant concentrations (Sanpui et al. 2008). Ag/chitosan nanoformulation has significant antifungal activity against the screened fungi such as *A. flavus*, *A. alternata* and *R. solani*. Therefore, it can be successfully used against diverse plant phytopathogenic fungi to protect the various crop plants and their products, in preference to using the commercially available synthetic fungicides, which show higher toxicity to humans and agri-ecosystems (Kaur et al. 2015). The synergistic antimicrobial activity of a CS-Ag nanoparticle composite in the presence of molecular iodine was increased (Banerjee et al. 2010). The application of silver nanocomposite and chitosan as antifungal effect in *C. militaris* (Cm) and *A. cinnamomea* (Ac) is uncommon. It was showed a selective inhibition by chitosan and silver nanoparticles–chitosan composite, suggesting that the cellular wall composition or modification of Cm and Ac is distinct. With the promotion of antifungal capability of chitosan and silver nanoparticles–chitosan composite, it could offer a new alternative to traditional antibiotics and a great prospect to develop the next-generation of antibiotics (Fig. 13.2) (Wang et al. 2015).

Notes: Droplets were pushed out from pin tip of a pump-driven syringe and then dropped into NaOH solution, which was used for silver nanoparticles' reduction and chitosan solidification. Silver nanoparticles–chitosan composite spheres were formed in 15 min. The distance between solidifying liquid surface and the tip was 1 cm. The diameter of the needle of the syringe was 8.73 mm.

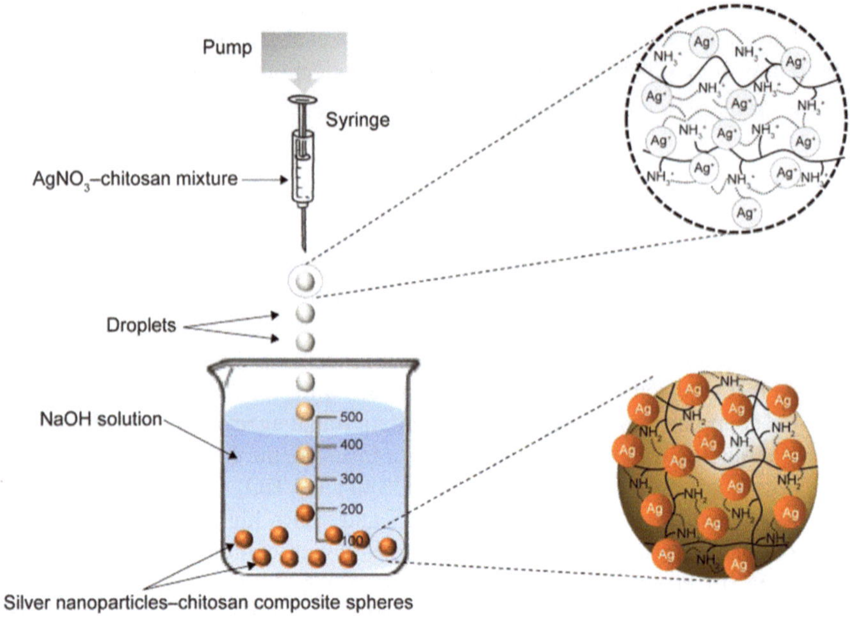

Fig. 13.2 Schematic drawing of the silver nanoparticles-chitosan composite spheres synthesis (Wang et al. 2015)

Green synthesized colloidal stable Cu-chitosan NPs with a size range of 20–30 nm have been successfully in the presence of a biopolymer. It was indicate that high antimicrobial activity of copper NPs for fighting pathogenic microorganisms. This formulation may be appropriate for novel kinds of antimicrobial agents for both of pharmaceutical and biomedical application (Manikandan and Sathiyabama 2015). The selection of chitosan as stabilizer of the Cu-NPs is because of its capacity to chelate metals, which makes a perfect candidate for metal NP synthesis. In general, the use of bio-polymers as stabilizers for the synthesis of Cu-NPs is gaining momentum because of their availability, biocompatibility and low toxicity (Hardy et al. 2004). Chitosan nanoparticles and copper-loaded nanoparticles could reduce the growth of different pathogens markedly and exhibit higher antibacterial activity than chitosan itself or doxycycline (Qi et al. 2004). The antimicrobial properties of metallic copper nanoparticles synthesized in chitosan polymer medium were evaluated (Usman et al. 2013). Copper-loaded chitosan nanoparticles have been synthesized and exhibit their biological activities due to the unique characters of nanoparticles, including the small size and quantum size effects (Du et al. 2008; Qi et al. 2004; Varma et al. 2004).

The ZnO–CS NPs showed significant antimicrobial activity and biofilm inhibition activity against *M. luteus* and *S. aureus*. It was showed the promising potential of ZnO–CS NPs as antimicrobial and biofilm inhibition agents (Dhillon et al. 2014). The advantages of combining chitosan with ZnONPs as an alternative to the widely used Ag NPs reside in their low cost, lack of color and UV-blocking properties (Abou-Okeil and El Shafei 2011). The combination of CS and Zn on the antimicrobial activity of the coated textile surface is discussed (Perelshtein et al. 2013). Biopolymer films, like chitosan, modified with CuNPs were evaluated to be effective in inhabiting of two microorganisms affecting food quality and therefore could be used to enhance food quality and extending the shelf-life of foods (Cárdenaz et al. 2009). Biopolymer chitosan one of the best plant resistance inducers (Youssef and Roberto 2014). The antifungal properties of some inorganic nanoparticles like S, Ag, CuO, MgO and ZnO were discovered alone or combined with biopolymer in very recently published articles (Brunel et al. 2013; Mohan et al. 2011). In vitro assay, The chemical combination of chitosan nanoparticles were found most efficient at 0.1 % concentration and showed 89.5 %, 63.0 % and 60.1 % growth reductions of *Alternaria alternate* (Saharan et al. 2013).

The colloidal Cu-nanoparticles-chitosan composite films were obtained using MVS (Cárdenaz et al. 2009). The antimicrobial activity of films against *Staphylococcus aureus* and *Salmonella enterica* serovar Typhimurium, were also tested. The composite film was effective in alteration of cell wall and reduction of microbial concentration in liquid culture for both bacteria tested. Nowadays, the dermal biomimetic scaffolds are widely used in regenerative medicine. Collagen-chitosan scaffold one of these materials possesses antibacterial activity, good compatibility with living tissues and has been already used as a wound-healing material. Collagen-chitosan scaffolds modified with Ag and Au nanoparticles have been synthesized using the metal-vapor synthesis (Rubina et al. 2016). The nanocomposite materials are characterized by XPS, TEM, SEM and synchrotron radiation-based X-ray techniques. According to XRD data, the mean size of the nanoparticles (NPs) is 10.5 and 20.2 nm in Au-Collagen-Chitosan (Au-CollCh) and Ag-Collagen-Chitosan (Ag-CollCh) scaffolds, respectively in fair agreement with the TEM data. SAXS analysis of the composites reveals an asymmetric size distribution peaked at 10 nm for Au-CollCh and 25 nm for Ag-CollCh indicative of particle's aggregation. According to SEM data, the metal-carrying scaffolds have layered structure and the nanoparticles are rather uniformly distributed on the surface material. XPS data indicate that the metallic nanoparticles are in their unoxidized/neutral states and dominantly stabilized within the chitosan-rich domains.

A high synergistic effect between chitosan and copper in reducing fungal growth of *Fusarium graminearum* was examined. The combined copper (II) chitosan colloids as a new generation of copper-based bio-pesticides (Brunel et al. 2013). Micro chitosan and chitosan based microparticles combined with copper ions was screened for fungicidal potency against toxicogenic fungi *F. solani* (Vokhidova et al. 2014). Chitosan biopolymer has been used to coat different type of fruits and vegetables because of its antimicrobial activities, and producing a variety of plasticized chitosan films for protection of food products due to the created-mediated modified atmosphere (Miranda et al. 2007).

4.4 Cellulose Metal Antimicrobial

Metal-cellulose had an excellent antimicrobial activity. The excellent antimicrobial property and binding durability of the nanoparticles for the metal cellulose make them effective in biological applications ranging from environmental materials such as air or water purification system to biomedical materials like wound dress. These robust metal-cellulose nanohybrids are quite attractive in environmental and biomedical technologies because of their excellent antimicrobial activity and low possibility for human tissue damage by enhanced binding durability of metal nanoparticles with fabric substrate (Park et al. 2012).

Cellulose fibers have found a broad application in medical textile field owing to the unique characteristic, such as high moisture and liquids' adsorption, low impurity content, antistatic behavior, and fine mechanical properties. Nevertheless, cellulose fibers offer an excellent surface for microbial growth. Because of their molecular structure and a large active surface area, cellulose fibers, may be an ideal matrix for the design of bioactive, biocompatible, and smart materials (Belyaev 2000; Stashak et al. 2004; Vigo 2001). According to the literature cellulose fibers are one of the most polymer materials for antimicrobial functionalization. The surface modification of the cellulose fibers is currently considered to be the best route for obtaining modern functionality on textiles for the use in medical applications (Gao and Cranston 2008).

In a suspension of cellulose nanocrystal and metallic salts, most of the metallic particles can absorb on the surface cellulose nanocrystal (CNC) due to electrostatic interactions between oxygen of polar hydroxyl and metallic particles (He et al. 2003). This effect controls sizes the size of metallic particles by avoiding particles agglomeration. Porous nanocomposites based on polysaccharides are widely used in cosmetic and medical fields (Sescousse et al. 2011). Materials from cellulose are of special interest due to abundance and renewability of this natural polymer. Modification of this natural polymer with metal nanoparticles gives special antibacterial properties to the material (Maneerung et al. 2008).

The cellulose/copper composites was generated by physical methods and showed a high antifungal effect. The microbial growth of *S. cerevisiae* was reduced in contact with the cellulose/copper composites, but the effectiveness depended on the procedures applied to synthesis the composite material, and on the obtainable concentration of copper ions the feasibility of the cellulose/copper composites to inhibit the growth of pathogens in contact with dripped fruit juices was tested under controlled conditions (Llorens et al. 2012a, b). The sensitivity of some strains of *Fusarium*, *Penicillium* and *Aspergillus* to a copper surface, and found resilient spores of *Aspergillus niger*. The minimal inhibitory concentrations of copper sulphate of about 4.7 mmol kg^{-1} for *P. expansum*, and around 8 mmol kg^{-1} for *B. cinerea*, some fungal isolates are resistant to copper (Judet-Correia et al. 2011).

Silver ions from cellulose/silver nanocomposites are effective to reduce microbial growth in contact with meat or fruit exudates. Conversely, proteins counteract the inhibitory action of silver ions against spoilage-related pathogens, being the fruit drips a mainly favorable substrate to realize the desired antimicrobial activities of silver in food preservation applications (Llorens et al. 2012b; Lloret et al. 2012).

Cellulose nanocrystal can offer high antimicrobial effect for ZnO nanoparticles (Azizi et al. 2008). The cellulose-antibiotics metal complexes were found to be highly active against *Escherichia coli*, *Pseudomonas aeruginosa*, *Staphylococcus aureus*, *Candida albicans* and *Aspergillus flavus* than the parent Cloxacillin, tetracycline and cellulose. The fabricated cellulose Rod-like nanocrystal/polyrhodanine (CNC/PR) core–sheath nanoparticles exhibited promising antimicrobial properties against *Escherichia coli* and *Bacillus subtilis* (Tang et al. 2015). The mesoporous hybrid material was prepared from microcrystalline cellulose modified with gold nanoparticles by metal-vapor synthesis (Vasil'kov et al. 2015). The BET surface area has been measured as 40 $m^2\ g^{-1}$.

4.5 Starch Metal Antimicrobial Activity

Synthesis of starch stabilized silver nanoparticles was approved using β-D-glucose as a nontoxic reducing agent (Raveendran et al. 2003, 2006). Corn starch stabilized silver nanoparticles synthesized quickly in aqueous system by the assistance of microwave Irradiation. Result also showed that AgNPs can reduce the growth of *A. niger* (Nnemeka et al. 2014).

The antifungal activity of silver nanoparticles against *Candida* spp. was evaluated (Panacek et al. 2009). Also, the antifungal activity of the silver nanoparticles against clinic isolates and ATCC strains of *Trichophyton mentagrophytes* and *Candida* spp, was investigated (Kim et al. 2008). The starch-stabilized silver NPs showed that a good antifungal activity against *C. albicans* (Raji et al. 2012). The fungicidal activity of some antifungals such as amphotericin B (MFC _2–16 mg=L), posaconazole (MFC _8 mg=L), itraconazole and voriconazole (MFC >10 mg=L) less than starch-stabilized silver NPs (Panacek et al. 2009).

ZnO nanoparticles loaded in polymer matrices such as soluble starch are a good example of functional nanostructures with potential for applications such as UV-protection ability in textiles, sunscreens, and antibacterial finishes in medical textiles and inner wears (Bhuyan et al. 2015). The stabilization of these nanoparticles was realized by the presence of soluble starch in the reaction medium.

The quick helical form of the soluble starch to protect and prevent the ZnO nanoparticles for agglomeration by action of steric hindrance. The average size of the ZnO nanoparticles was estimated to be 38±3 nm using a TEM (Vigneshwaran et al. 2006).

The biopolymer starch confirmed to be a well-organized matrix for stabilization of uniform, monodispersed metallic and bimetallic nanoparticles of copper, silver in aqueous medium. The biopolymer matrix gives the special advantage for use in biomedical applications and the nanoparticles solutions exhibited excellent antibacterial agent (Valodkar et al. 2012). The non-cytotoxic green Cu–starch conjugate offers a rational approach towards antimicrobial application and for integration to biomedical devices (Valodkar et al. 2012).

4.6 Polylactic Acid Metal Antimicrobial

A polylactic acid (PLA) nanobiocomposite was combined with silver nanoparticles and cellulose nanocrystals obtaining an antimicrobial film with improved barrier properties (Fortunati et al. 2013). Ag/PLA-NC films possessed a strong antibacterial activity with the increase in the percentage of Ag-NPs in the PLA. Thus, Ag/PLA-NC films can be used as an antibacterial scaffold for tissue engineering and medical application (Shameli et al. 2010).

Copper antimicrobial activity has been established over a broad spectrum of human pathogens such as *Escherichia coli* O157:H7, *Staphylococcus aureus*, *Enterobacter aerogenes*, *Pseudomonas aeruginosa*, *Listeria monocytogenes* (Wilks et al. 2006), *Salmonella entérica*, *Campylobacter jejuni* (Faundez et al. 2004), *Klebsiella pneumoniae*, *Mycobacterium tuberculosis*, *Candida albicans* (Mehtar et al. 2008), *Clostridium difficile* (Wheeldon et al. 2008), *Aspergillus flavus*, *Aspergillus niger*, *Pénicillium chrysogenum* (Weaver et al. 2010) and *Saccharomyces cerevisiae* (Quaranta et al. 2011). The antimicrobial effect of modified montmorillonites $MtCu^{2+}$/PLA nanocomposites, would be given by the presence of Cu^{2+}, because these ions have a strong antimicrobial activity removing electrons from the cell walls and cell membranes, thereby causing the output of the cytoplasm and the oxidation of the core cell, with the cell death, to *L. innocua* (Wilks et al. 2006) and *E. coli* (Nan et al. 2008). Alternatively, the antimicrobial activity of films $MtCu^0$/PLA nanocomposites can be attributed to copper nanoparticles which can be transformed into different chemical species with high antimicrobial activity, such as Cu^+, Cu^{+2} and CuO (Ren et al. 2009). The PLGA/CuO hybrid nanofiber scaffolds reduced both Gram-positive and Gram-negative bacterial growth. The mechanism of the antibacterial activity was accomplished to be based on the CuO nanoparticles and Cu^{++} ions release. Therefore, evaluated that PLGA/CuO hybrid nanofiber scaffolds can be a useful candidate for wound dressing (Haider et al. 2015). Copper-containing nanomaterials have been evaluated as antimicrobial agents for food packaging applications. This nano-antimicrobial was composed of copper nanoparticles inserted in poly-lactic acid, which has been tested as a biodegradable polymer matrix (Longano et al. 2013). PLA/ZnO nanocomposites are potential candidates for applications in packaging and in medical applications (Murariu et al. 2011).

4.7 Nanogels

Bioactive polysaccharides, such as chitosan (CS), dextran, starch, cyclodextrin, cellulose, pullulan and poly (amino acids) could be modified with various reactants to prepare nanogels (Namazi and Dadkhah 2008; Namazi and Mosadegh 2011; Ragauskas et al. 2007). The synthesis of nanogels can be achieved by diverse crosslinking techniques, like crosslinking copolymerization, chemical cross linking and functional group cross-linking (Yallapu et al. 2007). Nanogels in common have high capacity, low cytotoxicity and consistency. Encapsulation of antimicrobial essential oils within nano organogel has other benefits such as controlled and sustained release of a certain amount of oils from the carrier e.g. nanogel as well. This would allow the nutrients to be exposed to specific and adequate amounts of essential oils for a longer time (Herrero et al. 2014; Raemdonck et al. 2009). The encapsulation of *Mentha piperita* essential oil in chitosan–cinnamic acid nanogel was prepared, the extract showed high antifungal potency against *A. flavus*. The minimum inhibitory concentration of free and encapsulated *M. piperita* essential oils against *A. flavus* were at 2100 and 500 ppm, respectively (Beyki et al. 2014). The encapsulation of the thyme essential oils using chitosan and benzoic acid-made (CS-BA) nanogel in order to investigate their possible synergistic property in removal of *A. flavus* was screened. A high significantly antifungal potency of the encapsulated oils at concentrations over 700 mg/l was observed (Khalili et al. 2015).

5 Toxicity of Polymer Metal Nanocomposites

The use of engineered nanoparticles in the environment as a consequence of the development of nanotechnology is a serious case of concern of environmental biologists worldwide. However, a few studies have already demonstrated the toxic effects of NPs on various organisms, including mammals. Nanotechnology is still in discovery phase in which novel materials are first synthesized in small scale in order to identify new properties and further applications (Amara et al. 2009; Arora et al. 2012; Blaser et al. 2007; PCÆFVDK and Hofmann 2008; Toksha et al. 2008). The wide application of engineered NPs and their entry into the environment, the study of their impact on the ecosystem and a growing concern in society regarding the possible adverse effects of manufactured nanoparticles has been raised in recent years (Amara et al. 2009; Arora et al. 2012; Blaser et al. 2007; Innovation and Island 2010; Maynard 2006; PCÆFVDK and Hofmann 2008; Simonet and Valcárcel 2009; Tiede et al. 2009; Toksha et al. 2008). Therefore, it is required to study their release, uptake, and mode of toxicity in the organisms. Furthermore, to understand the long-term effect of NPs on the ecosystem, substantial information is required regarding their persistence and bioaccumulation.

Nanoparticles are increasing used in a wide range of applications in science, technology and medicine. Unfortunately, a material that has been shown to be safe and biocompatible at bulk level may turn out to be toxic at nano-level. The need to bring out these out comes is the aim of nanomaterial toxicity study. This would give us a better understanding of the variety of nanomaterial being used as well as help us design new material with lesser toxic effect.

Phytotoxicity of different types of nanomaterials has been widely studied in recent years (Dietz and Herth 2011; Khodakovskaya et al. 2013; Yang and Watts 2005). Under certain condition, any nano-sized material can be toxic for plants. The age and type of plants used for nano-toxicological experiments are particularly significant parameters for specific plant response. The level of toxicity will also dependent on the major properties of material, including chemical structure (toxic metals), size, shape, aggregation, concentration, functional groups or attachments, method of delivery, and purity of nanomaterial used (Rico et al. 2011). Silver nanoparticles (AgNPs) show strong inhibitory and antibacterial effects and limited toxicity to mammalian cells (Hollinger 1996; Mahapatra and Karak 2008). Chitosan inhibits the growth of a wide variety of bacteria and fungi, showing broad spectra of antibacterial activity, high killing rate, and low toxicity toward mammalian cells (Fang et al. 1994; Kim et al. 1997; Liu et al. 2001). Silver has a long history as an antimicrobial agent, especially in the treatment of wounds. Silver nanoparticles (AgNPs) show strong inhibitory and antibacterial effects and limited toxicity to mammalian cells (Hollinger 1996; Mahapatra and Karak 2008).

The level of the phytotoxicity depends on the type of nanoparticles and how they are applied. Lin and Xing (2007) studied the phytotoxicity of different nanoparticles (MWCNTs, Al_2O_3, ZnO, Al, and Zn) and their influence on the germination rates of radish, canola, ryegrass, lettuce, corn, and cucumber and reported that higher concentrations (2000 mg L^{-1}) of Zn and ZnO inhibited germination in all the plants tested. The toxicity of NPs depends on different physicochemical properties such as size distribution, state of dispersion, shape, agglomeration and aggregation, surface chemistry, surface charge, interaction with other chemicals in aqueous media, concentration, and porosity (Freese et al. 2012; Jiang et al. 2009; Mu et al. 2012).

6 Safety of Polymer-Metal Nanocomposites

The environmental safety of materials, which consist of or contain nanosize components, becomes one of the most important emerging topics of the Nanotechnology within the last few years. The main concerns dealing with the rapid development and commercialization of various nanomaterials are associated with (Ain 2007; Cushen et al. 2011; Siegrist et al. 2007):

- The approved higher toxicity of many nanomaterials (NMs) in comparison with their larger counterparts,
- The absence of the adequate analytical techniques for detection of NMs in the environment
- The absence of the legislation normative for permitted levels of various NMs in water and air. In this regard the increase of the safety of NMs is of particular importance. One way to prevent risk is the development of the environmentally-safe polymer-metal nanocomposite materials that consist in a functional polymer with immobilized MNPs distributed mainly by the surface of the polymer with a higher stability to prevent release of the MNPs. The material represents what makes them maximally accessible for the bacteria to be eliminated. Core-shell MNPs contain a superparamagnetic core coated with the functional metal shell, which provides the maximal bactericide activity. The MNPs are strongly captured inside the polymer matrix that prevents their escape into the medium under

treatment. The superparamagnetic nature of MNPs provides an additional level of the material safety as MNPs leached from the polymer matrix can be easily captured by the magnetic traps to completely prevent any post-contamination of the treated medium.

7 Mode of Action

Silver nanoparticles have a high specific surface area and a high fraction of surface atoms that lead to high antimicrobial activity compared to bulk silver metal (Cho et al. 2005). AgNPs biocidal action against microbial has been proposed to be that Ag interacts with the -SH groups of proteins on the cell walls, thereby blocking respiration and causing death (Prasad 2014; Sepideh et al. 2012). Other reports have that "pits" are formed in the cell wall of bacteria, thereby causing permeability and resulting in death (Joseph and Mathew 2014). Yet another explains silver ions form metal-organic complexes and insoluble compounds with the sulphhydryl groups in cell walls of bacteria and fungi, inhibiting metabolism and electron transport by making essential enzymes dysfunctional (Sondi and Sondi 2004). Silver nanoparticles may directly attach to the cell membrane and penetrate to kill spores, although penetration of silver nanoparticles into microbial cell membranes is not completely understood (Morones et al. 2005).

ZnO nanoparticles increased intensity of lipid and protein bands in *E. coli* (Liu et al. 2009). The antimicrobial activity of chitosan is assigned to the amino groups, which in acidic media form ammonium salts (Ravi-Kumar 2000). Mechanisms proposed for the antifungal activity of chitosan focused mainly on its effect on fungal cell wall (Allan and Hadwiger 1979) and cell membrane (Zakrzewska et al. 2005).

8 Applications and Future Trends

In this chapter a review of the most important and recent researches on development characterization and application of biodegradable polymer inorganic nanocomposites. Materials scientists, peptide chemists, and applied microbiologists should interact to develop these novel antimicrobial polymers effectively. Biomedical applications are still at the laboratory level. Additional studies are necessary before we can expect clinical applications and commercialization of chitin and chitosan-based nanofibers. With the emphasis on sustainability, nanocomposite technology may be applied to the development of biopolymers as viable packaging materials. The combination of active technologies such as antimicrobials and nanotechnologies such as nanocomposites can synergistically lead to bioplastic formulations. Researchers have reported various aspects of nanoparticle formulation, characterization, effect of their characteristics, and their applications in management of plant diseases (Alghuthaymi et al. 2015). Nanoparticles by themselves find many biomedical/biological applications.

In order to enhance their applicability these nanoparticles are conjugated with nanoparticles has been practiced since decades (Baekeland 2014). Nanomaterials are useful in developing diagnostic tool, drug delivery system, sunscreens formulation, antimicrobial bandages, disinfectants, nanobiosensors, as catalyst for greater efficiency in current manufacturing process by minimizing the use of toxic materials and an alternative energy production (Aziz et al. 2015; Prasad et al. 2012, 2014, 2015; Prasad and Swamy 2013; Prasad 2014; Swamy and Prasad 2012; Yumak et al. 2011). Chitosan-metal complexes could be promising candidates for novel antimicrobial agents in cosmetic, food and textile industries (Kaur et al. 2015). Previous studies provided the evidence of applicability of silver for controlling plant pathogenic fungi such as *Bipolaris sorokiniana, Magnaporthe grisea* (Jo et al. 2009), *Golovinomyces cichoracearum* or *Sphaerotheca fusca* (Lamsal et al. 2011) and *Raffaelea* sp. (Kim et al. 2009). Chitosan has been used as both a reducing agent and stabilizer to form AgNPs (Sanpui et al. 2008). The binding interaction between chitosan and the silver nanoparticles results in stabilization of the chitosan-AgNP composite. Once materials are prepared in the form of very small particles, they change significantly their physical and chemical properties. In fact in nano-dimension, percentage of surface molecule compare to bulk molecule is high and this enhances the activity of the particle in nano dimension and therefore, the normal properties of the particle like heat treatment, mass transfer, catalytic activity, etc are all increases. But compare to non-metal nanoparticles, metal nanoparticles have more industrial application. Nanoparticles offer many new developments in the field of biosensors, biomedicine and bio nanotechnology-specifically in the areas of drug delivery, diagnostic tools, and cancer treatment agent (Prasad et al. 2015). Nanoparticles and nanostructure are becoming a part in human medical application, including imaging or the delivery of therapeutic drugs to cell, tissues and organs. Drug loaded nanoparticles interact organ and tissues and are taken up by cells. Several studies have shown that the tissue, cell and even cell organelle distribution (Alexiou et al. 2000; Savic et al. 2003) of drugs may be controlled and improved by their entrapment in colloidal nanomaterials, mainly of the micellar structure, such as nanocontainer. Magnetic nanoparticles have been receiving considerable attention because of their wide range of applications, such as the immobilization of the proteins and enzymes, bioseparation, immunoassays, drug delivery, and biosensors (Chen and Liao 2002). Nanoparticles of ferromagnetic materials are of importance because of their reduced sizes that can support only single magnetic domains. The recent synthesis of arrays of 4 nm diameter FePt nanoparticles with an extremely narrow size distribution has promoted a significant research effort in this area, due to their potential technological application as recording media (Varlan and Sansen 1996).

For agricultural applications, polymer nanocomposites are ecofriendly (i.e., biodegradable and starch nanocomposites are commonly used). In this direction, the development of mulch films can be useful for farmers to retain moisture and control weeds (Ray 2013). Nanomaterials are including polymeric nanoparticles, iron oxide nanoparticles, gold nanoparticles, and silver ions have been exploited as pesticides. The biggest benefit of nanopesticides is that they're effective at protecting crops. They're effective at fighting pests and increasing crop yields, but they achieve that efficacy with added benefits, targeting with no residues left behind and with additional worker

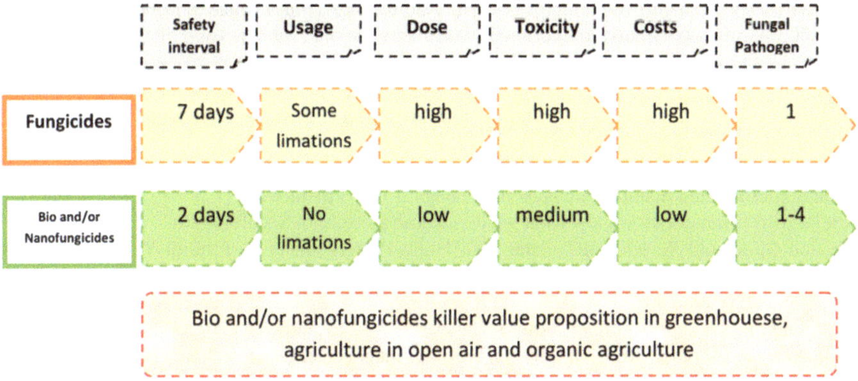

Fig. 13.3 Comparison between fungicides and bio and/or nanofungicides regarding safety interval, usage, dose, side effects and costs and number of targeted pathogens and their effects on value proposition in green house, agriculture in open air and organic agriculture

safety benefits. Comparison between fungicides and bio and/or nanofungicides regarding safety interval, usage, dose, side effects and costs and number of targeted pathogens and their effects on value proposition in greenhouse, agriculture in open air and organic agriculture was shown in Fig. 13.3.

Bio and/or safe nanofungicides could in no way penetrate mainstream agriculture without the big players leading this penetration. Development and commercialization of these products requires highly technical product development, market development and market education. This can come primarily from the big players for rapid growth. We must add to these factors also the macro trends of "going green nanotechnology" and of lack of unique products in the industry. All these together are reasons for the rapid development. That biopesticides are effective is the biggest selling point, but on top of that, they offer things which the marketplace, whether driven by regulations or consumers, is demanding.

Acknowledgments Unit of Excellence in Nano-Molecular Plant Pathology Research, Agricultural Research Center (ARC), Egypt for funding this study. This work was partially funded by the Science and Technology Development Fund (STDF), Egypt (STDF-ERJCG), Grant ID. 13791 to Kamel Abd-Elsalam. Also, this work was partially funded by Russian Foundation for Basic Research grant (RFBR-15-53-61030).

References

Abd-Elsalam KA, Khokhlov AR (2015) Eugenol oil nanoemulsion: antifungal activity against *Fusarium oxysporum* f. sp. *vasinfectum* and phytotoxicity on cotton seeds. Appl Nanosci 5:255–265

Abd-Elsalam KA, Alghuthaymi MA (2015) Nanobiofungicides: are they the next-generation of fungicides? J Nanotechol Mater Sci 2:1–3

Abou-Okeil A, El Shafei A (2011) ZnO/carboxymethyl chitosan bionano-composite to impart antibacterial and UV protection for cotton fabric. Carbohydr Polym 83:920–925

Abu-Laiwi FA, Ahmed MB, Ibrahim NA, Ab. Rahman MZ, Md. Dahlan KZ, Wan Yunus WMZ (2003) Graft copolymerization of methyl methacrylate onto rubber-wood fiber using H_2O_2 and Fe^{2+} as an initiator system. J Appl Polym Sci 88:2499–2503

Ahmad Z, Pandey R, Sharma S, Khuller GK (2006) Alginate nanoparticles as antituberculosis drug carriers: formulation development, pharmacokinetics and therapeutic potential. Indian J Chest Dis Allied Sci 48:171–176

Ahn S, Vang MS, Yang HS et al (2009) Histologic evaluation and removal torque analysis of nano- and microtreated titanium implants in the dogs. J Adv Prosthodont 1:75–84

Ain P (2007) Nanotechnologies: what we do not know. Technol Soc 29:43–61

Alexiou C, Arnold W, Klein RJ, Parak FG, Hulin P, Bergemann C, Erhardt W, Wagenpfeil S, Lubbe AS (2000) Locoregional cancer treatment with magnetic drug targeting. Cancer Res 60:6641–6648

Alghuthaymi MA, Almoammar H, Rai M, Galiev ES, Abd-Elsalam KA (2015) Myconanoparticles: synthesis and their role in phytopathogens management. Biotechnol Biotechnol Equip 29(2):221–236. doi:10.1080/13102818.2015.1008194

Alix S, Mahieu A, Terrie C, Soulestin J, Gerault E, Feuilloley MGJ, Gattin R, Edon V, Ait-Younes T, Leblanc N (2013) Active pseudo-multilayered films from polycaprolactone and starch based matrix for food-packaging applications. Eur Polym J 49:1234–1242

Allan CR, Hadwiger LA (1979) The fungicidal effect of chitosan on fungi of varying cell wall composition. Exp Mycol 3:285–287

Alonso A (2012) Development of polymeric nanocomposites with enhanced distribution of catalytically active or bactericide nanoparticles. Ph.D. Thesis, Universitat Autònoma de Barcelona

Al-Sagheer FA, Merchant S (2011) Visco-elastic properties of chitosan–titania nano-composites. Carbohydr Polym 85:356–362

Alt V, Bechert T, Steinrucke P, Wagener M, Seidel P, Dingeldein E, Domann E, Schnettler R (2004) An *in vitro* assessment of the antibacterial properties and cytotoxicity of nanoparticulate silver bone cement. Biomaterials 25:4383–4391

Amara D, Felner I, Nowik I, Margel S (2009) Physicochemical and engineering aspects synthesis and characterization of Fe and Fe_3O_4 nanoparticles by thermal decomposition of triiron dodecacarbony. Colloids Surf A Physicochem Eng Asp 339:106–110

Arora S, Rajwade JM, Paknikar KM (2012) Nanotoxicology and in vitro studies: the need of the hour. Toxicol Appl Pharmacol 258:151–165

Arul EJ, Asselin A, Benhamou N (1992) Antifungal activity of chitosan on postharvest pathogens induction of morphological and cytological alterations in *Rhizopus stolonifer*. Mycol Res 96:769–779

Aziz N, Faraz M, Pandey R, Sakir M, Fatma T, Varma A, Barman I, Prasad R (2015) Facile algae-derived route to biogenic silver nanoparticles: synthesis, antibacterial and photocatalytic properties. Langmuir 31:11605–11612

Azizi S, Ahmad MB, Mahdavi M, Abdolmohammadi S (2008) Preparation, characterization, and antimicrobial activities of ZnO nanoparticles/cellulose nanocrystal nanocomposites. Bioresources 8:1841–1851

Baba Y, Yamashita T, Kawano Y, Uchida Y (1996) Antifungal activity of aqueous soluble chitosan derivatives on *Fusarium* and *Verticillium*. Nippon Kagaku Kaishi 1:48–53

Baekeland LH (2014) Bakelite, a new composition of matter: its synthesis, constitution, and uses. Sci Am 68(Suppl 1909):322

Banerjee M, Mallick S, Paul A, Chattopadhyay A, Ghosh S (2010) Heightened reactive oxygen species generation in the antimicrobial activity of three component iodinated chitosan-silver nanoparticle composite. Langmuir 26:590–5908

Belyaev EY (2000) Drug synthesis methods and manufacturing technology, New medical materials based on modified polysaccharides (Review). Pharm Chem J 34:607–612

Belyakova OA, Shulenina AV, Zubavichus YV, Veligzhanin AA, Naumkin AV, Vasil'kov AY (2013) Diagnostics of gold containing surgical dressing materials with X-Ray and synchrotron radiation. J Surf Invest X-Ray Synchrotron Neutron Tech 7:509–514

Berger TJ, Spardaro JA, Bierman R, Chapin SE, Becker RD (1976) Antifungal properties of electrically generated metallic ions. Antimicrob Agents Chemother 10:856–860

Beyki M, Zhaveh S, Khalili S, Rahmani-Cherati T, Abollahi A, Bayat M, Tabatabaei M, Mohsenifar A (2014) Encapsulation of *Mentha piperita* essential oils in chitosan–cinnamic acid nanogel with enhanced antimicrobial activity against *Aspergillus flavus*. Ind Crop Prod 54:310–319

Bhuyan T, Mishra K, Khanuja M, Prasad R, Varma A (2015) Biosynthesis of zinc oxide nanoparticles from *Azadirachta indica* for antibacterial and photocatalytic applications. Mater Sci Semicond Process 32:55–61

Blaser SA, Scheringer M, Macleod M, Hungerbühler K (2007) Estimation of cumulative aquatic exposure and risk due to silver: contribution of nano-functionalized plastics and textiles. Sci Total Environ 390:396–409

Brunel F, Gueddari NE, Moerschbacher BM (2013) Complexation of copper (II) with chitosan nanogels: toward control of microbial growth. Carbohydr Polym 92:1348–1356

Cárdenaz G, Díaz JV, Meléndrez MF, Cruzat CC, Cancino AG (2009) Colloidal Cu nanoparticles/chitosan composite film obtained by microwave heating for food package applications. Polym Bull 62:511–524

Chen DH, Liao MH (2002) Preparation and characterization of YADH-bound magnetic nanoparticles. J Mol Catal B Enzym 16:283–291

Chien PJ, Sheu F, Yang FH (2007) Effects of edible chitosan coating on quality and shelf life of sliced mango fruit. J Food Eng 78:225–229

Cho KH, Park JE, Osaka T, Park SG (2005) The study of antimicrobial activity and preservative effects of nanosilver ingredient. Electrochim Acta 51:956–960

Cioffi N, Torsi L, Ditaranto N, Sabbatini L, Zambonin PG, Tantillo G, Ghibelli L, Sabbatini T, Bleve-Zacheo M, Zambonin PG, D'Alessio M, Traversa E (2004) Antifungal activity of polymer-based copper nanocomposite coatings. Appl Phy Lett 85:2417–2419

Cushen M, Kerry J, Morris M, Cruz-romero M, Cummins E (2011) Nanotechnologies in the food industry recent developments, risks and regulation. Trends Food Sci Technol 24:30–46

Dhillon GS, Kaur S, Brar SK (2014) Facile fabrication and characterization of chitosan-based zinc oxide nanoparticles and evaluation of their antimicrobial and antibiofilm activity. Int Nano Lett 4:107–117

Di Y, Li Q, Zhuang X (2012) Antibacterial finishing of Tencel/cotton nonwoven fabric using Ag nanoparticles-chitosan composite. J Eng Fibers Fabr 7:24–29

Dietz K, Herth S (2011) Plant nanotoxicology. Trends Plant Sci 16:582–589

Du WL, Xu YL, Xu ZR, Fan CL (2008) Preparation, characterization and antibacterial properties against *E. coli* K88 of chitosan nanoparticle loaded copper ions. Nanotechnology 19:085707. doi:10.1088/0957-4484/19/8/085707

Fang SW, Li CF, Shih DYC (1994) Antifungal activity of chitosan and its preservative effect on low-sugar candied kumquat. J Food Prot 57:136–140

Faundez G, Troncoso M, Navarrete P, Figueroa G (2004) Antimicrobial activity of copper surfaces against suspensions of *Salmonella enterica* and *Campylobacter jejuni*. BMC Microbiol 4:1–7

Fortunati E, Peltzer M, Armentano I, Jiménez A, Kenny JM (2013) Combined effects of cellulose nanocrystals and silver nanoparticles on the barrier and migration properties of PLA nanobio-composites. J Food Eng 118:117–124

Freese C, Uboldi C, Gibson MI, Unger RE, Weksler BB, Romero IA et al (2012) Uptake and cytotoxicity of citrate-coated gold nanospheres: comparative studies on human endothelial and epithelial cells. Part Fibre Toxicol 9:23. doi:10.1186/1743-8977-9-23

Friedman M, Juneja VK (2010) Review of antimicrobial and antioxidative activities of chitosans in food. J Food Prot 73:1737–1761

Gao Y, Cranston R (2008) Recent advances in antimicrobial treatments of textiles. Text Res J 87:60–72

Geranio L, Heuberger M, Nowack B (2009) The behavior of silver nano textiles during washing. Environ Sci Technol 43:8113–8118

Gong P, Li H, He X, Wang K, Hu J, Zhang S, Yang X (2007) Preparation and antibacterial activity of Fe3O4@Ag nanoparticles. Nanotechnology 28:604–611

Gu H, Ho PL, Tong E, Wang L, Xu B (2003) Presenting vancomycin on nanoparticles to enhance antimicrobial activities. Nano Lett 3:1261–1263

Haider A, Kwak S, Gupta KC, Kang IK (2015) Antibacterial activity and cytocompatibility of PLGA/CuO hybrid nanofiber scaffolds prepared by electrospinning. J Nanomater 2015:832762. doi:10.1155/2015/832762

Hardy JJE, Hubert S, Macquarrie DJ, Wilson AJ (2004) Chitosan-based heterogeneous catalysts for suzuki and heck reactions. Green Chem 6:53–56

He J, Kunitake T, Nakao A (2003) Facile in situ synthesis of noble metal nanoparticles in porous cellulose fibers. Chem Mater 15:4401–4406

He L, Liu Y, Mustapha A, Lin M (2011) Antifungal activity of Zinc oxide nanoparticles against *Botrytis cinerea* and *Penicillium expansum*. Microbiol Res 166:207–215

Herrero AM, Carmona P, Jiménez-Colmeneroa F, Ruiz-Capillas C (2014) Polysaccharide gels as oil bulking agents: technological and structural properties. Food Hydrocoll 36:374–381

Hollinger MA (1996) Toxicological aspects of topical silver. Crit Rev Toxicol 26:255–260

Honary S, Ghajar K, Khazaeli P, Shalchian P (2011) Preparation, characterization and antibacterial properties of silver-chitosan nanocomposites using different Molecular weight grades of chitosan. Trop J Pharm Res 10:69–74

Innovation N, Island R (2010) Ion release kinetics and particle persistence in aqueous nano-silver colloids. Environ Sci Technol 21:69–75

Jain RA (2000) The manufacturing techniques of various drug loaded biodegradable poly (lactide-co-glycolide) (PLGA) devices. Biomaterials 2:2475–2490

Jiang J, Oberdorster G, Biswas P (2009) Characterization of size, surface charge, and agglomeration state of nanoparticle dispersion for toxicological studies. J Nanopart Res 11:77–89

Jo Y K, Kim BH, Jung GH (2009) Antifungal activity of silver ions and nanoparticles on phytopathogenic fungi. Plant Dis 93:1037–1043

Joselito D, Soytong K (2014) Construction and characterization of copolymer nanomaterials loaded with bioactive compounds from Chaetomium species. J Agric Technol 10:823–831

Joseph S, Mathew B (2014) Microwave assisted biosynthesis of silver nanoparticles using the rhizome extract of *Alpinia galanga* and evaluation of their catalytic and antimicrobial activities. J Nanopart 2014:1–9

Judet-Correia D, Charpentier C, Bensoussan M, Dantigny P (2011) Modeling the inhibitory effect of copper sulfate on the growth of *Penicillium expansum* and *Botrytis cinerea*. Lett Appl Microbiol 53:558–564

Kalita M, Basel MT, Janik K, Bossmann SH (2009) Optical and electronic properties of metal and semiconductor nanostructures. In: Nanoscale materials in chemistry. Wiley, London, pp 537–578. doi:10.1002/9780470523674.ch16

Kaur P, Thakur R, Barnela M, Chopra M, Manujaand A, Chaudhury A (2015) Synthesis, characterization and *in vitro* evaluation of cytotoxicity and antimicrobial activity of chitosan-metal nanocomposites. J Chem Technol Biotechnol 90:867–873

Kavitha K, Sutha S, Rajendran V, Prabhu M, Jayakumar T (2013) In situ synthesized novel biocompatible titania-chitosan M nanocomposites with high surface area and antibacterial activity. Carbohydr Polym 93:731–739

Khalili ST, Mohsenifar A, Beyki M, Zhaveh S, Rahmani-Cherati T, Abdollahi A, Bayat M, Tabatabaei M (2015) Encapsulation of thyme essential oils in chitosan-benzoic acid nanogel with enhanced antimicrobial activity against *Aspergillus flavus*. LWT Food Sci Technol 60:502–508

Khodakovskaya MV, Kim BS, Kim JN, Alimohammadi M, Dervishi E, Mustafa T, Cernigla CE (2013) Carbon nanotubes as plant growth regulators: effects on tomato growth, reproductive system, and soil microbial community. Small 9:115–123

Kim CH, Choi JW, Chun HJ, Choi KS (1997) Synthesis of chitosan derivatives with quaternary ammonium salt and their antibacterial activity. Polym Bull 38:387–393

Kim KJ, Sung WS, Moon WS, Choi JS, Kim JG, Lee DG (2008) Antifungal effect of silver nanoparticles on dermatophytes. J Microbiol Biotechnol 18:1482–1490

Kim SW, Kim KS, Lamsal K, Kim YJ, Kim SB, Jung M, Sim SJ, Kim HS, Chang SJ, Kim JK, Lee YS (2009) An *in vitro* study of the antifungal effect of silver nanoparticles on oak wilt pathogen Raffaelea sp. J Microbiol Biotechnol 19:760–764

Kong M, Chen XG, Xing K, Park HJ (2010) Antimicrobial properties of chitosan and mode of action: a state of the art review. Int J Food Microbiol 144:51–63

Lamsal K, Kim SW, Jung JH, Kim YS, Kim KS, Lee YS (2011) Application of silver nanoparticles for the control of *Colletotrichum* species *in vitro* and pepper Anthracnose disease in field. Mycobiology 39:194–199

Lee HJ, Yeo SY, Jeong SH (2003) Antibacterial effect of nanosized silver colloidal solution on textile fabrics. J Mater Sci 38:2199–2204

Lee HY, Park HK, Lee YM, Kim K, Park SB (2007) A practical procedure for producing silver nanocoated fabric and its antibacterial evaluation for biomedical applications. Chem Commun 28:2959–2961

Li LH, Deng JC, Deng HR, Liu ZL, Xin L (2010) Synthesis and characterization of chitosan/ZnO nanoparticle composite membranes. Carbohydr Res 345:994–998

Li P, Wang Y, Peng Z, She F, Kong L (2011) Effects of starch nanocrystal on structure and properties of waterborne polyurethane-based composites. Carbohydr Polym 85:698–703

Lin D, Xing B (2007) Phytotoxicity of nanoparticles: inhibition of seed germination and root growth. Environ Pollut 150:243–250

Liu XF, Guan YL, Yao KD (2001) Antibacterial action of chitosan and carboxymethylated chitosan. J Appl Polym Sci 79:1324–1335

Liu Y, He L, Mustapha A, Li H, Hu ZQ, Lin M (2009) Antibacterial activities of zinc oxide nanoparticles against *Escherichia coli* O157:H7. J Appl Microbial 107:1193–1201

Llorens A, Lloret E, Picouet P, Fernandez A (2012a) Study of the antifungal potential of novel cellulose/copper composites as absorbent materials for fruit juices. Int J Food Microbiol 158:113–119

Llorens A, Lloret E, Picouet PA, Trbojevich R, Fernandez A (2012b) Metallic-based micro and nanocomposites in food contact materials and active food packaging. Trends Food Sci Technol 24:19–29

Lloret E, Picouet P, Fernåndez A (2012) LWT—Matrix effects on the antimicrobial capacity of silver based nanocomposite absorbing materials. Food Sci Technol 49:333–338

Longano D, Ditaranto N, Cioffi N, Niso FD, Sibillano T, Mezzapesa FP, Ancona A, Conte A, Del Nobile MA, Sabbatini L, Torsi L (2013) One- vs two-step preparation of antimicrobial coatings composed of laser ablated copper nanoparticles and polylactic acid. MRS Proc 1453, mrss12-1453-gg11-03. doi:10.1557/opl.2012.1044.

Mahapatra SS, Karak N (2008) Silver nanoparticle in hyperbranched polyamine: synthesis, characterization and antibacterial activity. Mater Chem Phys 112:1114–1119

Maneerung T, Tokura S, Rujiravanit R (2008) Impregnation of silver nanoparticles into bacterial cellulose for antimicrobial wound dressing. Carbohydr Polym 72:43–51. doi:10.1016/j.carbpol.2007.07.025

Manikandan A, Sathiyabama M (2015) Green synthesis of copper-chitosan nanoparticles and study of its antibacterial activity. J Nanomed Nanotechnol 6:1–5

Maynard AD (2006) Nanotechnology: risks. Nano Toady 1:22–33

McKenna KP (2009) Unique bonding in nanoparticles and powders. In: Nanoscale materials in chemistry. Wiley, London, pp 15–36. doi:10.1002/9780470523674.ch2

Mehtar S, Wiid I, Todorov SD (2008) The antimicrobial activity of copper and copper alloys against nosocomial pathogens and *Mycobacterium tuberculosis* isolated from healthcare facilities in the Western Cape: an *in-vitro* study. J Hosp Infect 68:45–51

Mikos A, Lyman M, Freed L, Langer R (1994) Wetting of poly (L-lactic acid) and poly (D, L-lactic-co-glycolic acid) foams for tissue culture. Biomaterials 15:55–58

Min JS, Kim KS, Kim SW, Jung JH, Lamsal K, Kim SB, Jung M, Le YS (2009) Effects of colloidal silver nanoparticles on sclerotium-forming phytopathogenic fungi. Plant Pathol J 25:376–380

Miranda SP, Miranda E, Serrano J (2007) Chitosan and gamma irradiated chitosan against *Aspergillus flavus* in maize. Asian Chitin J 3:37–47

Mohan R, Shanmugharaj AM, Hun RS (2011) An efficient growth of silver and copper nanoparticles on multiwalled carbon nanotube with enhanced antimicrobial activity. J Biomed Mater Res B 96:119–126

Morones JR, Elechiguerra JL, Camacho A, Holt K, Kouri JB, Ramírez JT, Yacaman JM (2005) The bactericidal effect of silver nanoparticles. Nanotechnology 16:2346–2353

Mu Q, Hondow NS, Krzemiński Ł, Brown AP, Jeuken LJ, Routledge MN (2012) Mechanism of cellular uptake of genotoxic silica nanoparticles. Part Fibre Toxicol 9:29. doi:10.1186/1743-8977-9-29

Murariu M, Doumbia A, Bonnaud L, Dechief AL, Paint Y, Ferreira M, Campagne C, Devaux E, Dubois P (2011) High-performance polylactide/ZnO nanocomposites designed for films and fibers with special end-use properties. Biomacromolecules 12:1762–1771

Najafzadeh H, Ghorbanpour M, Hekmati-Moghaddam SH, Karimiyan A (2015) Antifungal effect of magnesium oxide, zinc oxide, silicon oxide and copper oxide nanoparticles against *Candida albicans*. Zahedan J Res Med Sci 17, e2179

Namazi H, Dadkhah A (2008) Surface modification of starch nanocrystals through ring-opening polymerization of epsilon-caprolactone and investigation of their micro structures. J Appl Polym Sci 110:2405–2412

Namazi H, Mosadegh M (2011) Bio-nanocomposites based on naturally occurring common polysaccharides chitosan, cellulose and starch with their biomedical applications. In: Tiwari A (ed) Recent developments in bio-nanocomposites for biomedical applications. pp 379–397

Nan L, Yang W, Liu Y, Hui X, Ying L, Manqi L, Ke Y (2008) Antibacterial mechanism of copper-bearing antibacterial stainless steel against *E. coli*. J Mater Sci Technol 24:197–201

Nikitin LN, Vasil'kov AY, Banchero M, Manna L, Naumkin AV, Podshibikhin VL, Abramchuk SS, Buzin MI, Korlyukov AA, Khokhlov (2011) Composite materials for medical purposes based on polyvinylpyrrolidone modified with ketoprofen and silver nanoparticles, Russian J. of Phys Chem A 85, 1190–1195.

Nnemeka I, Sule M, Friday A, Philbus D, Godwin EU, Shola O, Moses O, Rufus S (2014) Rapid synthesis of silver nano particles capped in starch and its anti-mold activity. Int J Innov Sci Res 9:16–25

Palza H (2015) Antimicrobial polymers with metal nanoparticles. Int J Mol Sci 16:2099–2116

Panacek A, Kolar M, Vecerova R, Prucek R, Soukupova J, Krystof V, Hamal P, Zboril R, Kvitek L (2009) Antifungal activity of silver nanoparticles against *Candida* spp. Biomaterials 30:6333–6340

Park HJ, Kim SH, Kim HJ, Choi SH (2006) A new composition of nanosized silica-silver for control of various plant diseases. Plant Pathol J 22:295–302

Park SY, Chung JW, Priestley RD, YeopKwak S (2012) Covalent assembly of metal nanoparticles on cellulose fabric and its antimicrobial activity. Cellulose 19:2141–2215

Park T, Cohen S, Langer R (1992) Poly (L-lactic acid)/pluronic blends: characterization of phase separation behavior, degradation, morphology and as protein releasing matrices. Macromolecules 25:116–122

PCÆFVDK Æ, Hofmann MBÆT (2008) Nanoparticles: structure, properties, preparation and behaviour. Environ Media 17:326–343

Perelshtein I, Applerot G, Perkas N, Guibert G, Mikhailov S, Gedanken A (2008) Sonochemical coating of silver nanoparticles on textile fabrics (nylon, polyester and cotton) and their antibacterial activity. Nanotechnology 19:245705–245711

Perelshtein I, Ruderman E, Perkas N, Tzanov T, Beddow J, Joyce E, Mason TJ, Blanes M, Molla K, Patlolla A, Frenkele AI, Gedanken A (2013) Chitosan and chitosan–ZnO-based complex nanoparticles: formation, characterization, and antibacterial activity. J Mater Chem B 1:1968–1976

Petrak K (1990) Polymers for use in drug delivery. Br Polym J 22:213–219

Poosti M, Ramazanzadeh B, Zebarjad M, Javadzadeh P, Naderinasab M, Shakeri MT (2013) Shear bond strength and antibacterial effects of orthodontic composite containing TiO_2 nanoparticles. Eur J Orthod 35:676–679

Prasad R, Swamy VS, Varma A (2012) Biogenic synthesis of silver nanoparticles from the leaf extract of *Syzygium cumini* (L.) and its antibacterial activity. Int J Pharm Bio Sci 3(4):745–752

Prasad R, Swamy VS (2013) Antibacterial activity of silver nanoparticles synthesized by bark extract of *Syzygium cumini*. J Nanopart 2013, 431218. doi:10.1155/2013/431218

Prasad R (2014) Synthesis of silver nanoparticles in photosynthetic plants. J Nanopart 2014, 963961. doi:10.1155/2014/963961

Prasad R, Kumar V, Prasad KS (2014) Nanotechnology in sustainable agriculture: present concerns and future aspects. Afr J Biotechnol 13(6):705–713

Prasad R, Pandey R, Barman I (2015) Engineering tailored nanoparticles with microbes: *quo vadis*? WIREs Nanomed Nanobiotechnol 8(2):316–330. doi:10.1002/wnan.1363
Prasad R, Pandey R, Varma A, Barman I (2016) Polymer based nanoparticles for drug delivery systems and cancer therapeutics. In: Kharkwal H, Janaswamy S (eds) Natural polymers for drug delivery. CABI, UK (in Press)
Qi L, Xu Z, Jiang X, Hu C, Zou X (2004) Preparation and antibacterial activity of chitosan nanoparticles. Carbohydr Res 339:2693–2700
Quaranta D, Krans T, Espirito Santo C, Elowsky CG, Domaille DW, Chang CJ, Grass G (2011) Mechanisms of yeast contact-killing on dry metallic copper surfaces. Appl Environ Microbiol 77:416–426
Rabea EI, Badawy MET, Stevens CV, Smagghe G, Steurbaut W (2003) Chitosan as antimicrobial agent: applications and mode of action. Biomacromolecules 4:1457–1465
Raemdonck K, Demeester J, De Smedt S (2009) Advanced nanogel engineering for drug delivery. J Soft Mater 5:707–715
Ragauskas AJ, Zhang AJ, Elder TJ, Pu YQ (2007) Facile synthesis of spherical cellulose nanoparticles. J Carbohydr Polym 69:607–611
Saharan V, Mehrotra A, Khatik R, Rawal P, Sharma SS, Pal A (2013) Synthesis of chitosan based nanoparticles and their in vitro evaluation against phytopathogenic fungi. Int J Bio Macromol 62:677–683
Raghupathi KR, Koodali RT, Manna AC (2011) Size-dependent bacterial growth inhibition and mechanism of antibacterial activity of zinc oxide nanoparticles. Langmuir 27:4020–4028
Raji V, Chakraborty M, Parikh PA (2012) Synthesis of starch-stabilized silver nanoparticles and their antimicrobial activity. Particulate Sci Technol 30:565–577
Raveendran P, Fu J, Wallen SL (2003) Completely green synthesis and stabilization of metal nanoparticles. J Am Chem Soc 125:13940–13941
Raveendran P, Fu J, Wallen SL (2006) A simple and green method for the synthesis of Au, Ag, and Au-Ag alloy nanoparticles. Green Chem 8:34–38
Ravi-Kumar MNV (2000) A review of chitin and chitosan applications. React Funct Polym 46:1–27
Ray SS (2013) Environmentally friendly polymer nanocomposites: types, processing and properties. Woodhead Publishing, Cambridge
Reddy MVB, Arul J, Ait-Barka E, Angers P, Richard C, Castaigne F (1998) Effect of chitosan on growth and toxin production by *Alternaria alternata* f. sp. *lycopersici*. Biocontrol Sci Technol 8:33–43
Ren GG, Hu D, Cheng EWC, Vargas-Reus MA, Reip P, Allaker R (2009) Characterisation of copper oxide nanoparticles for antimicrobial applications. Int J Antimicrob Agents 33:587–590
Rico CM, Majumdar S, Duarte-Gardea M, Peralta-Videa J, Gardea-Torresdey JL (2011) Interaction of nanoparticles with edible plants and their potential interactions in the food chain. J Agric Food Chem 59:3485–3498
Rinaudo M (2006) Chitin and chitosan: properties and applications. Prog Polym Sci 31:603–632
Romanska D, Mazur M (2003) Electrochemical preparation of thiol-coated silver nanostructures on highly oriented pyrolytic graphite. Langmuir 19:4532–4534
Rubina MS, Kamitov KK, Zubavichus YV, Peters GS, Naumkin AV, Suzer S, Vasil'kov AY (2015) Collagen-Chitosan Scaffold modified with Au and Ag nanoparticles: synthesis and structure. Appl. Surf. Sci, 2016, DOI:10.1016/j.apsusc.2016.01.107
Sanpui P, Murugadoss A, Prasad PVD, Ghosh SS, Chattopadhyay A (2008) The antibacterial properties of a novel chitosan-Ag-nanoparticle composite. Int J Food Microbiol 124:142–147
Sautista-Banos S, Hernandez-Lauzardo AN, Velazquez-Del Valle MG, Hernandez-Lopez M, Barka E, Bosquez-Molina E, Wilson CL (2006) Chitosan as a potential natural compound to control pre and postharvest diseases of horticultural commodities. Crop Prot 25:108–118
Savic R, Luo L, Eisenberg A, Maysinger D (2003) Micellar nanocontainers distribute to defined cytoplasmic organelles. Science 300:615–618
Sawai J, Yoshikawa T (2004) Quantitative evaluation of anti-fungal activity of metallic oxide powders (MgO, CaO and ZnO) by an indirect conductimetric assay. J Appl Microbiol 96:803–812

Schmid G, Simon U (2005) Gold nanoparticles: assembly and electrical properties in 1–3 dimensions. Chem Commun 6:697–710

Sebastien F, Stephane G, Copinet A, Coma V (2006) Novel biodegradable films made from chitosan and poly(lactic acid) with antifungal properties against mycotoxinogen strains. Carbohydr Polym 65:185–193

Sepideh H, Ghaseminezhad SM, Shojaosadati SA, Shokrollahzadeh S (2012) Comparative study on silver nanoparticles properties produced by green methods. Iran J Biotechnol 10:191–197

Sescousse S, Gavillon R, Budtova T, (2011) Aerocellulose from cellulose-ionic liquid solutions: preparation, properties and comparison with cellulose-NaOH and srllulose-NMMO routes. Carbohydrate Polym 83, 1766–1774

Shameli K, Ahmad MB, Yunus WMZ, Ibrahim NA, Abdul Rahman R, Jokar M, Darroudi M (2010) Silver/poly (lactic acid) nanocomposites: preparation, characterization, and antibacterial activity. Int J Nanomed 5:573–579

Shanmugam S, Viswanathan B, Varadarajan TK (2006) A novel single step chemical route for noble metal nanoparticles embedded organic–inorganic composite films. Mater Chem Phys 95:51–55

Shih CM, Shieh YT, Twu YK (2009) Preparation and characterization of cellulose/chitosan blend films. Carbohydr Polym 78:169–174

Siegrist M, Cousin-eve M, Kastenholz H, Wiek A (2007) Public acceptance of nanotechnology foods and food packaging: the influence of affect and trust. Appetite 49:459–466

Simonet BM, Valcárcel M (2009) Monitoring nanoparticles in the environment. Anal Bioanal Chem 393:17–21

Smiechowicz E, Kulpinski P, Niekraszewicz B, Bacciarelli A (2011) Cellulose fibers modified with silver nanoparticles. Cellulose 18:975–985

Sondi I, Sondi B (2004) Silver nanoparticles as antimicrobial agent: a case study on *E. coli* as a model for Gram-negative bacteria. J Colloid Interface Sci 275:177–182

Soytong K, Charoenporn C, Kanokmedhakul S (2013) Evaluation of microbial elicitors to induce plant immunity for tomato wilt. Afr J Microbiol Res 7:1993–2000

Sriram MI, Kanth SB, Kalishwaralal K, Gurunathan S (2010) Antitumor activity of silver nanoparticles in Dalton's lymphoma ascites tumor model. Int J Nanomed 5:753–762

Stashak ST, Farstvedt E, Othis A (2004) Update on wound dressings: indication and best use. Clin Tech Equine Pract 3:148–163

Sundaresan K, Sivakumar A, Vigneswaran C, Ramachandran T (2012) Influence of nano titanium dioxide finish, prepared by sol-gel technique, on the ultraviolet protection, antimicrobial, and self-cleaning characteristics of cotton fabrics. J Ind Text 41:259–277

Supp AP, Neely AN, Supp DM, Warden GD, Boyce S (2005) Evaluation of cytotoxicity and antimicrobial activity of acticoat burn dressing for management of microbial contamination in cultured skin substitutes grafted to athymic mice. J Burn Care Rehabil 26:238–246

Swamy VS, Prasad R (2012) Green synthesis of silver nanoparticles from the leaf extract of *Santalum album* and its antimicrobial activity. J Optoelectron Biomed Mater 4(3):53–59

Tang J, Song Y, Tanvir S, Anderson WA, Berry RM, Ta KC (2015) Polyrhodanine coated cellulose nanocrystals: a sustainable antimicrobial agent. ACS Sustain Chem Eng 3:1801–1809

Taylor MS, Daniels AU, Andriano KP, Heller J (1994) Six bioabsorbable polymers: in vitro acute toxicity of accumulated degradation products. J Appl Biomater 5:151–157

Tiede K, Hassellöv M, Breitbarth E, Chaudhry Q, Boxall ABA (2009) Considerations for environmental fate and ecotoxicity testing to support environmental risk assessments for engineered nanoparticles. J Chromatogr A 1216:503–509

Toksha BG, Shirsath SE, Patange SM, Jadhav KM (2008) Structural investigations and magnetic properties of cobalt ferrite nanoparticles prepared by sol-gel auto combustion method. Solid State Commun 147:479–483

Tripathi S, Mehrotra GK, Dutta PK (2009) Physicochemical and bioactivity of cross-linked chitosan–PVA film for food packaging applications. Int J Biol Macromol 45:372–376

Usman MS, El Zowalaty ME, Shameli K, Zainuddin N, Salama M, Ibrahim NA (2013) Synthesis, characterization, and antimicrobial properties of copper nanoparticles. Int J Nanomed 8:4467–4479

Valodkar M, Rathore PS, Jadeja RN, Thounaojam M, Devkar RV, Thakore S (2012) Cytotoxicity evaluation and antimicrobial studies of starch capped water soluble copper nanoparticles. J Hazard Mater 201:244–249
Varlan AR, Sansen W (1996) Nondestructive electrical impedance analysis in fruit: normal ripening and injuries characterization. Electromagn Biol Med 15:213–227
Varma AJ, Deshpande SV, Kennedy JF (2004) Metal complexation by chitosan and its derivatives: a review. Carbohydr Polym 55:77–93
Vasil'kov AY, Rubina MS, Gallyamova AA, Naumkin AV, Buzin MI, Murav'eva JP, (2015) Mesoporic material from microcrystalline cellulose with gold NPs: new approach to obtain metal-carrying polysaccharides. Mendeleev Comm 25, 358–360
Vigneshwaran N, Nachane RP, Balasubramanya RH, Varadarajan PV (2006) A novel one-pot 'green' synthesis of stable silver nanoparticles using soluble starch. Carbohydr Res 341:2012–2018
Vigo TL (2001) Antimicrobial polymers and fibers: retrospective and prospective in bioactive fibers and polymers. Am Chem Soc 11:175–200
Vokhidova NR, Sattarov ME, Kareva ND, Rashidova SS (2014) Fungicide features of the nanosystems of silkworm (*Bombyx mori*) chitosan with copper ions. Microbiology 83:751–753
Wang LS, Wang CY, Yang CH, Hsieh CL, Chen SY, Shen CY, Wang JJ, Huang KS (2015) Synthesis and anti-fungal effect of silver nanoparticles-chitosan composite particles. Int J Nanomed 10:2685–2696
Weaver L, Michels HT, Keevil CW (2010) Potential for preventing spread of fungi in air-conditioning systems constructed using copper instead of aluminium. Lett Appl Microbiol 50:18–23
Wheeldon LJ, Worthington T, Lambert PA, Hilton AC, Lowden CJ, Elliott TSJ (2008) Antimicrobial efficacy of copper surfaces against spores and vegetative cells of *Clostridium difficile*: the germination theory. J Antimicrob Chemother 62:522–525
Wilks SA, Michel HT, Keevil CW (2006) Survival of *Listeria monocytogenes* Scott A on metal surfaces: implications for cross-contamination. Int J Food Microbiol 111:93–98
Xue J, Luo Z, Li P, Ding Y, Cui Y, Wu Q (2014) A residue-free green synergistic antifungal nanotechnology for pesticide thiram by ZnO nanoparticles. Sci Rep 4:5408. doi:10.1038/srep05408
Yallapu MM, Reddy MK, Labhasetwar V (2007) Nanogels: chemistry to drug delivery. In: Labhasetwar V, Pelecky LDL (eds) Biomedical applications of nanotech. Wiley, New York, pp 131–172
Yamamoto O (2001) Influence of particle size on the antibacterial activity of zinc oxide. Int J Inorg Mater 3:643–649
Yang FL, Li XG, Zhu F, Lei CL (2009) Structural characterization of nanoparticles loaded with garlic essential oil and their insecticidal activity against *Tribolium castaneum* (Herbst) (Coleoptera:Tenebrionidae). J Agric Food Chem 57:10156–10162
Yang L, Watts DJ (2005) Particle surface characteristics may play an important role in phytotoxicity of alumina nanoparticles. Toxicol Lett 158:122–132
Yang YJ, Chuang CC, Yang HB, Lu CC, Sheu BS (2012) *Lactobacillus acidophilus* ameliorates H. pylori-induced gastric inflammation by inactivating the Smad7 and NFkB pathways. BMC Microbiol 12:38–45
Youssef K, Roberto SR (2014) Applications of salt solutions before and after harvest affect the quality and incidence of postharvest gray mold of 'Italia' table grapes. Postharvest Biol Technol 87:95–102
Yu L, Deana K, Li L (2006) Polymer blends and composites from renewable resources. Prog Polym Sci 31:576–602
Yumak T, Kuralay F, Muti M, Sinag A, Erdem A, Abaci S (2011) Preparation and characterization of zinc oxide nanoparticles and their sensor applications for electrochemical monitoring of nucleic acid hybridization. Colloids Surf B 86:397–403
Zakrzewska A, Boorsma A, Brul S, Hellingwerf KJ, Klis FM (2005) Transcriptional response of *Saccharomyces cerevisiae* to the plasma membrane-perturbing compound chitosan. Eukaryot Cell 4:703–715

Chapter 14
Advances in Bio-Botanicals Formulations with Incorporation of Nanotechnology in Intensive Crop Management

Pinki Bhandari, Megha Pant, P.K. Patanjali, and S.K. Raza

Abstract With changing environmental conditions, biotic and abiotic stresses with direct impact on pest-diseases complexes, the usage of chemical pesticides has increased tremendously in recent times. However, the increasing awareness on environmental and human health safety along with food security concerns has opened up the path for an array of botanical based pesticide formulations. The plant based products with insecticidal activity are well used as active ingredients in different types of new generation formulations as developed at Institute of Pesticide Formulation Technology (IPFT). Various researches on new molecules to combat the different Isotypes of pest population in different geography created a need to strengthen the Formulation technology now days, although various agrochemical companies are indulge to enhance and improvise their products and add significant value to the intensive crop management. Based on research and development at IPFT with use of nanotechnology certain commercially viable factor has been discovered and developed to develop new eco-friendly effective nano based pesticide formulations for the safety of environment and human kind.

1 Introduction

Being a developing country, India exhibits a high resource potential in agriculture as evident with attained status among top three global producers in major crops like wheat, rice and sugarcane. It is acquiescent in respect of chemical pesticides that lead to agriculture revolution has attributed successfully to fulfill food demand for decades. The continuous development and discovery of new chemical molecules acted as backbone to protect the plant health and delivered as high yielder. Chemical protectants and Indigenous knowledge or technologies from rural agriculture has worked together for many years to boost the Indian Agriculture. In attention to pest ecology system, various scientific efforts has been made to reveal the huge plant protectant potential of Biological pesticides and botanicals.

P. Bhandari (✉) • M. Pant • P.K. Patanjali • S.K. Raza
Institute of Pesticide Formulation Technology, Sector-20, Udyog Vihar, Opp. Ambience Mall, Gurgaon, Haryana, India
e-mail: drpbhandari.ipft@gmail.com

R. Prasad (ed.), *Advances and Applications Through Fungal Nanobiotechnology*, Fungal Biology, DOI 10.1007/978-3-319-42990-8_14

1.1 Biopesticides and Intensive Crop Management

The ever increasing problems of bio-accumulation of toxic components, food chain poisoning, increased toxicity levels thereby significantly increased environmental pollution and immense contamination of agricultural commodities with excessive use of pesticides, ill-effects of the pesticide residues to human health have caused to the adoption of better pest management strategies. Hence, the use of plant-based products appeared as an alternative to the synthetic chemical based pesticides (Dayan et al. 2009).

The plant based products fall under the category of biopesticides (Koul and Walia 2009) which include microbial pesticides consisting of bacteria, entomopathogenic fungi or viruses (and sometimes includes the metabolites that bacteria or fungi produce), biochemical pesticides or herbal pesticides are naturally occurring substances that control (or monitor in the case of pheromones) pests and microbial diseases and plant-incorporated protectants (PIPs) have genetic material from other species incorporated into their genetic material (i.e. GM Crops).

The process of mixing a pesticide with other materials to give it certain advantageous properties makes up its "formulation" which includes processes as: physical treatment including grinding and liquefaction, combination with inert materials like talc, silica, diatomaceous earth etc., addition of adjuvants like wetters, spreaders, emulsifiers, stickers, deflocculators and stabilizers for unique properties and combination with other toxic or non-toxic chemicals to increase their effectiveness (synergistic action). The formulations are mainly categorized as conventional and new generation. 'Conventional', refer to 'old technology' or 'classical' or 'traditional' because of their higher dose rate or repeated applications to get desired bioefficacy (Copping and Sugavanam 2000). These higher doses and repeated applications lead to accumulation of pesticide residues in food commodities along with environmental pollution. Conventional formulations, because of their characteristics i.e. dustiness and use of volatile organic solvents (VOCs) in their preparation maximize several problems like pesticide residues in food and finished products etc (Foy and Pritchard 1996). The former are replaced by the water based new generation formulations with a reduced dosage of active ingredients too. They are user and environment friendly, target specific, low mammalian toxicity, pests do not develop resistance to them easily etc (De et al. 2013).

Formulation technology improve operator safety, reduce dose rate and wastage of pesticides applied to crops. Also, it helps in reducing environmental impact and increasing food safety. Some of the possible areas in agrochemical formulation technology are: Water based formulation technology, Floating tablet/spreading formulation technology, Microemulsion gel technologies, Microemulsion technology in mixed formulation, Controlled release/ZW/nano-encapsulation formulation technology/Floating tablet/ spreading formulation technology etc.

2 Biological Pesticide Formulations in India

The total world production of biopesticides have made biopesticide market of 3.6 billion USD in year 2014, which is expected to grow as 6.9 billion USD by year 2019 with 13.9 % CAGR. In India, biopesticide consumption has shown its increased use over the time. In 2005–2006, consumption of biopesticides in India stands at 1920 MT. India has a vast potential for biopesticides. However, its adoption by farmers in India needs education for maximizing gains. Biopesticides represent only 2.89 % (as on 2005) of the overall pesticide market in India and is expected to exhibit an annual growth rate of about 2.3 % in the coming years (Thakore 2006).

In India, bio pesticide market of 0.16 Billion USD with was reported by FICCI in 2014 which is expected to grow upto 0.35 billion USD with 17.3 % CAGR. The major contributing formulation in Indian biological pesticide market is Bt, NPV, neem based pesticides, etc. which have already been registered under Central Insecticide Board, India.

Biopesticides registered under Insecticide Act, 1968 are *Bacillus thuringiensis* var. *israelensis*, *Bacillus thuringiensis* var. *kurstaki*, *Bacillus thuringiensis* var. *galleriae*, *Bacillus sphaericus*, *Trichoderma viride*, *Trichoderma harzianum*, *Pseudomonas fluorescens*, *Beauveria bassiana*, NPV of *Helicoverpa armigera* and *Spodoptera litura*, Neem based pesticides and *Cymbopogon*.

Baculoviruses with rod shaped nucleocapsid and enveloped and contains a single large covalently closed, double stranded DNA genome ranging between 80–180 Kbps (King et al. 2011). As represented by two different phenotypes: the budded viruses (BV) and the occluded virus (OV) (Jehle et al. 2006). Later classification of family baculoviridae further divided into two sub families: the Eubaculovirinae and the Nudibaculovirinae. Burand further classified the Eubaculovirinae into two genera viz. Nuclear Polyhedrosis Virus (NPV) and Granulovirus (GV) while Nudibaculovirinae contained only genus non-occluded baculovirus (NOB); based on lack of occlusion body formation and virion morphology.

These baculoviruses existing with main difference of type I NPVs which contain the fusion protein GP64, actually required for virus entry into the cell and for cell-to-cell transmission, whereas type II NPVs lack GP64 and instead contain a generic fusion protein, called F protein, with the same role as GP64. This protein can also be found in some vertebrate viruses (Beas-Catena et al. 2014). Such a vast array of potential viruses with particle sizes of 1.5–2.00 µm/600 nm may act as a potential tool for the management of most damaging insects through the application of suitable nano- encapsulation of virus particles and application of nanotechnology in form of suitable nano delivery systems for biological viruses to be used in agriculture (Fig. 14.1).

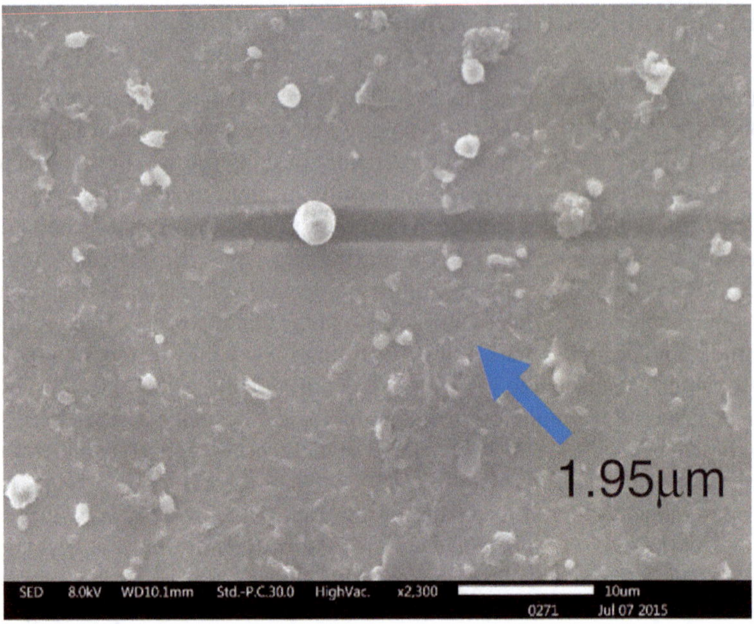

Fig. 14.1 High potential strain of *Spodoptera litura* identified from IPFT field. Attributed to Bhandari et al. (2015)

2.1 Plant Based Products as Green Pesticides Under Development for Nano Application

2.1.1 Biodiesel by-Products

De-oiled seed cake of jatropha showed molluscicidal, insecticidal, fungicidal, and nematicidal activity. De-oiled seed cake of karanja also possesses insecticidal and nematicidal activity (Sharma et al. 2011). Neem and karanja oil cakes in combination also showed insecticidal properties against mosquito vectors (Shanmugasundaram et al. 2008). Aqueous extract of Karanja and Jatropha seedcakes had a significant effect on the termite (Odontotermes obesus) (Sharma et al. 2011). The main constituents of Karanja and Jatropha responsible for insecticidal activity are karanjin and phorbol esters which was 0.18 % and 0–0.8 %, respectively. But there is no relevant information regarding the use of these aqueous extracts to formulate an essential oil to increase their shelf life and effectiveness.

The biodiesel production in India utilizes two main crops, karanja (*Pongamia pinnata*) and jatropha (*Jatropha curcas*) as major source of non-edible oils. Production of biodiesel generates 10 % biodiesel waste liquid by volume approximately as in general 10 kg of crude glycerol is generated as a by-product for every 100 kg of biodiesel produced. The process involves transesterification of vegetable oils with methanol catalyzed by KOH. The technology of biodiesel production consumes only extracted vegetable oil from non-edible seeds and leaves behind a large amount of unutilized mass as seed cake. Majority of industrial products employ only purified glycerol as a raw material, and therefore bio diesel waste liquid is often discarded as a waste product. This liquid biodiesel waste creates disposal problems and in future, liquid biodiesel waste is likely to be produced in large amounts, with detrimental effects on the environment. Hence, there is an urgent need to convert crude glycerol into more valuable products. The self-decomposition of seed cake in the open atmosphere by the action of various micro-organisms generated the gases CH_4, N_2O, H_2S, NH_3 and CO_2 volatile organic compounds (VOCs). The approach of using of non-edible cakes as "Biomass" resources instead of disposing as "waste" made possible as to meet energy and economic benefits and also environmental benefits. Many strategies pertaining to this have been worked out. Commonly, the generation of biogas from these cakes is figured out as a best solution for its efficient utilization.

Biodiesel production produces three major biodiesel waste products: glycerin, methanol, and (sometimes) water (Fig. 14.2).

Composition of biodiesel by product -Glycerol 40–50 %, alcohol 0.2–1 %, Mono-, di- & tri-glycerides 1.5–3 %, Fee fatty acid 1–2 %, Triglyceride ester 0.1–0.2 %, water 0.1–0.3 %. The composition works as an emulsifier in the formulations and helps in better mixing of the ingredients used. Glycerol helps in phase stabilization whereas alcohol acts as a co-surfactant. Mono and di-glycerides function as emulsifiers for oil in water emulsions and tri-glycerides for cream formulations. Free fatty acids are used as emulsion stabilizers and the triglyceride esters as solvents.

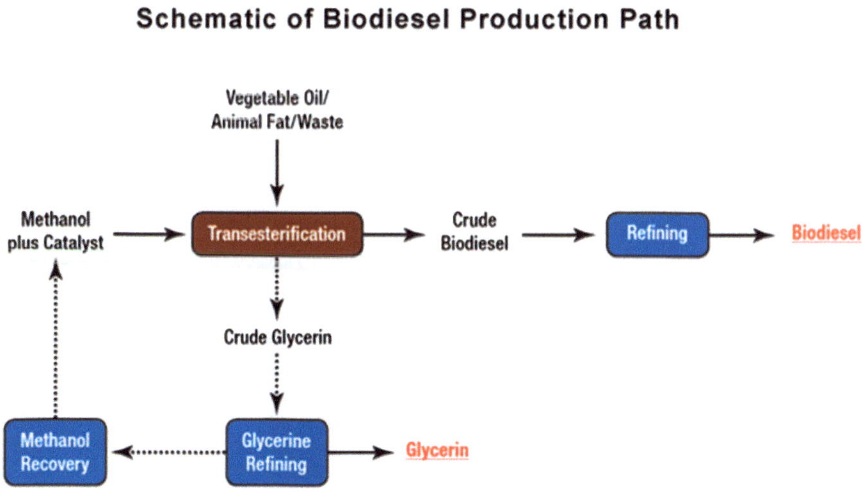

Fig. 14.2 The biodiesel production process

2.1.2 Essential Oils

Essential oils (Fig. 14.3) and their major components are emerging as potential pest control agents because of their insecticidal and repellant properties.

Eucalyptus Oil

Eucalyptus oil has been placed under the GRAS (generally regarded as safe) category by The U.S. Food and Drug Administration and classified as non-toxic (USEPA 1993; Mahboubi et al. 2013). Even the Council of Europe has approved use of eucalyptus oil as a flavoring agent in foods (>5 mg kg^{-1}), candies, and confectionary items (<15 mg kg^{-1}). Vilela et al. (2009) also studied the effect of eucalyptus essential oil by contact and head space volatile assay on food-spoiling fungi. The marker compound of eucalyptus oil responsible for its insecticidal property, which makes 75–80% content of the oil along with some other components (Table 14.1).

The values of the oral and acute LD50 of eucalyptus oil and 1,8-cineole for rats is 4440 mg kg^{-1} body weight (BW) and 2840 mg kg^{-1} (BW), respectively (Regnault-Roger 1997), rendering it much less toxic than pyrethrins (with LD50 values of 350–500 mg kg^{-1} BW) (USEPA 1993) and even technical-grade pyrethrum (LD50 value 1500 mg kg^{-1} BW) (Batish et al. 2008).

Fig. 14.3 Naturally occurring sources of essential oils. *Source*: Google Images

Table 14.1 Composition of Eucalyptus oil (*Eucalyptus globulus*)

Compound identified	Percentage composition
cis-Ocimen	21.33
Camphene	0.21
β-Pinen	1.23
β-Myrcene	1.00
1,8-Cineole*	66.28
4-Terpineol	0.52
α-Terpineol	1.73
α-Terpinol acetate	3.39
Aromadendrene	2.85
Globulol	1.43

*1,8-Cineole is the main marker compund in eucalyptus oil responsible for insecticidal activity

Citronella Oil

Citronella oil is one of the essential oils obtained from the leaves and stems of different species of Cymbopogon (lemongrass). The oil is used extensively as a source of perfumery chemicals such as citronellal, citronellol and geraniol. Citronella oil is also a plant-based insect repellent, and has been registered for this use in the United States since 1948. The United States Environmental Protection Agency considers oil of citronella as a biopesticide with a non-toxic mode of action. Research also shows that citronella oil has strong antifungal properties, and has even been used as a successful spray-on deterrent against pets destroying household items including effectiveness in repelling *Aedes aegypti* (Kim et al. 2005). The US Environmental Protection Agency states that citronella oil has little or no toxicity when used as a topical insect repellent, with no reports of adverse effects of concern over a 60 year period. The US Food & Drug Administration considers citronella oil as generally recognized as safe (GRAS) (USEPA 1993).

Dillapiole

Dillapiole is a phytochemical, monophenols, essential oil, commonly extracted from Dill weed, but can also be found in variety of other plants, *Piper aduncum*, Fennel root etc. (Shulgin and Sargent 1967). Dillapiole has been found to have a significant synergistic action with insecticide like DDT, Carbaryl and Pyrethrin. Dillapiole also shows anti-inflammatory, anti-bacterial, anti-housefly and larvicidal activity against mosquitoes.

3 Biopesticides Preference Over Synthetic Chemical Based Pesticides for Application of Nano Delivery Systems

- As compared with the persistent and highly toxic synthetic chemical pesticides, the bio-pesticides are biodegradable in nature and do not leave any harmful residue.
- These are ecofriendly in contrast to the synthetic ones, which are hazardous to the environment and users.
- These are non-phytotoxic with greater selectivity towards the targeted pest unlike the synthetic pesticides having phyto and mammalian toxicity and a broad spectrum of action.
- The biopesticides have an array of constituents responsible for the insecticidal action, so the chances of pests developing resistance is relatively less, whereas in case of synthetic chemical pesticides, pest resistance and resurgence are quite common.
- Availability of biological components (derived from naturally occurring plants) is relatively easy as compared to the synthetic chemicals.
- The biopesticidal products have a great economic feasibility as low cost infrastructure (equipments, raw materials etc.) is used in their production in contrast to the synthetic products which use high cost machines and other ingredients.
- Bio pesticides are mostly target specific and susceptible to varying environment conditions which mainly attribute to affect the efficacy of these molecules therefore the fast and accurate nano delivery systems may act as boon for eco- friendly bio pesticides.
- Micro-emulsions, floating granules, surface spreading oils, combination formulation with synergistic action and nano-encapsulations will be the future formulation in biological pesticides.

4 Different Types of Plant Based Agrochemical Formulations

4.1 Conventional Formulations

4.1.1 Granules

A granular formulation is a product with a size range from 16 to 60 British Standard BS mesh (250–1000 μm) with at least 90 % of the granules within the specified mesh size range. Granules (Fig. 14.4A) are, therefore, the largest of the solid pesticide formulations (apart from tablets) and their large size virtually eliminates drift leading to much less loss of pesticide than with powder and liquid formulations. Granular formulations are often used as pre-emergence herbicides or as soil insecticides for direct broadcasting to the field. The active ingredient concentration is usually from 1 to 40 % and the granules should be free flowing and should disintegrate in the soil to release the active ingredient (Agrow 1995).

4.1.2 Wettable Powders

Wettable powders (Fig. 14.4B) are finely-divided solid pesticide formulations which are applied after dilution and as a suspension in water. The powders contain dry surfactants as powder wetting and dispersing agents and inert carriers or fillers with the concentration of active up to 50 %. Wettable powders contain many particles of less than 5 μm and all the particles should pass through a 45 μm screen (350 BS Mesh or 325 ASTM Mesh) (Foy et al. 1998).

4.1.3 Dustable Powder

It consists of active ingredients (botanical powders) along with inerts and carriers like china clay. Active ingredient either solid or liquid is gradually added in china clay. After complete addition of the active ingredient it is ground in a mixer to get a uniform composition. It is easy to formulate and use of inhalation during application. Dust formulations (Fig. 14.4C) of spinosad have been found to be effective for the control of some stored pests (Bonjour and Opit 2010). Dustable powder formulations have been developed by IPFT, Gurgaon for the control of stored grain pests and a patent has been filed.

4.1.4 Emulsifiable Concentrate (EC)

They are formulated by dissolving the active ingredient with emulsifying surfactants in an organic solvent. EC (Fig. 14.4D) formulations are easy to use and, when diluted in water, should give a stable "milky" emulsion with very little creaming and no oil separation. Emulsion droplets up to about 10 μm are formed when the product is diluted in water in the spray tank (Knowles 2008).

4.2 Drawbacks of Conventional Formulations

- Granules are considered to be the low strength products and require specialized applicator equipments.
- Hazardous effects of dust and wettable powder during manufacture and application.
- These are difficult to mix in spray tanks and have poor compatibility with other formulations.
- Emulsion stability problems may arise after dilution in case of EC. They may be phytotoxic to crops and use of solvent may harm the rubber/plastic in spray applicators.

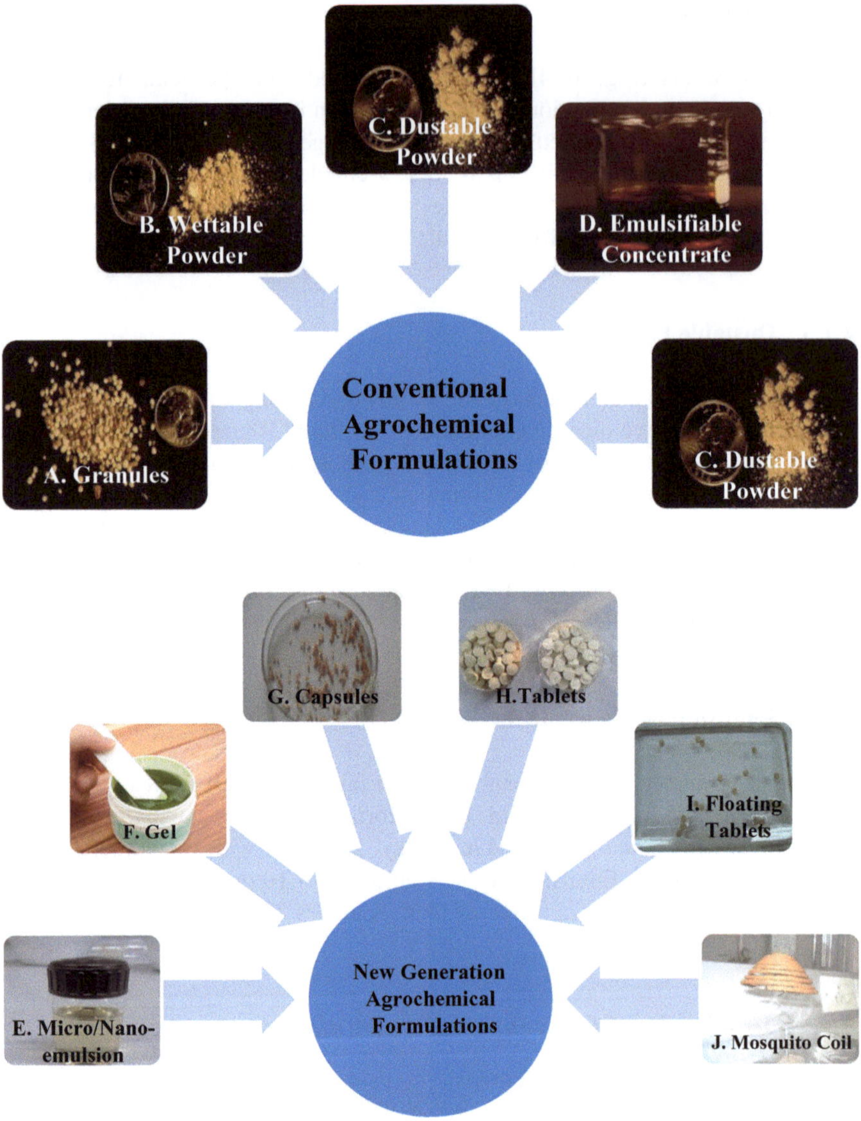

Fig. 14.4 Plant based conventional and nanotechnology based new generation formulations developed at IPFT

5 New Generation Formulations

5.1 *O/W Emulsions (EW)*

Oil-in-water emulsions are now receiving considerable attention because of the need to reduce or eliminate volatile organic solvents (VOCs) for safer handling. A solid active may be dissolved in a water–immiscible solvent. Oil-in-water emulsions (EWs) consist of a dispersion of oil droplets in a continuous aqueous medium (Tadros 1995).

5.2 *Microemulsions*

Microemulsions (Fig. 14.4E) are thermodynamically stable transparent dispersions of two immiscible liquids and are stable over a wide temperature range. They have a very fine droplet size of less than 0.05 μm (50 nm) and consist of three components, namely: oily liquid or solid dissolved in organic solvent, water and surfactant/co-surfactant system These components form a single phase containing relatively large "swollen micelles" (Tadros 1995). Neem oil based microemulsions were successfully developed at IPFT, Gurgaon, India (Singla and Patanjali 2013).

5.3 *Nanoemulsions*

Nanoemulsions are fine oil-in-water dispersions having droplet size ranging from 100 to 600 nm (Solans et al. 2003). They are thermodynamically and kinetically stable with a natural oil and water in combination with a surfactant (Bouchernal et al. 2004). Nanoemulsions using eucalyptus oil were successfully developed against *Tribolium castaneum*, red flour beetle at IPFT, Gurgaon (Pant et al. 2014).

5.4 *Novel Gel Technology*

Gel (Fig. 14.4F) formulations are innovative products, which can be described as thickened ECs packed in water-soluble bags. Gelation of liquid formulations can be brought about by thickening agents such as polyacrylic acids, xanthan gum, silicas, clays, surfactants and combinations thereof.

5.5 Controlled Release Formulations (CRF), Microencapsulation

This technology allows the controlled release of pesticide active ingredients and can reduce product toxicity appreciably as well as reducing leaching from the soil (Ribeiro et al. 2007). The polymer membrane or microencapsulation technique has become popular in recent years which use the principle of interfacial polymerization (Knowles 1998). Interfacial polymerization occurs at the interface where the active component gets encapsulated. The rate of release of the active ingredient can be controlled by adjusting the droplet size, the thickness of the polymer membrane and the degree of cross-linking or porosity of the polymer (Fernández-Pérez 2007). Microcapsules (Fig. 14.4G) using neem seed oil and karanja oil were developed against larvae of *Aedes aegypti* at IPFT, Gurgaon (Pant et al. 2012).

5.6 Tablet Formulations

Tablets (Fig. 14.4H) are emerging type of dry formulations especially useful with compounds that are effective at grams per hectare application rate. The active ingredients are compressed into a solid mass i.e. tablet with the help of a tablet machine. The Basic components of pesticide tablet are active materials, diluents or filler, binder, lubricants and wetting/dispersing agent. Floating tablets (Fig. 14.4I) are used for aquatic insects and pests like mosquito larvae. Effervescent tablets can be developed causing self-dispersion of pesticide. Botanical tablets were successfully developed against *Periplaneta americana* (american cockroach) at IPFT, Gurgaon, India. The technology was patented (Patanjali et al. 2014; Application No.: 2705/DEL/2012).

5.7 Coil Formulations

Coil formulation comprises of botanical active ingredients, inerts: burning materials and binders and preservatives in the proportion: Saw dust (binder) 30–60 %, Jigat (binder) 20–30 %, Guar gum (binder) 5–7 %, Potassium Nitrate (burning) 2–3 %, Sodium Benzoate (preservative) 0.5–1.0 %. These ingredients are properly dried, powdered, mixed well and finally extruded through a coil machine to get the product i.e. coil. Preservatives are used to preserve the repellant properties for a longer period and to increase their shelf life. Mosquito coils (Fig. 14.4J) against the adults of *Aedes aegypti* were successfully developed at IPFT, Gurgaon, India. The technology is patented (Patanjali et al. 2012; Application No.: 365/DEL/2010).

6 Advantages of New Generation Formulations

- Replacement of toxic solvents by water in micro/nanoemulsions and selection of safer solvents.
- Use of safer surfactant components with low toxicity, low skin irritation and enhanced biodegradability in formulations like micro-emulsions, gels, o/w emulsions etc.
- Broad spectrum inerts/fillers can be used in the novel formulations viz. tablets and coils.
- Controlled release formulations can be used with triggered release of active for better performance and shelf life.
- Better surface coverage and target specificity is achieved in case of nanotechnology based pesticide formulations due to size reduction.
- Majority of these formulations have compatibility in tank mixes.
- Sustained and residual bio-efficacy along with increased shelf life.

7 Conclusion

Increasing attention is being paid to the development of safer, effective and more environmentally friendly pesticide formulations to combat the alarming problems of pest-resistance, pest resurgence pesticide residues and contamination of biosphere. This has led to the development of water-based liquid formulations regarded as 'new technology'. This chapter describes the new possible technological developments in pesticide formulations using nano- technology with details on research and development done at Institute of Pesticide Formulation Technology which has evolved new technologies that resulted into improving the shelf life of the product/formulation and increasing its bio-efficacy against the targeted insect pest.

The developed world has progressed substantially in this regard to develop eco-friendly formulations which are safer to food and the environment. Application of nanotechnology in formulation product development will be definitely helpful in decreasing the load of pesticide by decreasing its dose required for pest control/crop protection.

The work carried out at IPFT, greatly emphasizes on the development of environment and user-friendly pesticide formulations, also biodegradable, with the incorporation of latest technologies, and also on their commercialization.

References

Agrow (1995) Methyl Bromide. Environmental Health Criteria 166 International Programme on chemical safety, UNEP/ILO/WHO, World health Organization, Geneva, 324 p

Batish DR, Singh HP, Kohli RK, Kaur S (2008) Eucalyptus essential oil as a natural pesticide. For Ecol Manag 256:2166–2174

Beas-Catena A, Sanchez-Miron A, Garcia-Camacho F, Contraras-Gonez A, Molina-Grima E (2014) Baculovirus biopesticide; an overview. J Anim Plant Sci 24(2):362–373

Bhandari P, Chaudhary S, Nayana ED (2015) Evaluation of viral potential of a new baculovirus strain against cabbage worm. Indian Journal of Entomology (Accepted)

Bonjour EL, Opit GP (2010) Efficacy of dust formulations of spinosad for controlling insects infesting stored wheat. In: 10th International working conference on stored product protection, 425:815–818, Julius-Kuhn-Archiv

Bouchernal K, Brianeon S, Perrier E, Fessi H (2004) Nano-emulsion formulation using spontaneous emulsification, solvent oil and surfactant optimization. Int J Pharmacol 280:241–251

Copping LG, Sugavanam B (eds) (2000) Crop protection chemicals—present developments and future prospects in the 21st century. China Agriculture Press, Beijing

Dayan FE, Cantrell CL, Duke SO (2009) Natural products in crop protection. Bioorg Med Chem 17:4022–4034

De A, Bose R, Kumar A, Mozumdar S (2013) Targeted delivery of pesticides using biodegradable polymeric nanoparticles. Springer Science & Business Media, New Delhi

Fernåndez-Pérez M (2007) Controlled release systems to prevent the agro-environmental pollution derived from pesticide use. J Environ Sci Health B 42:857–862

Foy CL, Pritchard DW, Beestman GB (1998) Formulation Science, vol 1. Proceedings of Formulations Forum'97, Las Vegas, NV, Published by The Association of Formulation Chemists, USA

Foy CL, Pritchard DW (1996) Pesticide formulation and adjuvant technology. CRC Press, Boca Raton, FL

Jehle JA, Blissard GW, Bonning BC, Cory JS, Herniou EA, Rohrmann GF, Theilmann DA, Thiem SM, Vlak JM (2006) On the classification and nomenclature of baculoviruses: a proposal for revision. Arch Virol 151:1257–1266

Kim JK, Kang CS, Lee JK, Kim YR, Han HY, Yun HK (2005) Evaluation of repellency effect of two natural aroma mosquito repellent compounds, citronella and citronellal. Entomol Res 35:117–120

King AMQ, Adams MJ, Carstens EB, Lefkowitz EJ (2011) Virus taxonomy ninth report of the International Committee on Taxonomy of Viruses. Elsevier Academic Press, London, San Diego

Knowles A (2008) Recent developments of safer formulations of agrochemicals. Environmentalist 28:35–44

Knowles DA (1998) Chemistry and technology of agrochemical formulations. Kluwer Academy Publishers, Boston, pp 158–175

Koul O, Walia S (2009) Comparing impacts of plant extracts and pure allelochemicals and implications for pest control. CAB Rev Perspect Agric Vet Sci Nutr Nat Res 4(049):1–30

Mahboubi M, Kazempour N, Valian M (2013) Antimicrobial activity of natural respitol-B and its main components against poultry microorganisms. Pak J Biol Sci 16:1065–1068

Pant M, Dubey S, Patanjali PK, Naik SN, Sharma S (2014) Insecticidal activity of eucalyptus oil nanoemulsion with karanja and jatropha aqueous filtrates. Int Biodeterior Biodegrad 91:119–127

Pant M, Saurabh D, Raza SK, Patanjali PK (2012) Encapsulation of neem and karanja oil mixture for synergistic as well as larvicidal activity for mosquito control. J Sci Ind Res 71:348–352

Patanjali PK, Aggarwal A, Dubey S, Chauhan M, Babu R (2012) A novel synergistic mosquito repellent composition for preparation of mosquito coils. Patent Filed, Application No: 365/DEL/2010

Patanjali PK, Dubey S, Pant M, Raza SK, Naik SN (2014) Insecticidal compositions for controlling household pests. Patent Filed, Application No: 2705/DEL/2012

Regnault-Roger C (1997) The potential of botanical essential oils for insect pest control. Integr Pest Manag Rev 2:25–34

Ribeiro ML, Lourencetti C, Pereira SY, Marchi MRR (2007) Groundwater contamination by pesticides: preliminary evaluation. Quim Nova 30:688–694

Shanmugasundaram R, Jeyalakshmi T, Dutt MS, Murthy PB (2008) Larvicidal activity of neem and karanja oil cakes against mosquito vectors, *Culex quinquefasciatus* (say), *Aedes aegypti* (L.) and *Anopheles stephensi* (L.). J Environ Biol 29:43–50

Sharma S, Verma M, Prasad R, Yadav D (2011) Efficacy of non-edible oil seed cakes against termites, *Odontotermes obesus*. J Sci Ind Res 70:1037–1041

Shulgin AT, Sargent T (1967) Psychotropic phenylisopropylamines derived from apiole and dillapiole. Nature 215:1494–1495

Singla M, Patanjali PK (2013) Phase behaviour of neem oil based microemulsion formulations. Ind Crop Prod 44:421–426

Solans C, Esquena J, Forgiarini A, Uson N, Morales D, Izquierdo P, Azemar N, Garcia-Celma MJ (2003) Nano-emulsions: formation, properties, and applications. In: Mittal KL, Shah Dinesh O (eds) Adsorption and aggregation of surfactants in solution. Marcel Dekker, New York, NY, pp 525–554

Tadros TF (1995) Surfactants in agrochemicals. Marcel Dekker, New York, NY

Thakore Y (2006) The biopesticide market for global agricultural use. Ind Biotechnol 2(3):194–208

USEPA (United States Environment Protection Agency) (1993) R.E.D. FACTS: flower and vegetables oil. http://www3.epa.gov/pesticides/chem_search/reg_actions/reregistration/fs_G-114_1-Dec-93.pdf

Vilela GR, de Almeida GS, D'Arce MABR et al (2009) Activity of essential oil and its major compound, 1,8-cineole, from *Eucalyptus globulus* Labill., against the storage fungi *Aspergillus flavus* link and *Aspergillus parasiticus* Speare. J Stored Prod Res 45:108–111

Chapter 15
Nano-Biofungicides: Emerging Trend in Insect Pest Control

Atanu Bhattacharyya, Prabha Duraisamy, Marimuthu Govindarajan, Abdul A. Buhroo, and Ram Prasad

Abstract Application of pesticides, which encompasses insecticides, fungicides, herbicides, nematicides etc., being used for plant defence mechanisms, embarrasses pest organisms. Although, pesticides eliminate the problem of pests, the use of synthetic pesticides has resulted in unsustainable management of our soil resources. This can be explained as due to the development of resistance by the pest organisms on continuous exposure to the pesticides, thus posing a challenge which leads to development of new classes of pesticides. These pesticides, apart from targeting the pest organisms, causes undesirable effects to all matrices of the environment-viz, soil, water, air, biota etc., Hence, the need for the development of ecofriendly pesticides becomes immediate inevitability. However, there is no one single method for efficient command on insect pests. Among the various classes of pesticides, fungicides form a major group of domineering plant diseases of fungal origin, either by inhibiting the growth of the fungi or by complete biocidal activity. The significance of fungicides is due to the fact that fungal diseases stands first in crop thrashing world wide. Currently, apart from the existing synthetic fungicides in the market, biofungicides occupies an unique position in controlling target diseases of fungal

A. Bhattacharyya (✉)
Department of Biomedical Engineering, Nanotechnolgy Section, Rajiv Gandhi Institute of Technology and Research Centre, Cholanagar, R.T. Nagar Post, Hebbal, Bangalore 560032, India
e-mail: atanubhatt@rediffmail.com

P. Duraisamy
Department of Environmental Sciences, Bharathiar University, Coimbatore 641046, Tamilnadu, India

M. Govindarajan
Unit of Vector Control, Phytochemistry and Nanotechnology, Department of Zoology, Annamalai University, Annamalai Nagar, Chidambaram, Tamil Nadu 608002, India

A.A. Buhroo
Section of Entomology, P.G. Department of Zoology, University of Kashmir, Hazratbal, Srinagar 190006, India

R. Prasad (✉)
Amity Institute of Microbial Technology, Amity University, Noida 201303, Uttar Pradesh, India
e-mail: rprasad@amity.edu

R. Prasad (ed.), *Advances and Applications Through Fungal Nanobiotechnology*, Fungal Biology, DOI 10.1007/978-3-319-42990-8_15

origin. An advanced technology in the field of biopesticides is the employment of engineered nanomaterials. These nanomaterials are more reactive can conjugate with biofungicides forming covalent bonds. This unique property of nanomaterials are exploited to manage the plant-pest chain. Therefore, this manuscript focus on nanobiofungicides as a powerful alternative for eco-friendly management of insect pests, in the coming decade.

1 Introduction

Biofungicides denotes the class of fungicides which have been derived from a biological organism inclusive of, but not constrained to bacteria, fungi, animal or plant and plant products. A consortium of beneficial fungi and bacteria that can colonize and defeat plant pathogens can be very effective against plant pathogens, thereby thwarting the diseases they cause. As these microorganisms are found in the soil, they offer an eco-friendly way of suppressing insect pests. The mechanism of biocidal activity of the biofungicides are different depending on the microorganisms employed-viz, rhizosphere competence, parasitism, antibiosis, inducing metabolic changes, stimulating plant growth etc. (Saraf et al. 2014; Shrivastava et al. 2014). Although integrated pest management, which includes processes like, crop rotation, selection of disease-tolerant or disease-resistant cultivars (cultivars genetically less susceptible than other cultivars) (Francis and Keinath 2010) is considered efficient in inspecting plant diseases, fungicides also play a significant role in plant disease supervision as fungal infection has been shown to be the major cause of economic loss in agriculture and crops of economic importance (Francis and Keinath 2010; Gawai 2015). *Bacillus pumilus*, *Bacillus subtilis*, *Pseudomonas syringe*, *Coniothyrium minitans*, *Streptomyces griseoviridis*, *Trichoderma harzianum*, *Streptomyce lydicus*, *Gliocladium virens*, are a few examples of microorganisms that play an ingredient in plant disease supervision (Francis and Keinath 2010). A wide variety of chemical compounds are being used as antimicrobial agents to inhibit pathogenic plant fungi (Gawai 2015). In view of the increased hazard put forward by the synthetic pesticides, thus nanotechnology is being explored for their possible controlling phenomena of numerous plant diseases. The role of nanotechnology in plant and soil systems demonstrates that nanomaterials may help in the controlled release of agrochemicals for sustenance and defence against insect pests and pathogens of plants (Sekhon 2014) (Fig. 15.1).

2 Biopesticides

Biopesticides refers to certain types of pesticides that may have their origin from any biological form like animals, plants, bacteria etc. their non-toxic mechanism of action projects them as a viable alternative towards sustainable insect pest control

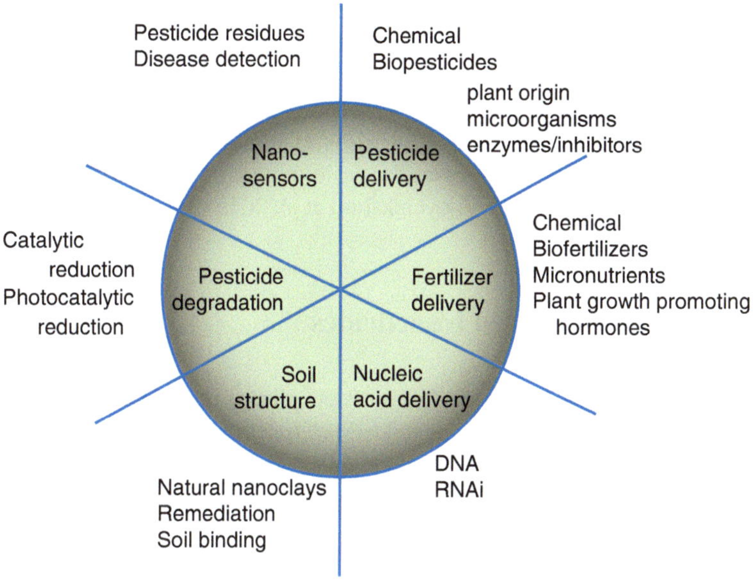

Fig. 15.1 Applications of nano-biotechnology in plant protection and nutrition (adopted from Ghormade et al. 2011)

(Usta 2013). Microbial bio-pesticides are designed based on the pathogenicity of a microorganism towards a target pest organism. The antimicrobial metabolites of some species of bacteria belonging to the genus *Bacillus* sp. and *Pseudomonas* sp. have been demonstrated to be effective bio-control agents. Russia, Australia and the United States, etc. are marching towards new fungicide development. However, only a few entomic-pathogens have been developed as bio-control agents. *Trichoderma* species is a potential candidate which act as a safe microbial pesticide. China, Russia, Belarus and to a lesser extent India and Thailand, is now becoming an imperative producer of *Bacillus thuringiensis* products which are used at length. Several studies have demonstrated the insecticidal action of Bt. delta-endotoxin Cry1C, which carry out bacterium bio-pesticides potency (Leng et al. 2011; Prasad et al. 2014).

3 Biopesticide and Its Mode of Action

Bio-pesticides are microorganisms and their substances produced by microorganisms. The ingredient of microorganisms performing as a biopesticide which should be registered and can be used as a crop protection product. Plant stimulates microorganisms by the ascent of sap procedure. Furthermore, it is very essential to know that the different modes of action of the bio-pesticide at the time of fungi-host interaction. This process exists a competition with several microorganisms. The effect of a bio-pesticide is an amalgamation of different modes of functions. *Trichoderma harzianum*, for example exhibits a competition with antibiosis and there by induces resistance in the plants. The current scenario of January, 2015 can be demonstrated that most of the tomato and sweet pepper growers used integrated control for most insect pests and plant diseases, almost year-round. It may be consider that eight new bio-pesticides were used as test cases for easier registration in a 'Green Deal process' (Gupta and Dikshit 2010; Vijayalakshmi et al. 2015).

4 Nanomaterials-Based Insecticides

Nanotechnology is a promising field of interdisciplinary research extending a broad spectrum of opportunities in various fields like insecticides, pharmaceuticals, electronics, agriculture etc (Bhattacharyya et al. 2015; Prasad et al. 2014). The potential uses and benefits of nanotechnology are enormous which also includes supervision of insect pests through formulation of nanomaterials-based insecticides. Conventional strategies like integrated pest management used in agriculture are inadequate and hence solicit for prevention of chemical pesticide application which have adverse impacts on animals and mammalian system (Bhattacharyya et al. 2015). Therefore, it is considered that nanotechnology can provide a green and efficient alternative for the management of insect pests in agriculture without disturbing the natural harmony (Ragaei and Sabry Al-Kazafy 2014). The pathogenic fungi such as *Rhizoctonia solani*, *Fusarium* spp., *Phythophora* spp., infects whole plants whereas species such as *Botrytis cinerea* infects only the green parts and fruit tissues (Abd-Elsalam and Alghuthaymi 2015). Development of nanopesticide, however, is still in initial stage. Expansion option of eco-safe antifungal agent like bio-based nanomaterials is the immediate need of the hour to defend fungal pathogens. In this aspect, nanotechnology coupled with biological mechanisms can prove to be an effective technology in agricultural pest management. The expression nano-fungicides is used to explain any fungicide formulation which includes, organic ingredients like, active ingredients, polymer-based inorganic silica nanoparticles, titanium dioxide, nanoemulsions and nanoclays in various forms (Lee et al. 2013; Nuruzzaman et al. 2016). Further, the self regulating ability of the carriers aids in delivering the required dose of pesticides into the plant tissue, thus minimizing pesticide use (Sharon et al. 2010). Therefore, formulation of nano-biofungicide is under process which enables the microorganism to exhibit its biocidal activity on the pathogen, only when the biopesticide is within the target organism. This is made possible by the self-regulation capacity of the nanomaterial (Nuruzzaman et al. 2016).

The commercially available "Diyarex Gold" can be shown as a best example for biofungicide, developed and processed through molecular nanotechnology by R.V. Agri Corporation. Diyarex Gold has been demonstrated to be an efficient bactericide and fungicide which are attributed as due to the natural components that make up the product. It has been certified that the product can be used as an eco-friendly bio-fungicide that leaves no residue in the plants or their tissues (Certified by Biocert India) thus rendering safety to humans and the environment, especially non toxic to economically important insect communities like bees etc. Diyarex Gold is the new generation of bio-fungicides used in the control of powdery mildews, downy mildews, rust and early and late blight diseases in vegetables, herbs, grapevines and orchards etc.

4.1 Nanomaterials in Agriculture

Application of biosynthesized nanoparticles in agriculture aims towards sustainable development (Fig. 15.2). Nanomaterials are carrier of agrochemicals (Bhattacharyya 2009a), which facilitating targeted delivery of essential nutrients, thus enhancing growth and yield of the host. They also act as nano-biosensors for crop protection (Singh et al. 2015a). Moreover, much importance is given to the recent development in plant science pertaining to nano-biotechnology that focuses on agricultural practice, plant growth and yield etc. (Bhattacharyya et al. 2009b, 2015). Bio-reduction of nanomaterials- may be obtained from either in vitro or in vivo process. Enzymes, proteins, sugars, and phytochemicals like flavonoids, phenolics, terpenoids, cofactors, mostly act as reducing and stabilizing agents for synthesis of nanoparticles (Prasad 2014). Synthesis of TiO_2-NPs (36–38 nm,

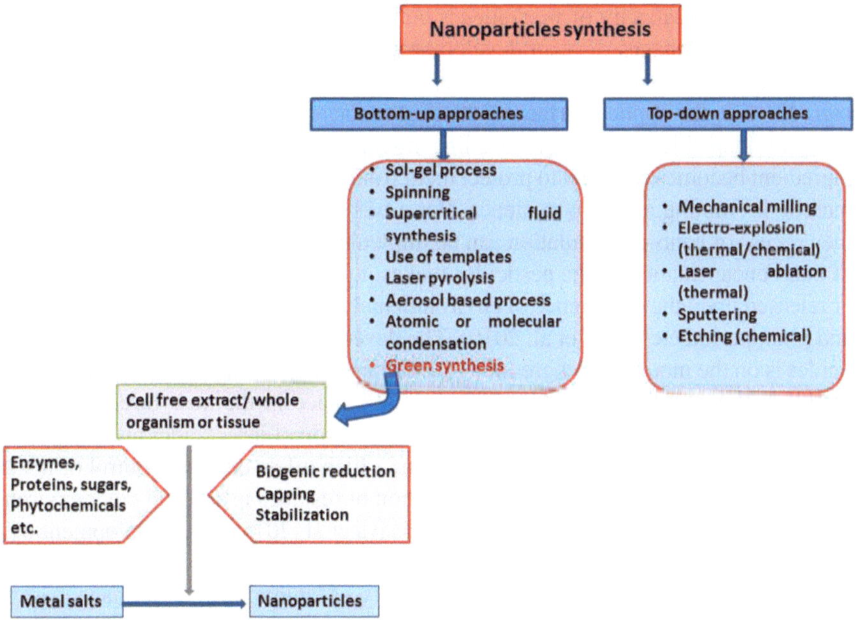

Fig. 15.2 Schematic representation of synthesis of nanoparticles through biological/green method (adopted from Singh et al. 2015a)

spherical shape) using a leaf extract of *Eclipta prostrata* which clearly denotes that nanomaterials, specially, titanium hydroxide may organize at room temperature (Rajakumar et al. 2012). The reduction was endorsed due to presence of carboxyl group (COOH) stretch and amine (N-H) group (other several secondary metabolites) present in the extract of the plant. It can also be considered that the Cu ions present in the plant that helps to reduce the biosynthesis of CuO NPs (5–10 nm, spherical size) and they were found to be active against both Gram positive and Gram negative bacteria (Awwad et al. 2015; Acharyulu et al. 2014; Philip 2009; Vardhana and Kathiravan 2015). Moreover, the produced nanoparticles exhibit greater antibacterial activity against *Bacillus subtilis* in comparison to ampicillin. CeO_2 NPs (5 nm, spherical shape) with antibacterial properties were successfully synthesized from *Gloriosa superba* leaf extract (Kargara et al. 2015; Arumugama et al. 2015; Singh et al. 2015a). The use of nanoparticles for release of anti-microbiological or drug molecules will be highly demanding task in near expectations for conduct of all pathological torment plants. The possible reimbursement of nanotechnology for agriculture and food need to be impartial alongside of concern as the soil, water and environment are related in this process (Bhattacharyya et al. 2010; Khot et al. 2012; Singh et al. 2015a, b).

4.2 Nano-Pesticide

The persistence capacity of pesticides is considered to be of significant importance during the early stages of plant development as it helps in reduces pest population thus have an effective control over pests for longer phase. Hence, the use of active ingredients in the applied surface of the host remnants is one of the most cost-effective and adaptable means of controlling insect pests. Encapsulation of the active ingredient becomes essential to protect them from adverse environmental conditions, thereby promoting a long persistence. For a biopesticide to act efficiently and thus the process of nano-encapsulation can be implement successfully. Since, the process of nano-encapsulation of the pesticides makes it possible for time controlled release or released upon the occurrence of environmental triggers like temperature, humidity and photoperiod etc. (Nair et al. 2010). The development of nano-encapsulated pesticides is on the mounting progression (OECD and Allianz 2008). However, its availability in the commercial market is yet to take place in the near future, although recent researches have proposed that the encapsulation of insecticides and fungicides will help in improving the nano-formulations which offers efficient control of insect pests, simultaneously preventing accumulation of residues in soil and other environmental matrix (Manimaran 2015; Channabasava et al. 2015). The development of a nano-encapsulated pesticide formulation that has slow releasing resources with enhanced solubility, permeability, and stability (Manimaran 2015; Channabasava et al. 2015). These assets are mainly attained through either protecting the encapsulated active ingredients from early degradation or increasing their insect pest control

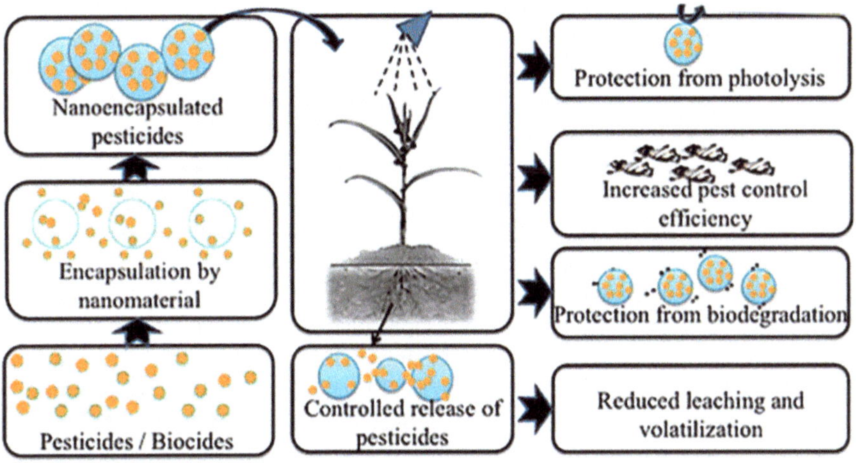

Fig. 15.3 Schematic diagram of nanoencapsulation in pesticide applications (reprinted with permission from reference Nuruzzaman et al. (2016), copyright 2016 American Chemical Society)

efficacy for a longer period. Nano-encapsulated pesticide formulation is able to reduce the dosage of pesticides and thus no human hazard will expose. Moreover, it may consider as high quality of eco-friendly material for crop protection. However, be short of knowledge of the mechanism of synthesis and there is no appropriate information of a cost-benefit analysis of nano-encapsulation materials delayed their application process in relation to pesticide delivery (Fig. 15.3). The scientific research deliver fundamental and critical information for technocrats in the field of nanotechnology, especially in the use of nano-encapsulation techniques in relation to pesticides deliver (Nuruzzaman et al. 2016).

4.3 *Nanobiofungicide*

Review of literature reveals extensive application of biosynthesized silver nanoparticles as antifungal agents against potential plant pathogenic fungi (Jo et al. 2009; Lee et al. 2013; Gopinath and Velusamy 2013; Mishra et al. 2014; Adil et al. 2015). Microorganisms like bacteria, algae, yeast, fungi, actinomycetes and virus are being used in biomediated synthesis of nanometals (Prasad 2014; Prasad et al. 2014, 2015; Aziz et al. 2015). In the near the beginning, copper nanoparticles were dissolved in water and used as fungicide for controlling diseases of grape trees and other fruit trees (Lee et al. 2013; Bhattacharyya et al. 2015). Seeds coated with silver nanoparticles have shown enhanced germination potential (Parveen and Rao 2015). However, no significant difference has been observed when the silver nanoparticles were used for seed germination in comparison with fungicides process. Therefore, it can be an option to use conventional fungicides for protecting the seeds against fungal infection. The interest in nanopesticides appears as a centre part predominantly. The different formulation process like, polymer-based nanoformulations, inorganic nanoparticles such as silica and titanium dioxide, and nanoemulsions are the predominant one (Vijayalakshmi et al. 2015). Application of these formulations, compared to existing formulations results in the discharge of active ingredient in a slow and targeted manner to the host, thus protecting the seeds against degradation (Vijayalakshmi et al. 2015). Nano encapsulated pesticide requires only a small quantity which can be used for plant pathogen control. This pesticide is effectively used without causing much harm to the environment. Several nano-pesticides have been developed which are bio-safe and has a molecular communication with plant, soil and environment on a dose dependent manner, when dispersed onto the soil (Nair et al. 2010; Abd-Elsalam, 2013; Prasad et al. 2014). It is expected that large number of nano-pesticide formulations will be introduced in agricultural fields commercially in near future which move towards sustainability process (Vijayalakshmi et al. 2015). Fungi are associate with the plant diseases and thus subsequently the most important to economic loss. Numerous genera of plant pathogenic fungi, are widespread in nature. To cite a few examples like, *Fusarium* sp., *Phoma* sp., *Aspergillus* sp., *Phytopthora* sp., *Phyllosticta* sp., are considered the most effective pathogens of plants (Rai and Ingle 2012). It has been demonstrated that most of the pathogens can be managed by nanomaterials or by nano-biofungicide (Ingle et al. 2014; Singh et al. 2015a; Yadav et al. 2015). Preliminary studies have shown the efficiency of nanomaterials in improving seed germination and development. The strengthening of natural fibres from *Cocos nucifera* and *Agave sisalana* is made possible by employing carbon nano fibres (Khot et al. 2012; Abd-Elsalam and Alghuthaymi 2015).

It may be proposed that the several nanoparticles have been formulated with pesticide colloidal suspensions or in powder form with nano or micro scale. These formulations are effective in improving the stability of the active organic composite (UV, thermal, hydrolysis, etc.), foliar settling, reduction in foliar leaching, systemic action, synergism, specificity and other several forms. Collectively these mechanisms results in minimum dose of pesticide application. Frequency of application of pesticides thus mitigating the impacts of these pesticides on human and other environmental exposures. These formulations may not be restricted to synthetic insecticides but can also be extended to natural products like herbal or other plant extracts for controlling insect pests (Joselito and Soytong 2014). Efficient green synthesis of silver nanoparticles by extracellular processes of microorganisms have been reported (Lamsal et al. 2011a, b). Extracellular synthesis of nanoparticles mediated through the fungi- *Pestaloptiopsis pauciseta* and the endophytic fungus *Pestaloptiopsis* sp. are exploited for use as antimicrobial agents either individually or in combination with antibiotics. Silver nanoparticles synthesized extracellularly by the bacterium *Serratia* sp. BHU-S4 has been demonstrated to be effective against the fungus *Bipolaris sorokiniana*, a causative agent of Spot blotch disease in wheat (Lamsal et al. 2011b). Further investigations in that meadow can lead to the improvement of the methods for the treatment of microbial infections (Kah et al. 2013; Abd-Elsalam 2013; Vardhana and Kathiravan 2015).

4.4 Fungi as Efficient Mycosystems

Mycosystems serve as an invaluable resource rendering immense provision in various fields including but not limited to biotechnology, bioremediation of contaminated lands, biomining, bioleaching, biomineralization and biocorrosion etc., (Klaus-Joerger et al. 2001; Korbekandi et al. 2009; Narayanan and Sakthivel 2010). Among the various biological agents that are harnessed for the synthesis of metal nanoparticles, fungi are used predominantly due to their high metal tolerance potential and their ability to bioaccumulate metals (Prasad et al. 2015). *Fusarium oxysporum*, *Aspergillus fumigates*, *Verticillum* spp., *Phytophthora infestans* are few examples of mycosynthesis of metal nanoparticles (Rai and Duran 2011). The fungal system possesses high cell wall binding capacity and also intracellular metal uptake capability (Sanghi and Verma 2008; Hemath et al. 2010; Ingle et al. 2014). The fungus, *Verticillium* sp., for example, showed efficient reduction of silver ions leading to the synthesis of silver nanoparticles below the surface of the fungal cells (Mukherjee et al. 2001). Furthermore, Fungi can be easily cultured therefore synthesis of nanoparticles is also very simple (Chan and Mashitah 2012). Furthermore, it is an economically feasible technology on a large scale (Prasad et al. 2015). Fungi secrete a large amount of extracellular enzymes required for synthesis and higher yield of nanoparticles (Alani et al. 2012; Birla et al. 2009; Kumar et al. 2007; Narayanan and Sakthivel 2010). Moreover, the nanoparticles precipitated extracellular and hence are devoid of cellular components and therefore, can be directly used for different applications (Narayanan and Sakthivel 2010). It is also important to mention here that fungi requires simple nutrition for chemo-organotrophs (Siqueira et al. 2002). The majority of these fungi grow on land and derive their food from dead organic matter. Some fungi also grow as parasites (Volesky and Holan 1995). Fungi generally feed by secreting enzymes that digest their food in extracellular process and the remaining food is then absorbed and completely digested internally (Frisvad et al. 2008; Alghuthaymi et al., 2015). As we know that the secondary metabolite is a chemical compound produced by a limited number of fungal species. Several secondary metabolites consists of all the different compounds of fungus that can produce on a given substratum and includes toxins, antibiotics and other several compounds (Frisvad et al. 2008). As denoted earlier, the fungal cell wall is a dynamic structure that provides the cells with perfunctory potency to tolerate changes in osmotic pressure and environmental stress (Valentine et al. 2002; Durán and Nombela 2004; Bowman and Free 2006).

5 Conclusion

Nanoformulations are viewed to be a safer and environment friendly option for plant disease management. Nano-bio-fungicide may evolve as the next generation pesticide owing to its properties like high effectiveness, durability and less dose of active ingredient required in this process. Nano-bio-fungicides can be equipped in a simple and cost-effective manner for formulating hybrid nanocides. The component can be used as a eco-friendly antimicrobial agent to different fungal pathogens in different plants and also may act as insect pest control of the host plant. Fungi are used as bio-manufacturing unit which will offer an additional advantage in existence form trouble-free use as compared to other microbes. The non-pathogenic nature of some fungal species in synergism with other fungi helps in mass production of silver nanoparticles. In view of safety of nanobiofungicides to public health and to the environment as a whole, nanobiofungicides, similar to any new substance should be subjected to toxicological evaluation before being made commercially available. This is essential, as because, human uptake of nanomaterials through food chain by consumption of plants and animals or their products treated with nanomaterials during agriculture or farming practice which comprise a chief transmit source of exposure.

References

Abd-Elsalam KA (2013) Nanoplatforms for plant pathogenic fungi management. Fungal Genom Biol 2, e10

Abd-Elsalam KA, Alghuthaymi MA (2015) Nanobiofungicides: are they the next-generation of fungicides? J Nanotech Mater Sci 2:1–3

Acharyulu NPS, Dubey RS, Swaminadham V, Kalyani RL, Kollu P, Pammi SV (2014) Green synthesis of Cuo nanoparticles using *Phyllanthus amarus* leaf extract and their antibacterial activity against multidrug resistance bacteria. Int J Eng Res Technol 3:639–641

Adil SF, Assal ME, Khan M (2015) Biogenic synthesis of metallic nanoparticles and prospects toward green chemistry. Dalton Trans 44:9709–9717

Alani F, Moo-Young M, Anderson W (2012) Biosynthesis of silver nanoparticles by a new strain of Streptomyces sp. compared with *Aspergillus fumigatus*. World J Microbiol Biotechnol 28(3):1081–1086

Alghuthaymi MA, Almoammar H, Rai M, Said-Galiev E, Abd-Elsalam KA (2015) Myconanoparticles: synthesis and their role in phytopathogens management. Biotechnol Biotechnol Equip 29(2):221–236

Arumugama A, Karthikeyan C, Hameed ASH, Gopinath K, Gowri S, Karthika V (2015) Synthesis of cerium oxide nanoparticles using *Gloriosa superba* L. leaf extract and their structural, optical and antibacterial properties. Mater Sci Eng 49:408–415

Awwad AM, Albiss BA, Salem NM (2015) Antibacterial activity of synthesized copper oxide nanoparticles using *Malva sylvestris* leaf extract. SMU Med J 2:91–100

Aziz N, Faraz M, Pandey R, Sakir M, Fatma T, Varma A, Barman I, Prasad R (2015) Facile algae-derived route to biogenic silver nanoparticles: synthesis, antibacterial and photocatalytic properties. Langmuir 31:11605–11612

Bhattacharyya A (2009) Nanoparticles from drug delivery to insect pest control. Akshar 1(1):1–7

Bhattacharyya A, Bhaumik A, Nandi M, Viraktamath S, Kumar R, Nayak A, Chowdhury MU, Mandal S (2009) Nano—a new frontier in present century. Advances Life Sci 3:18–23

Bhattacharyya A, Bhaumik A, Usha Rani P, Mandal S, Epidi TT (2010) Nano-particles—a recent approach to insect pest control. Afr J Biotechnol 9(24):3489–3493

Bhattacharyya A, Antoney PU, Raja Naika H, Reddy SJ, Adeyemi MM, Omkar (2015) Nanotechnology and butterflies: a mini review. J Appl Biosci 41: 27–32

Birla SS, Tiwari VV, Gade AK, Ingle AP, Yadav AP, Rai MK (2009) Fabrication of silver nanoparticles by *Phoma glomerata* and its combined effect against *Escherichia coli*, *Pseudomonas aeruginosa* and *Staphylococcus aureus*. Lett Appl Microbiol 48:173–179

Bowman SM, Free SJ (2006) The structure and synthesis of the fungal cell wall. Bioessays 28(8):799–808

Chan YS, Mashitah MD (2012) Optimization of process parameter in bioreduction of Ag nanoparticles by *Pycnoporus sanguineus*. In: Proceeding of international conference on advanced material engineering and technology (ICAMET 2012). Bayview Beach Resort, Penang, Malaysia, 28–30 Nov 2012

Channabasava A, Lakshman HC, Jorquera MA (2015) Effect of fungicides on association of arbuscular mycorrhiza fungus *Rhizophagus fasciculatus* and growth of Proso millet (*Panicum miliaceum* L.). J Soil Sci Plant Nutr 15:35–45

Durán A, Nombela C (2004) Fungal cell wall biogenesis: building a dynamic interface with the environment. Microbiology 150:3099–3103

Francis R, Keinath A (2010) Biofungicides and Chemicals for Managing Diseases in Organic Vegetable Production. Clemson Cooperative Extension Information Leaflet 88. (http://www.clemson.edu/psapublishing/PAGES/PLNTPATH/IL88.pdf)

Frisvad JC, Andersen B, Thrane U (2008) The use of secondary metabolite profiling in chemotaxonomy of filamentous fungi. Mycol Res 112:231–240

Gawai DU (2015) Antifungal activity of essential oil of *cymbopogon citratus* stapf against different *Fusarium* species. Bionano Frontier 8(2):186–189

Ghormade V, Deshpande MV, Paknikar KM (2011) Perspectives for nano-biotechnology enabled protection and nutrition of plants. Biotechnol Adv 29:792–803

Gopinath V, Velusamy P (2013) Extracellular biosynthesis of silver nanoparticles using Bacillus sp. GP-23 and evaluation of their antifungal activity towards *Fusarium oxysporum*. Spectrochim. Acta Mol Biomol Spectrosc 106:170–174

Gupta S, Dikshit AK (2010) Biopesticides: an ecofriendly approach for pest control. J Biopesticides 3:186–188

Hemath Naveen KS, Kumar G, Karthik L, Bhaskara Rao KV (2010) Extracellular biosynthesis of silver nanoparticles using the filamentous fungus *Penicillium* sp. Arch Appl Sci Res 2:161–167

Ingle A, Duran N, Rai M (2014) Bioactivity, mechanism of action and cytotoxicity of copper-based nanoparticles: a review. Appl Microbiol Biotechnol 98:1001–1009

Jo YK, Kim BH, Jung GH (2009) Antifungal activity of silver ions and nanoparticles on phytopathogenic fungi. Plant Dis 93:1037–1043

Joselito D, Soytong K (2014) Construction and characterization of copolymer nanomaterials loaded with bioactive compounds from *Chaetomium* species. J Agr Tech 10(4):823–831

Kah M, Beulke S, Tiede K (2013) Nano-pesticides: state of knowledge, environmental fate and exposure modelling. Crit Rev Environ Sci Technol 43(16):1823–1867

Kargara H, Ghasemi F, Darroudid M (2015) Bioorganic polymer-based synthesis of cerium oxide nanoparticles and their cell viability assays. Ceram Int 41:1589–1594

Khot LR, Sankaran S, Maja JM, Ehsani R, Schuster EW (2012) Applications of nanomaterials in agricultural production and crop protection: a review. Crop Prot 35:64–70

Klaus-Joerger T, Joerger R, Olsson E, Granqvist CG (2001) Bacteria as workers in the living factory: metal-accumulating bacteria and their potential for materials science. Trends Biotechnol 19:15–20

Korbekandi H, Iravani S, Abbasi S (2009) Production of nanoparticles using organisms. Crit Rev Biotechnol 29:279–306

Kumar SA, Abyaneh MK, Gosavi SW, Kulkarni SK, Pasricha R, Ahmad A (2007) Nitrate reductase-mediated synthesis of silver nanoparticles from $AgNO_3$. Biotechnol Lett 29(3):439–445

Lamsal K, Kim SW, Jung JH, Kim YS, Kim KS, Lee YS (2011a) Application of silver nanoparticles for the control of *Colletotrichum* species in vitro and pepper anthracnose disease in field. Mycobiol 39:194–199

Lamsal K, Kim SW, Jung JH, Kim YS, Kim KS, Lee YS (2011b) Inhibition effects of silver nanoparticles against powdery mildews on cucumber and pumpkin. Mycobiol 39:26–32
Lee KJ, Park SH, Govarthanan M (2013) Synthesis of silver nanoparticles using cow milk and their antifungal activity against phytopathogens. Mater Lett 105:128–131
Leng P, Zhang Z, Pan G, Zhao M (2011) Applications and development trends in biopesticides. Afr J Biotechnol 10(86):19864–19873
Manimaran M (2015) A review on nanotechnology and its implications in agriculture and food industry. Asian J Plant Sci Res 5:13–15
Mishra S, Singh BR, Singh A (2014) Biofabricated silver nanoparticles act as a strong fungicide against *Bipolaris sorokiniana* causing spot blotch disease in wheat. PLoS One 9(5), e97881
Mukherjee P, Ahmad A, Mandal D, Senapati S, Sainkar SR, Khan MI, Parishcha R, Ajaykumar PV, Alam M, Kumar R, Sastry M (2001) Fungus-mediated synthesis of silver nanoparticles and their immobilization in the mycelial matrix: a novel biological approach to nanoparticle synthesis. Nano Lett 1(10):515–519
Nair R, Varghese SH, Nair BG, Maekawa T, Yoshida Y, Kumar DS (2010) Nanoparticulate material delivery to plants. Plant Sci 179:154–163
Narayanan KB, Sakthivel N (2010) Biological synthesis of metal nanoparticles by microbes. Adv Colloid Interface Sci 156:1–13
Nuruzzaman M, Rahman MM, Liu Y, Naidu R (2016) Nanoencapsulation, nano-guard for pesticides: a new window for safe application. J Agric Food Chem 64:1447–1483
OECD and Allianz (2008) Report in cooperation with the OECD International Futures Programme. http://www.oecd.org/dataoecd/32/1/44108334.pdf
Parveen A, Rao S (2015) Effect of nanosilver on seed germination and seedling growth in *Pennisetum glaucum*. J Cluster Sci 26:693–701
Philip D (2009) Biosynthesis of Au, Ag and Au-Ag nanoparticles using edible mushroom extract. Spectrochim Acta A 73:374–381
Prasad R (2014) Synthesis of silver nanoparticles in photosynthetic plants. J Nanopart, Article ID 963961, doi:10.1155/2014/963961
Prasad R, Kumar V, Prasad KS (2014) Nanotechnology in sustainable agriculture: present concerns and future aspects. Afr J Biotechnol 13:705–713
Prasad R, Pandey R, Barman I (2015) Engineering tailored nanoparticles with microbes: quo vadis. WIREs Nanomed Nanobiotechnol. doi:10.1002/wnan.1363
Ragaei M, Sabry Al-Kazafy H (2014) Nanotechnology for insect pest control. Int J Sci Environ Technol 3:528–545
Rai M and Duran N (eds.) (2011) Metal nanoparticles in microbiology. Springer, Berlin/Heidelberg, doi:10.1007/978-3-642-18312-6_1
Rai M, Ingle A (2012) Role of nanotechnology in agriculture with special reference to management of insect pests. Appl Microbiol Biotechnol 94:287–293
Rajakumar G, Rahuman AA, Priyamvada B, Khanna VG, Kumar DK, Sujin PJ (2012) *Eclipta prostrata* leaf aqueous extract mediated synthesis of titanium dioxide nanoparticles. Mater Lett 68:115–117
Sanghi R, Verma P (2008) A method for preparing auto capped nanoparticles such as CdS in continuous flow columns. Indian Patent Application No. 1395/Del/2008
Saraf M, Pandya U, Thakkar A (2014) Role of allelochemicals in plant growth promoting rhizobacteria for biocontrol of phytopathogens. Microbiol Res 169:18–29
Sekhon BS (2014) Nanotechnology in agri-food production: an overview. Nanotechnol Sci Appl 7:31–53
Sharon M, Choudhary AK, Kumar R (2010) Nanotechnology in agricultural diseases and food safety. J Phytol 2(4):83–92
Shrivastava S, Prasad R, Varma A (2014) Anatomy of root from eyes of a microbiologist. In: Morte A, Varma A (eds) Root engineering, vol 40. Springer, Berlin/Heidelberg, pp 3–22
Singh A, Singh NB, Hussain I, Singh H, Singh SC (2015a) Plant-nanoparticle interaction: an approach to improve agricultural practices and plant productivity. Int J Pharm Sci Invent 4:25–40

Singh S, Singh BK, Yadav SM, Gupta AK (2015b) Applications of nanotechnology in agricultural and their role in disease management. Res Nanosci Nanotechnol 5:1–5
Siqueira JF, Rocas IN, Lopes HP, Elias CN, de Uzeda Milton (2002) Fungal infection of the radicular dentin. J Endod 28:770–773
Usta C (2013) Microorganisms in biological pest control—a review (bacterial toxin application and effect of environmental factors). In: Silva-Opps M (ed) Current progress in biological research. InTech, doi:10.5772/55786
Valentine NB, Wahl JH, Kingsley MT, Wahl KL (2002) Direct surface analysis of fungal species by matrix-assisted laser desorption/ionization mass spectrometry. Rapid Commun Mass Spectrom 16:1352–1357. doi:10.1002/rcm.721
Vardhana J, Kathiravan G (2015) Biosynthesis of silver nanoparticles by endophytic fungi *Pestaloptiopsis pauciseta* isolated from the leaves of *Psidium guajava* Linn. Int J Pharm Sci Rev Res 31:29–31
Vijayalakshmi C, Chellaram C, Kumar SL (2015) Modern approaches of nanotechnology in agriculture—a review. Biosci Biotechnol Res Asia 12:327–331
Volesky B, Holan ZR (1995) Biosorption of heavy metals. Biotechnol Prog 11:235–250
Yadav A, Kon K, Kratosova G, Duran N, Ingle AP, Rai M (2015) Fungi as an efficient mycosystem for the synthesis of metal nanoparticles: progress and key aspects of research. Biotechnol Lett 37:2099–2120

Chapter 16
Nanocellulose Production Using Cellulose Degrading Fungi

Nadanathangam Vigneshwaran and Prasad Satyamurthy

Abstract Nanocellulose, a novel material derived from cellulosic biomass, consists of cellulose having at least one dimension in the nano-size (<100 nm). Very high surface area to volume ratio (50–200 m^2/g), high tensile strength (1–10 GPa) and low density (1.45 g/cc) make nanocellulose an attractive material as reinforcement agents in high performance composites. Earlier, nanocellulose was produced by concentrated sulphuric acid hydrolysis that removed the amorphous region leaving behind highly crystalline nanocellulose whiskers. Though they are stable due to sulfation on surface, scaling up could not be achieved due to reasons related to handling of concentrated (64 %) sulphuric acid and effluent disposal. Recently, research effort is towards mechanical preparation of nanocellulose by high pressure homogenization process that could circumvent the effluent problem. But, here the bottleneck is very high energy consumption (30,000 kWh/tonne) for nanocellulose production and frequent clogging of the production system. Various pre-treatments methodologies are evolved to reduce energy consumption and to avoid clogging in homogenizer. One among them, cellulase enzyme pre-treatment, is very popular and highly researched due to eco-friendliness and efficacy. Apart from cellulase enzyme the cellulase secreting fungi as such are being used for ease of handling and to reduce the cost of enzyme processing. Well studied fungi include *Trichoderma* sp. and *Aspergillus* sp. for pre-treatment of cellulosic biomass before homogenization process for production of nanocellulose. Lately, controlled hydrolysis by fungi itself evolved for production of nanocellulose thereby bypassing the homogenization process step. This makes fungi a versatile organism for production of nanocellulose.

N. Vigneshwaran (✉) • P. Satyamurthy
ICAR-Central Institute for Research on Cotton Technology,
Adenwala Road, Matunga, Mumbai 400019, India
e-mail: nvw75@yahoo.com; Vigneshwaran.N@icar.gov.in

R. Prasad (ed.), *Advances and Applications Through Fungal Nanobiotechnology*, Fungal Biology, DOI 10.1007/978-3-319-42990-8_16

1 Nanocellulose

Cellulose, the most abundant biomass available on Earth, is a biodegradable homo -polymer of β-(1, 4) linked D-glucose units. Cellulose is a straight chain polymer consisting of multiple units of D-glucose linked together in a repeating, overlapping pattern, resulting in a high tensile strength polymer. Cellulose is the main structural component of the primary cell wall of plants, many forms of algae and fungi. For industrial use, cellulose is obtained from wood pulp, agro-biomass and cotton (Satyamurthy and Vigneshwaran 2013). Nanocellulose is a novel biomaterial derived from any cellulosic biomass by various processes viz., mechanical, chemical, biological and in combinations of them. They are very much interesting due to its renewable nature, anisotropic shape, excellent mechanical properties, good biocompatibility, tailorable surface chemistry, and interesting optical properties (Prasad et al. 2015; Abitbol et al. 2016). At least, any one dimension of nanocellulose has to fall in the region of nanometres (1–100 nm). They are classified as nanocrystalline cellulose (NCC) and nanofibrillated cellulose (NFC) according to its aspect ratio. NCC has the aspect ratio less than 100, and in general, called as nanowhiskers due to their elongated whisker shape. In case of NFC, the aspect ratio is more than 100, and in general, they are more than 1000 so that it forms a very long fibrillated structure. Further classifications can be based on the method of preparation, source of the raw material and intended application area. Figure 16.1 shows the overall classification of nanocellulose.

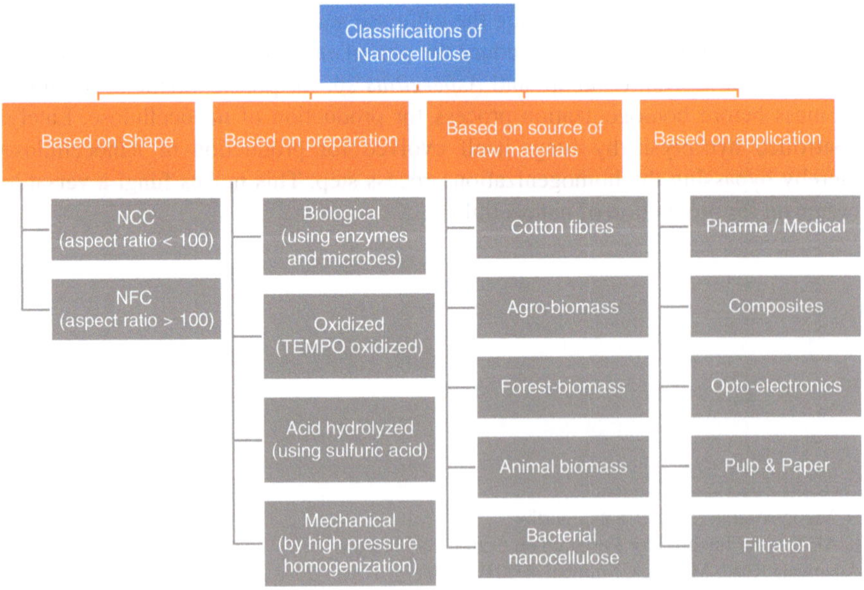

Fig. 16.1 Classifications of nanocellulose

The purest form of nanocellulose could be obtained from cotton fibres and bacterial cellulose, while other raw materials require extensive purification to remove lignin, hemicellulose and other impurities. Nanocellulose can be produced by top-down approach (mechanical/chemical/enzymatic degradation) or bottom-up approach (bacterial cellulose synthesis). The major areas of application of nanocellulose include pulp and paper, polymer film composites, paint and pigments, non-calorific food thickeners and drug delivery system. In spite of established application potential of nanocellulose, the major bottleneck encountered is the requirement of huge amount of energy in production of nanocellulose. An extensive review on this aspect was recently published by our research group (Bharimalla et al. 2015). The NCC and NFC isolated from pure rice straw cellulose via sulfuric acid hydrolysis, mechanical blending and TEMPO-mediated oxidation resulted in 16.9, 12 and 19.7 % yields, respectively. Sulfuric acid hydrolysis produced highly crystalline (up to 90.7 % CrI) rod-like (3.96–6.74 nm wide, 116.6–166 nm long) NCCs with negative surface charges (−67 to −57 mV); Mechanical defibrillated NFCs were 82.5 % crystalline and bimodally distributed in sizes (2.7 nm wide and 100–200 nm long; 8.5 nm wide and micrometers long); and TEMPO mediated oxidation liberated the most uniform, finest (1.7 nm) and micrometer long, but least crystalline (64.4 % CrI) NCCs (Jiang and Hsieh 2013). Figure 16.2 shows the various options of modifying nanocellulose for diversified applications (Dufresne 2013).

2 Cellulose Degrading Fungi

Cellulose is degraded by cellulase enzymes that are highly specific in nature and the product of hydrolysis is glucose. The utility cost of enzymatic hydrolysis is low as compared to acid or alkaline hydrolysis process since enzyme hydrolysis happens at relatively mild conditions viz., pH 4.8 and temperature 45 °C (Sun and Cheng 2002). In general, aerobic fungi and anaerobic bacteria are known to produce cellulase enzyme. In case of aerobic fungi, cellulases are produced as multi component enzyme system comprised usually of three components that act synergistically in the hydrolysis of cellulose; endoglucanases (EC 3.2.1.4), cellobiohydrolase (EC 3.2.1.91) and cellobiase (β-glucosidase, EC 3.2.1.91). In case of anaerobic bacteria, large multi protein complexes known as cellulosome are involved in degradation of cellulose and it has about 11 different enzymes aligned on the non-catalytic scaffolding protein that ensure a high local concentration, together with the correct ratio and order of the components. Figure 16.3 shows the action of cellulase enzyme on cellulosic substrate and Fig. 16.4 shows the action of cellulosome on cellulosic substrate (Beckham et al. 2011).

Cellulase has applications in diversified industries including agriculture (for enhanced plant growth and flowering), bioconversion (ethanol from cellulose), detergents (superior cleaning with fibre damage), fermentation (improved aroma of wines), food (clarification of fruit juices), pulp and paper (co-additive in pulp bleaching), and textile (biopolishing of textile fibers) (Kuhad et al. 2011; Salahuddin et al. 2012).

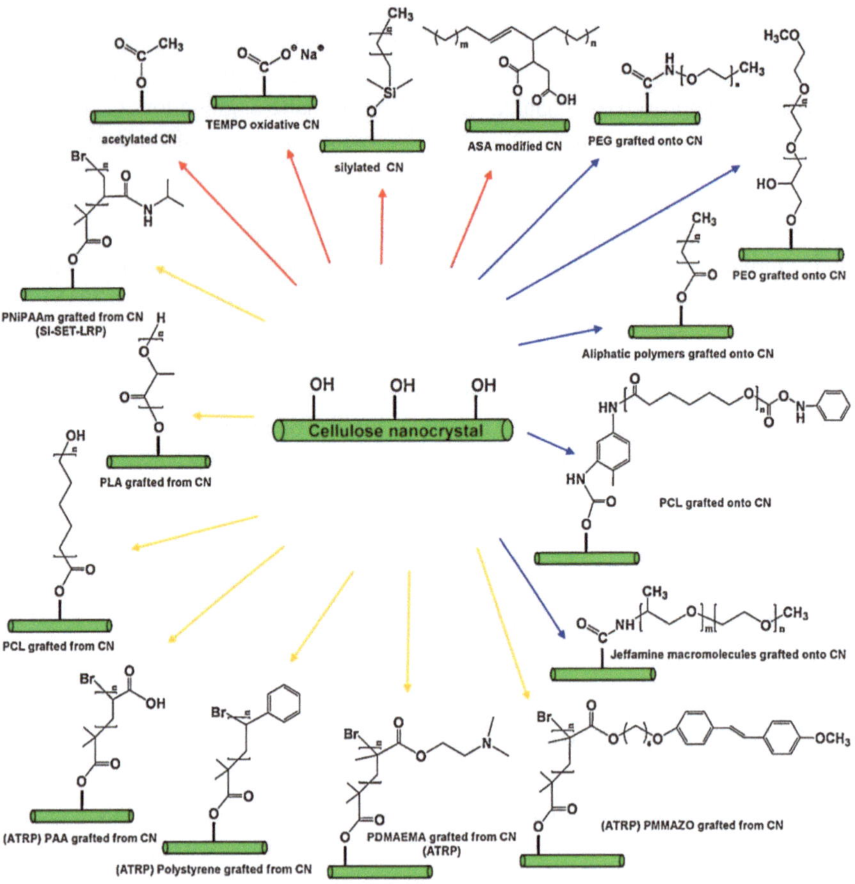

Fig. 16.2 Various modifications of nanocellulose for diversified applications. Reprinted with permission from Dufresne (2013)

3 Production of Nanocellulose by Cellulose Degrading Fungi

The enzymatic hydrolysis of cellulose, particularly hydrogen-bonded and ordered crystalline regions, is a very complex and slow process. Among the two major types of cellulose (algal-bacterial type rich in cellulose Iα crystalline region and cotton–ramie type rich in cellulose Iβ), algal–bacterial type is highly susceptible to cellulase enzyme. The cotton cellulose is recalcitrant due to the dominance of cellulose Iβ structure. In our work (Satyamurthy et al. 2011) we have explored a possibility of controlled hydrolysis of microcrystalline cellulose (MCC) using the fungus *T. reesei* with the yield of 22 %. The penetration of fungus into the ordered regions of MCC during incubation resulted in reduced crystallinity of nanocellulose prepared by microbial hydrolysis compared to that of acid hydrolysis. Figure 16.5 shows the AFM images of nanocellulose prepared by controlled microbial hydrolysis in comparison with that of sulfuric

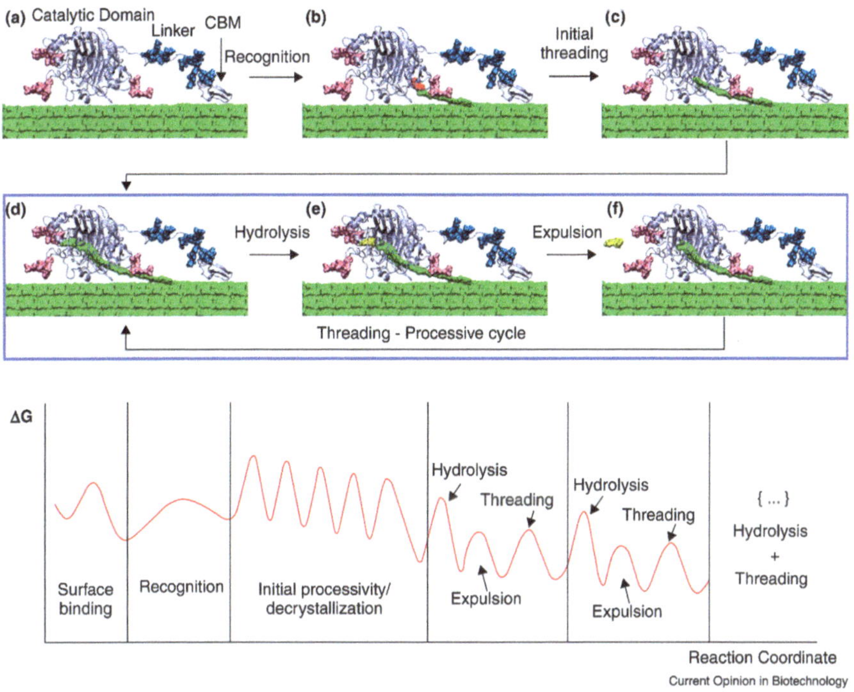

Fig. 16.3 The *Trichoderma reesei* Family 7 cellobiohydrolase (Cel7A) acting on cellulose. Cel7A is comprised of a 36-amino acid CBM, a linker domain with O-glycan (*dark blue*), and a large catalytic domain with N-linked glycan (*pink*) and a 50-Å tunnel for processing cellulose chains (*green*). The cellobiose product is shown in *yellow* (**e**) and (**f**). Here, the putative steps that Cel7A takes to deconstruct biomass and the hypothesized free energy surface for each elementary step (**a**–**f**) is shown. Reprinted with permission from Beckham et al. (2011)

acid hydrolysis process. The soft rot ascomycetes fungus *Trichoderma reesei* is utilized for industrial production of secreted enzymes, especially lignocellulose degrading enzymes. *T. reesei* uses several different enzymes for the degradation of plant cell wall-derived material, including nine characterized cellulases, 15 characterized hemicellulases and at least 42 genes predicted to encode cellulolytic or hemicellulolytic activities (Mari Häkkinen et al. 2014). As this fungus is being exploited for commercial use, nanocellulose production using this fungus also will add a new dimension. Earlier studies reported the successful production of NFC using the endoglucanase enzyme in combination with mechanical shearing and high-pressure homogenization (Henriksson et al. 2007; Pääkkö et al. 2007; Zhu et al. 2011), but, with a lot of energy input.

In our another work, the enriched anaerobic microbial consortium (for cellulase production) is proven to be efficient in hydrolyzing microcrystalline cellulose to produce nanocellulose in a span of 7 days with a maximum yield of 12.3 %. Nanocellulose prepared by this process has a bimodal particle size distribution (43 ± 13 and 119 ± 9 nm) (Satyamurthy and Vigneshwaran 2013). Figure 16.6 shows the AFM image of nanocellulose prepared by anaerobic microbial consortium.

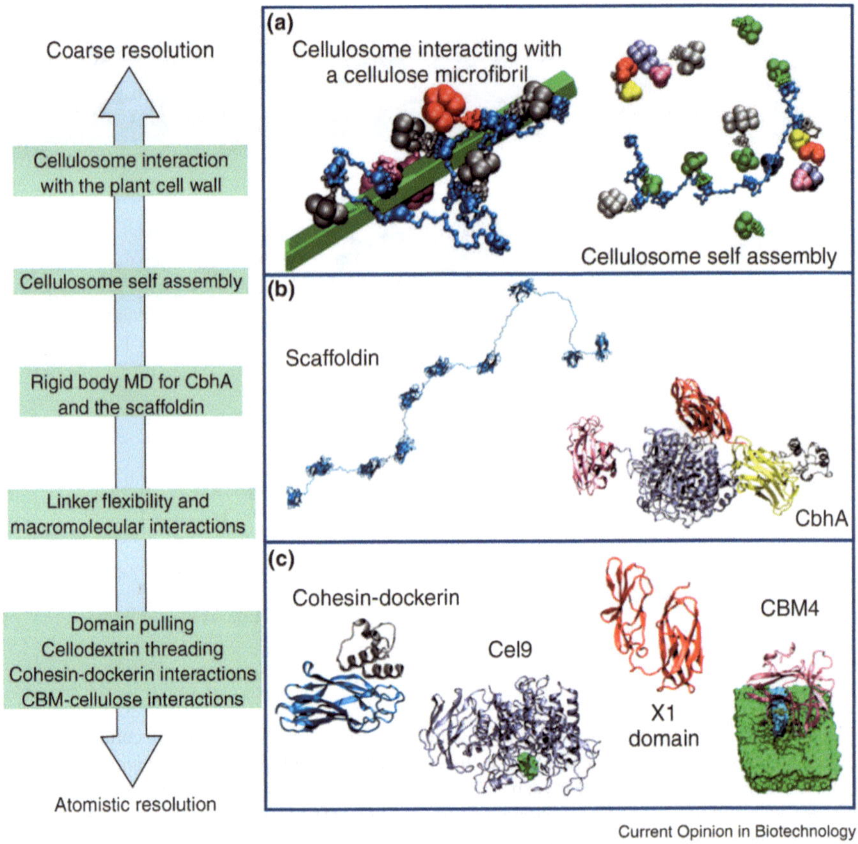

Fig. 16.4 Multi-scale modeling can aid in the understanding of the cellulosomal complex and enzyme–cellulose interactions in the cellulosome. Here are several open questions at various degrees of resolution together with methods to probe each question. (**a**) A simple coarse-grained model has been developed to study self-assembly of the entire cellulosome enzyme complex as a function of enzyme concentration and other relevant variables. (**b**) Rigid body MD enables calculation of solution behavior directly from simulation to compare with SAXS and FRET experiments of the large CbhA enzyme. (**c**) Multiple scientific questions exist at the atomistic scale that can be examined with methods such as rare event simulation to understand mechanisms of threading cellodextrin chains into cellulase tunnels, free energy perturbation methods for relative binding free energies and absolute binding free energies of carbohydrates to cellulases and CBMs, docking calculations to understand the non-covalent binding at the atomic scale, steered MD to understand the work to extend putatively flexible proteins, and REMD to understand intrinsic disorder. Reprinted with permission from Beckham et al. (2011)

4 Purification of Nanocellulose

Purification of nanocellulose becomes a bottleneck while dealing with the microbial process of nanocellulose production. Many of the broth components and fungal secretes are also fall in the nanometer size range that makes it difficult for separation of nanocellulose. The two different options of separation of nanocellulose from the fungal

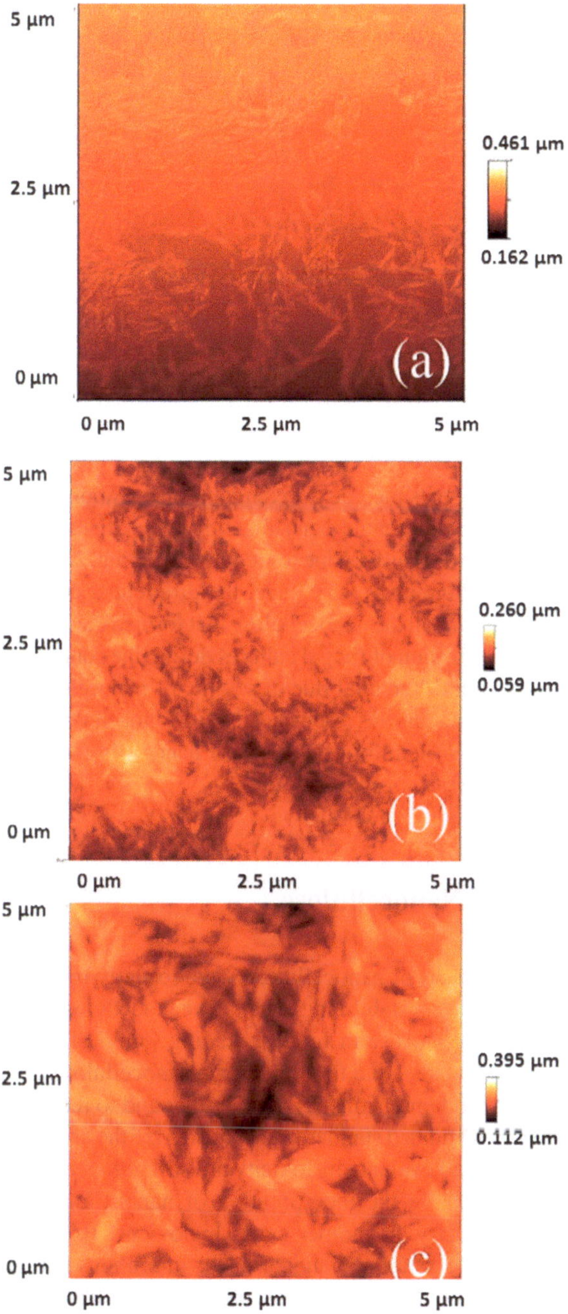

Fig. 16.5 AFM images of the nanocellulose prepared by acid (**a**) and fungal (**b**, **c**) hydrolysis. Reprinted with permission from Satyamurthy et al. (2011)

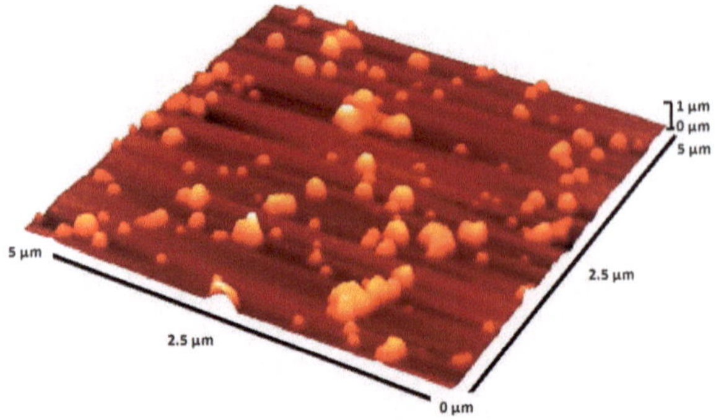

Fig. 16.6 AFM image of spherical nanocellulose prepared using anaerobic microbial consortium. Reprinted with permission from Satyamurthy and Vigneshwaran (2013)

culture and broth components include differential centrifugation and filtration through membrane filters. While differential centrifugation is a tedious and time consuming process, the filtration process suffers due to frequent blocking. Also, since the nanocellulose product size distribution is very wide, the differential centrifugation/filtration techniques could not purify all the nanocellulose that formed during the process. Newer ideas including immobilization of fungal cultures during the controlled hydrolysis process in combination with differential rate of settling of nanocellulose in the broth are being developed as alternate means of purification of nanocellulose. Figure 16.7 shows the overall fermentation system for production of nanocellulose by hydrolysis of cellulose using the fungus *Trichoderma reesei.*

5 Application of Nanocellulose

Three different applications of nanocellulose were recently reviewed (Gómez et al. 2016): (1) nanocellulose as a stabilizing agent, (2) nanocellulose as a functional food ingredient and (3) nanocellulose in food packaging. The last is the most common application of nanocellulose in the food industry. Nanocellulose has potential use as a stabilizing agent in food emulsions, as dietary fiber and to reduce the caloric value of food. Nevertheless, validated standards to characterize the produced nanostructure, quantify its properties and evaluate its toxicity are still required to answer safety and regulatory issues to achieve the incorporation of nanocellulose as a commercial product in the food industry. In another work (Ioelovich 2016), applications of five kinds of nanocellulose, crystalline nanoparticles, amorphous nanoparticles, nanofibrillated cellulose, bacterial nanocellulose, and cellulose nanoyarn that in various areas of care and cure were discussed. The crystalline nanoparticles are applied as multifunctional agents in cosmetic remedies and dentifrices. The amorphous nanoparticles can be used as an antibacterial and hemostatic nanoagent. Nanofibrillated cellulose is characterized by excellent thickening and gel-forming properties. Bacterial nanocellulose finds applications in diverse areas of personal care and biomedicine. Nanoyarn can be used to create new types of wound dressings.

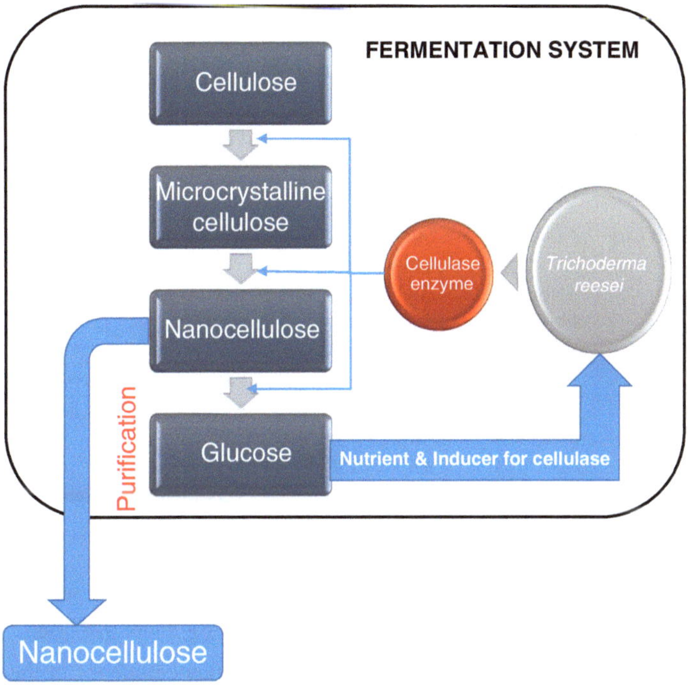

Fig. 16.7 Schematic representation of fermentation system for production of nanocellulose

The other main areas of nanocellulose research including photonics, films and foams, surface modifications, nanocomposites, and medical devices were reviewed in an another work (Abitbol et al. 2016). Nanocellulose, with its ability to form hydrogen bonds resulting in strong network makes it very hard for the molecules to pass through, suggesting excellent barrier properties associated with films made from these material (Nair et al. 2014). In most of the applications dealing with bio logical system, it is better to have the nanocellulose without any surface modification and without metallic contamination. While the chemically produced nanocellulose inherits the sulfated or carboxylated surface, mechanically produced nanocellulose suffers heavy metal contamination during processing in high energy refining and milling processing. In these circumstances, nanocellulose produced by fungal hydrolysis route offers an excellent alternative as they are bio-compatible and retains the cellulosic nature on its surface.

6 Challenges and Ways Ahead

Table 16.1 shows the comparative aspects of nanocellulose produced by different processes, viz., mechanical, chemical, enzymatic and microbial processes. Each and every process has its own merits and demerits and selection depends on the demand and potential application to be explored.

Table 16.1 Comparison of nanocellulose produced by different processes

Parameters of final product (Nanocellulose)	Methods of preparation			
	Mechanical	Chemical	Enzymatic	Microbial/fungal
Rate of formation	Fast	Fast	Very slow	Slow
Yield (%)	>80	40–60	30–40	10–25
Morphology	NFC	NCC	NCC	NCC
Surface chemistry	Not changed	Sulfated or carboxylated	Not changed	Not changed
Stability	Stable due to fibrillar structure	Highly stable due to very high surface charge	Relatively unstable	Stable due to bound protein and broth components
Ease of operation	Very easy	Difficult	Easy	Easy
Effluent	No effluent generated	Chemical effluent (acidic) generated with high COD	Effluent with high BOD generated	
Cost of production	High	Medium	Very high	High

The cellulose degrading fungi are found to have scope for large scale production of nanocellulose for commercial exploitation. The major challenges are:

(a) To control the size distribution of the product (nanocellulose) in the dynamic production system using cellulose degrading fungi
(b) To increase the yield as increase in substrate concentration act as a limiting factor in a fermentation system
(c) Purification of nanocellulose from the broth substances and the fungal biomass
(d) Efficient control of cellulose degradation process by the cellulase enzyme secreted by fungi.

By overcoming the above said challenges and as governed by the need for eco-friendly system of production, fungal based nanocellulose production could be of the future for diversified bio-based applications.

References

Abitbol T, Rivkin A, Cao Y, Nevo Y, Abraham E, Ben-Shalom T, Lapidot S, Shoseyov O (2016) Nanocellulose, a tiny fiber with huge applications. Curr Opin Biotechnol 39:76–88

Beckham GT, Bomble YJ, Bayer EA, Himmel ME, Crowley MF (2011) Applications of computational science for understanding enzymatic deconstruction of cellulose. Curr Opin Biotechnol 22(2):231–238

Bharimalla A, Deshmukh S, Patil P, Vigneshwaran N (2015) Energy efficient manufacturing of nanocellulose by chemo- and bio-mechanical processes: a review. World J Nano Sci Eng 5:204–212

Dufresne A (2013) Nanocellulose: a new ageless bionanomaterial. Mater Today 16(6):220–227

Gómez HC, Serpa A, Velásquez-Cock J, Gañán P, Castro C, Vélez L, Zuluaga R (2016) Vegetable nanocellulose in food science: a review. Food Hydrocoll 57:178–186
Häkkinen M, Valkonen MJ, Westerholm-Parvinen A, Aro N, Arvas M, Vitikainen M, Penttilä M, Saloheimo M, Pakula TM (2014) Screening of candidate regulators for cellulase and hemicellulase production in *Trichoderma reesei* and identification of a factor essential for cellulase production. Biotechnol Biofuels 7(14):1–21
Henriksson M, Henriksson G, Berglund LA, Lindström T (2007) An environmentally friendly method for enzyme-assisted preparation of microfibrillated cellulose (MFC) nanofibers. Eur Polym J 43(8):3434–3441
Ioelovich M (2016) Nanocellulose-fabrication, structure, properties, and application in the area of care and cure. In: Mihai GA (ed) Fabrication and self-assembly of nanobiomaterials. William Andrew Publishing, Amsterdam, pp 243–288
Jiang F, Hsieh Y-L (2013) Chemically and mechanically isolated nanocellulose and their self-assembled structures. Carbohydr Polym 95(1):32–40
Kuhad RC, Gupta R, Singh A (2011) Microbial cellulases and their industrial applications. Enzym Res 2011:10
Nair SS, Zhu J, Deng Y, Ragauskas AJ (2014) High performance green barriers based on nanocellulose. Sustain Chem Process 2(23):1–7
Pääkkö M, Ankerfors M, Kosonen H, Nykänen A, Ahola S, Österberg M, Ruokolainen J, Laine J, Larsson PT, Ikkala O, Lindström T (2007) Enzymatic hydrolysis combined with mechanical shearing and high-pressure homogenization for nanoscale cellulose fibrils and strong gels. Biomacromolecules 8(6):1934–1941
Prasad R, Pandey R, Barman I (2015) Engineering tailored nanoparticles with microbes: quo vadis. WIREs Nanomed Nanobiotechnol. doi:10.1002/wnan.1363
Salahuddin K, Prasad R, Gor SH, Visavadia MD, Soni VK, Hussain DM (2012) Biochemical characterization of thermostable cellulose enzyme for mesophilic strains of actinomycetes. Afr J Biotechnol 11(43):10125–10134
Satyamurthy P, Vigneshwaran N (2013) A novel process for synthesis of spherical nanocellulose by controlled hydrolysis of microcrystalline cellulose using anaerobic microbial consortium. Enzyme Microb Technol 52(1):20–25
Satyamurthy P, Jain P, Balasubramanya RH, Vigneshwaran N (2011) Preparation and characterization of cellulose nanowhiskers from cotton fibres by controlled microbial hydrolysis. Carbohydr Polym 83(1):122–129
Sun Y, Cheng J (2002) Hydrolysis of lignocellulosic materials for ethanol production: a review. Bioresour Technol 83(1):1–11
Zhu JY, Sabo R, Luo X (2011) Integrated production of nano-fibrillated cellulose and cellulosic biofuel (ethanol) by enzymatic fractionation of wood fibers. Green Chem 13(5):1339–1344

Index

A
Absorption, 202–203
Aedes aegypti, 297
Aerobic fungi, 323
Aflatoxins, 176
Ag NPs, 269
Agaricus bisporus, 4
Ag-Collagen-Chitosan (Ag-CollCh), 273
Agriculture, 170, 177, 180, 184, 185
  nanomaterials in, 311–312
Agro-ecological zone, 266
Alfalfa sprouts, 93
Allylamines, 207
Alternaria, 174
Alternaria alternaria, 80
Alternaria alternata, 248
Alzheimer's disease (AD), 228
Amorphous nanoparticles, 328
Amphotericin B, 207
Ampicillin (Amp), 247
Anaerobic fungi, 325
Anastomosis groups (AGs), 173
Anidulafungin, 209
Antifungal agent, 205–210
Antifungal azoles, 205
Antifungal effect, 274
Antimicrobial activity, 111–113, 172, 181
Antimicrobial agents
  antifungal agents and packaging systems, 199
  chemicals, 197–198
  natural, 198–199
  probiotics, 199–200
Antimicrobial effects, 170, 248, 266, 269
Antimicrobial food packaging, 195, 196
Antimicrobial mechanisms
  germicidal, 200
  microbial inhibitory, 201
  microorganisms, 200
Antimicrobial nano-packaging, 197
Antioxidative system, 11–15
Ascomycetes, 174–176
Ascorbic acid, 14
Ascospores, 174
Aspergillus, 176
  A. aculeatus, 81
  A. flavus, 29, 176, 181, 248, 277
  A. fumigatus, 3, 102, 248
  A. niger, 29, 52, 181, 246
Atomic force microscope (AFM), 268
Au-Collagen-Chitosan (Au-CollCh), 273
Aureobasidium pullulans, 7

B
Bacillus megaterium, 54
Bacillus subtilis, 275, 312
Bacillus thuringiensis, 309
Bacterial nanocellulose, 328
Bacterialcidal effects, 271
Baculoviruses, 293
Baker's yeast, 3
Basidiomycetes, 258, 261
Basidiomycetes fungi, 258
Basidiomycota, 173
Bimetallic nanocomposites, 267
Bio pesticides, 298
Bioactive polysaccharides, 276
Biochemical pesticides, 292
Biocompatible polymers, 267

R. Prasad (ed.), *Advances and Applications Through Fungal Nanobiotechnology*, Fungal Biology, DOI 10.1007/978-3-319-42990-8

Biodegradable polymers, 266, 268
Biodiesel, 294
  production process, 295
  waste products, 295
Biodiesel by-products, 294–295
Biofungicides, 308
Biopesticide market, 293
Biopesticides, 292, 293, 298, 308, 309, 312
Biopolymer, 275
Bioremediation, 254, 257, 260
Biosensors laccases, 260
Bio-separation, 231–232
Biosynthesis, 242, 245
  cellular level organization, 1
  exponential leap, 1
  laccase, 40
  metal/oxide nanoparticles, 5
  metallic nanoparticles, 39, 40
  primary decomposers, 1
Biosynthesized nanoparticles, 247
Biotechnological, 254, 256
Biotechnology, 254, 255, 261
Bipolaris sorokiniana, 170, 314
Botanical tablets, 302
Botrytis cinerea, 310
Brazil, 170
Bronw patch, 173
Budded viruses (BV), 293

C

Cadmium sulfide (CdS), 242
Cadmium sulphide (CdS), 179
  nanoparticles, 179–180
Cancer, 228–229
Candida albicans, 80, 248
Candida glabrata, 3, 242
Carbon nanotubes (CNTs), 220
Caspofungin, 209
Catalases, 11
Cellobiose, 325
Cellulase, 323, 324, 326, 330
Cellulose, 270, 322, 323, 325, 326
Cellulose degrading fungi, 323–326, 330
Cellulose fibers, 274
Cellulose Metal Antimicrobial, 274–275
Cellulose nanocrystal (CNC), 274
Cellulosome, 323, 326
Chemical pesticides, 291
Chitosan (CS), 152, 153, 182, 183, 185, 270–273, 278, 279
  application, 209–210
  spoilage microorganisms, 208
Chitosan nanoparticles, 272
Chitosan-Ag nanoparticles, 271
Chlamydomonas reinhardtii, 244
Cholesterol, 207
Citronella oil, 297
Coil formulation, 302
Collagen-chitosan scaffold, 273
Colletotrichum, 176, 177
Computed tomography (CT), 229
Controlled hydrolysis, 324, 328
Controlled Release Formulations (CRF), 302
Controlled release technology, 203–205
Conventional food packaging, 195
Copper, 181
Copper antimicrobial activity, 276
Copper nanoparticles, 183, 269, 272, 313
  cetyl trimethyl ammonium bromide, 154
  hydroextractor, 155
Crop management, 292
Crystalline nanoparticles, 328
CS/TCS-SA NPs, 154
Cu-chitosan nanoparticles, 183
CuO nanoparticles, 276
Cymbopogon flexuosus, 116

D

Debaromyces hansenii, 57
Dendrimers, 217–218
DeniLite® laccase, 261
De-oiled seed cake, 294
Desulfovibrio desulfuricans, 102
Diabetes, 229
Diabetes mellitus (DM), 229
Differential centrifugation, 328
Dillapiole, 297
Dimetallic nanoparticles, 260
Disk diffusion method, 248
Diyarex Gold, 310
Drug delivery, 232
Dustable powder, 299
Dynamic light scattering (DLS), 268

E

Echinocandins, 208
Eco-friendly, 7, 330
Eco-friendly nanofungicides, 180–183
Economic polymers, 268
Effervescent tablets, 302
Electron microscopy (EM), 268
Emulsifiable concentrate (EC), 299
Endophytic fungus, 179
Enzyme Commission (EC), 256
Enzyme-Linked Immuno Sorbent Assay (ELISA), 225
Epidermophyton floccosum, 54

Ergosterol, 207
Escherichia coli, 33, 101, 275, 276
Essential oils, 296–297
Eubaculovirinae, 293
Eucalyptus globulus, 297
Eucalyptus oil, 296
Extracellular chelation, 6–7
Extracellular mucilaginous materials (ECMM), 7

F
F. graminearum species complex (FGSC), 175
F. oxysporum f.sp. cubense (Foc), 178
F. oxysporum species complex (FOSC), 175
F. solani species complex (FSSC), 175
Filamentous fungi, 29–30, 177
Floating tablets, 302
Fluconazol, 206
Flucytosine, 206–207
Food packaging
  antimicrobial packaging systems, 201
  characteristics, 203
  chemical immobilization, 201
  description, 194
  functions, 194
  microbial agents, 201
  requirements, 194
  time period, performances and implications, 194
Formulation technology, 292
Formulations, 181, 292–299, 303
Fourier Transform Infrared (FT-IR) spectroscopy, 178, 268
Fungal diseases, 172–177, 180–185
Fungal nanotechnology, 233
Fungal plant pathogens
  phytopathogenic fungi, 163
  spore germination, 163
Fungal postharvest diseases, 176–177
Fungi, 173, 176, 177, 181, 241–249, 258, 314–316, 323–326
  green synthesis, 73–75, 78–80, 85
    advantages and disadvantages, 85
    Ag-MNPs synthesis, 75
    cell wall, 74
    enzymatic reduction, 75
    extracellular synthesis, 75, 80
    heterotrophic organisms, 73
    inactive cells, 79
    intracellular synthesis, 79
    MCF-7 cells, 78
    metal species, 74
    metal-binding, 74
    NADH-dependent reductase activity, 78
    NiO-MNPs synthesis, 79
  marine habitats, 50
Fungicides, 169, 170, 174, 198, 281, 308, 313
Fungus-mediated AgNPs
  antimicrobial activities, 54
  antimicrobial properties, 56
  Aspergillus candidus, 55
  culture filtrate, 52
  dark brownish coloration, 54
  extracellular biosynthesis, 52
  gentamicin, 54
  illumination microscopy, 53
  industrial effluents, 57
  light transmission experiment, 53
  marine sediment, 54
  marine-derived fungal strains, 57
  marine-derived fungus, 56
  nanoparticle synthesis, 57
  NCIM 3590 cells, 57
  NP biosynthesis, 57
  physico-chemical parameters, 53
  secondary metabolites, 56
  TEM image, 56
Fusarium, 3, 175, 176
  F. graminearum, 181
  F. oxysporum, 3, 34, 75, 106, 107, 178, 179, 244
  F. semitectum, 248
  F. solani, 171

G
Ganoderma lucidum, 37
Ganoderma neo-japonicum, 38, 79
Gelation, 301
Gene transfection, 230
Generally recognized as safe (GRAS), 296, 297
Geranium leaves, 179
Germicidal, 200
Gloriosa superba, 312
β-glucan, 183
Glucan nanoparticles (GNPs), 183
Glutaredoxins (Grx), 14
Glutathione (GSH), 6
Glutathione peroxidises (Gpx's), 14
Glutathione transferases (GSTs), 14
Glycerol, 295
GNPs biosynthesis, 57, 58, 61
Gold nanoparticles (GNPs), 30–33, 63, 179, 216, 245, 267, 275, 280
Gram-negative bacteria, 198
Granules, 298–299
Granulovirus (GV), 293
Green Deal process, 310

Green House Condition, 164
Green nanotechnology, 29, 281
Green processes, 253, 254
Green synthesis, 76–77, 82–84, 233
  Ag-MNPs synthesis, 81
  application, 72
  bioremediation, 81
  CdS-NPs, 81
  medical applications, 80, 81
  MNP
    fungi, 76–77
    predictability challenges, 83
    regulatory challenges, 84
    reproducibility, 82–83
    risk assessment and management, 84
    scaling-up, 83
  MNPs, 72
  nanotechnology, 81
Green Technology, 214

H
Herbal pesticides, 292
Herbicides, 169
Heterocyst polysaccharide layer (HEP), 104
HIV, 229
Hybrid nanofungicides, 182–183
Hydrogen peroxide (H_2O_2), 11
Hydrothermal method, 160

I
Immobilized laccase, 260
India, 291, 293–297
Inorganic nanoparticles, 314
Inorganic-based nanofungicides, 181–182
Insect pests, 308, 310, 312, 314
Insecticide Act, 1968, 293
Intracellular chelation, 8–10
Iron oxide, 145
Iron oxide NPs
  biological method, 157–158
  chemical precipitation, 156–157
  types, 156
Itraconazol, 206

J
Jatropha seedcakes, 294

K
Karanja oil, 294, 302

L
Laccase enzyme, 40
Laccase mediator system (LMS), 260
Laccases, 253–255, 258–261
  application, 259–261
    bioremediation, 260
    nanobiotechnology, 260
    pulp and paper industry, 260–261
    textile industry, 261
  production, 258
  properties, 256–261
  sources, 257–259
    bacteria and archea, 258
    fungi, 258
    insects, 259
Ligninolytic enzyme, 256
Liquid biodiesel, 295
Liquid crystals, 218
Lymphatic filariasis, 230

M
Magnaporthe grisea, 170
Magnesium, 160–163
Magnesium oxide nanoparticles, 248
Magnetic resonance images (MRI), 229, 232
Management, 169–172, 177, 180–185
Marine environment
  AgNPs and GNPs, 51
  marine-derived fungi, 51
Marine-derived fungal nanotechnology, 52
MDP design, 246
Mentha piperita, 277
Mesoporous silica nanoparticles (MSN), 171
Metal, 244, 266
Metal antimicrobial activity, 269–277
Metal nanoparticles (MNPs), 146, 177, 269, 278, 279
  biocontrol, 111
  bio-inspired synthesis, 98–100
  compositional difference, 116
  downstream steps, 108
  eco-friendly biosynthesis, 116
  electrophoresis, 109
  green synthesis, 73
  large-Scale Synthesis, 108
  microorganisms' control, 72
  plasmodial pathogens, 115
  platinum nanoparticles, 115
  scaling-up, 109–111
  silver nanoparticles, 114
  surface-to-volume ratio, 72
  zinc oxide nanoparticles, 115

Metal oxide nanoparticles, 269
Metal polymer nanocomposites, 268
Metal-microorganism, 214
Metal-polymer nanoparticles, 266, 267
Metal-vapor synthesis (MVS), 267
Methionine Sulfoxide Reductase, 12
Micafungin, 209
Microbial enzymes
  bottom up approaches, 92
  building blocks, 92
  cell factories, 94–95
  characterization methods, 95–98
  metal nanoparticles, 93
  microorganisms, 92
  top down approaches, 92
Microbial laccases, 257
Microbial NPs synthesis
  antimicrobial activity, 106
  biological agents, 105
  biomass, 107
  characterization technologies, 108
  fungal cells surface, 105
  hydrophobic channels, 107
  immobilization, 108
  NADPH-dependent nitrate, 106
  nitrate reductase, 106
  oxidoreductases, 105
  proteins/enzymes, 107
  resistance machinery, 106
  Sephacryl S-300, 107
Microcrystalline cellulose (MCC), 324
Microemulsion gel technology, 292
Microemulsions, 301
Microencapsulation, 302
Microorganism detection, 185
Microorganisms, 308, 309, 313
Minimum inhibitory concentration (MIC), 200
Monolithic system, 204
Mono-nanocomposites, 267
Morpholines, 207
Multiple Drug Resistant (MDR), 222
Myoellium, 182
Myconanotechnology (MNT), 214, 221, 233
Mycosystems, 315

N
Nano delivery systems, 298
Nano electro mechanical sensors (NEMS), 217
Nanoantimicrobial, 266
Nano-biofungicides, 183–184, 310, 313–316
Nano-biosensors, 184–185, 228, 261
Nano-biotechnology, 185, 213, 254–256, 260, 261, 309
  application, 255
  conventional methods, 29
  defined, 256
  description, 28
  electrospinning method, 29
  enzymes, 28
  fungi, 28
  laccase enzyme, 28
  nanometer size, 27
  nanoscale particles, 27
  silver nanoparticles, 28
Nanobots, 218–220
Nanocellulose, 322–325, 327, 328
  application, 328–329
  comparison, 330
  production, 324–326
  purification, 326–328
  representation of fermentation system, 329
Nanochitosans, 271
Nanocomposite materials, 273
Nanocomposites, 277–279
Nanocrystal, 219
Nanocrystalline cellulose (NCC), 322
Nanoelectomechanical sensors, 217
Nanoemulsions, 219, 230, 301
Nano-encapsulation, 312, 313
Nanofactories, 233
Nanofibers, 256, 279
Nanofibrillated cellulose (NFC), 322, 325
Nanofungicdes, 180, 266, 270, 281
Nanogels, 276–277
Nanomaterial synthesis methods, 214
Nanomaterials (NMs), 181, 183, 255–256, 269, 277, 278, 280, 308, 310–316
  advantages, 256
  defined, 256
Nanoparticle synthesis
  AgNPs, 36–38
Nanoparticles (NPs), 169–172, 177, 185, 242–249, 256, 266, 279, 280, 311, 312
  alumino-silicates, 171
  antimicrobial activity, 221–225
  bacterial diseases, 226
  biochemical mechanism, 214
  bioinformatic tools, 6
  biological approach, 49–50
  biological synthesis, 213
  bottom up approach, 48
  C60 and C70, 144
  carbon, 170
  cellular processes, 213
  chitosan, 171–172
  classification, 48

Nanoparticles (NPs) (*cont.*)
- desorption, 161–162
- discontinuity behavior, 144
- engineered, 144–145
- fungal synthesis, 216
- GNPs, 216
- gold, 47
- immunohistochemical analysis, 226
- living organisms, 141
- medical science, 223–224
- metabolic flux, 3
- miscellaneous, 216
- nanobiosensors, 228
- nano-DNA technology, 226–227
- nanoproteomic analysis, 227–228
- nanoscale materials, 2
- nano-science and technology, 2
- nano-sized scale, 2
- nanotransformation, 5
- organism types, 215
- physical and chemical methods, 213
- protozoan diseases, 226
- redox systems, 215
- RoHS and WEEE legislation, 2
- silica, 171
- SNPs, 47, 170–171, 215, 227
- strained-layer growth, 2
- surface effects, 144
- synthesis, 49
- top down approach, 48
- viral diseases, 225
- yeast cells, 4

Nano-pesticides, 312–314
Nanoprism, 256
Nano-remediation, 233
Nanoscale, 256
Nanoshells, 219–220
Nanosphere, 256
Nanostars, 218–219
Nanostructure, 280
Nanotechnology (NT), 47, 169, 177, 213, 253, 254, 256, 261, 277–279, 300, 303, 308, 310, 312
Nanotubes, 220, 221
Nanowires, 256
Nanoyarn, 328
Natural polymers
- properties, 268

Neem, 294
Neem seed oil, 302
New Generation Formulations, 300–302
- advantages, 303
- coil formulations, 302
- CRF, 302
- microemulsions, 301
- nanoemulsions, 301
- novel gel technology, 301
- O/W Emulsions (EW), 301
- tablet formulations, 302

Nickel oxide nanoparticles (NiO-MNPs), 81
Nitro aryl c compounds, 155–156
Noble metal nanoparticles (NPs), 266
Non-immigration system, 202–203
Non-occluded baculovirus (NOB), 293
Non-toxic polymers, 268
Non-volatile immigration, 201–202
Novel gel technology, 301
Nuclear Polyhedrosis Virus (NPV), 293
Nudibaculovirinae, 293

O

Occluded virus (OV), 293
Oil-in-water emulsions (EW), 301
Organic acids, 198
Organogel, 277
Organosol, 267
Oxidoreductases, 259

P

Paper Industry, 260–261
Paraconiothyrium variabile, 40
Parthenium hysterophorus, 181
Pathogens, 308–310, 314, 316
Pencillium, 3
- *P. brevicompactum*, 4, 7
- *P. atramentosum*, 178
- *P. fellutanum*, 51
- *P. funiculosum*, 34
- *P. notatum*, 29

Pepper, 177
Peptidoglycan, 98
Peroxiredoxins (Prx), 13
Pesticide detection, 184–185
Pesticides, 169, 308–310, 312–315
Phagokinetics, 232
Phanerochaete chrysoporium, 29, 36
Phenolic compounds, 260
Phoma glomerata, 248
Phoma herbarum, 248
Photon correlation spectroscopy (PCS), 268
Phytochelatins, 10
Phytohtora capsici, 181
Phytopathogenic fungi, 172, 271
Phytotoxicity, 277, 278
Pichia salicaria, 57
Pimenta dioica, 173
Plant-incorporated protectants (PIPs), 292
Pleurotus florida, 75

Pleurotus ostreatus, 37
Point of purchase (POP), 194
Poly aromatic hydrocarbons (PAHs), 260
Poly lactic acid (PLA), 266, 270
Polyaniline
  description, 129
  electrosynthesis, 131
  idealized oxidation states, 130
  structure and properties, 130
  voltammograms, 131
Polyene, 207
Polyethylene glycol (PEG), 266
Polylactic acid (PLA), 276
Polylactic acid metal, 276
Polymer, 270–271
Polymer metal
  safety, 278–279
  toxicity, 277–278
Polymer nanocomposites, 280
Polymerase chain reaction (PCR), 225
Porous nanocomposites, 274
Probiotics, 199–200
Proteus vulgaris, 54
Pseudomonas stutzeri, 100
Pycnoporus sanguineus, 39

Q
Quantum dots, 220–222

R
Reactive oxygen species (ROS), 181
Redox agents, 130
Rhizoctonia, 173–174
Rhizoctonia solani, 173
Rhizophora annamalayanna, 54
Rhizopus oryzae, 101, 104
Rhizospheric fungus, 249
Rhodotorula minuta, 57
Rhodotorula mucilaginosa, 81

S
Sac fungi, 174
Saccharomyces cerevisiae, 3, 5, 57
Scanning electron microscopy (SEM), 268
Scanning probe microscope (SPM), 268
Schizophyllum commune, 38
Schizosaccharomyces pombe, 10, 242
Sclerotia, 172, 173
Sclerotinia sclerotiorum, 174
Sclerotium, 173–174
Sclerotium rolfsii, 173
Shigella sonnei, 54
Signal transduction
  biosynthetic mechanism, 5
  regulation, 13
Silica (SiO_2) nanoparticles, 171
Silver nanocomposites, 274
Silver nanoparticles (SNPs), 33–34, 178, 181, 182, 215, 243, 244, 246–249, 269–272, 275, 276, 278–280, 313–316
  aggregation, 36
  antibacterial effect, 61–62
  antifungal effect, 62
  antiviral activity, 62, 63
  biosynthesis, 36
  biosynthesis, 149–151
  bottom up', 147
  classification, 142
  culture filtrate, 37, 38
  extracellular enzymes, 149
  first method, 148
  mycosynthesis, 37
  pH aqueous nitric acid, 148
  scientific fields, 147
  second method, 148
  spectral and electron microscopy, 38
  third method, 148
  top down' synthesis, 147
Silver-resistance system, 106
Single nucleotide polymorphisms (SNPs), 227
Sol-Gel Route, 159
Spodoptera litura, 294
Staphylococcus aureus, 33, 54
Starch, 270
Starch metal, 275
Sulfur NPs
  fungicide, 151
  S-NPs preparation, 151–152
Supercritical carbon dioxide (SC CO_2), 267
Superoxide dismutases (SODs), 12
Surface plasmon resonance (SPR), 179
Sustainability, 72, 73
Synergetic effect, 246, 249
Synergitic antifungal effect, 248

T
Tablet formulations, 302
Textile industry, 261
Thermogravimetric Analysis (TGA), 268
Thiolated chitosan (TCS)
  Ellman's method, 153–154
  synthesis, 153–154
Thioredoxins (TRX), 12

Tissue engineering, 230
Titania-chitosan nanocomposites, 270
Titanium, 162–163, 181
Titanium dioxide, 145
Titanium dioxide nanoparticles, 248
Tomato, 176
Tomato seeds, 170
Toxic metal ions tolerance
  bacterial community, 100
  cyanobacterium, 102
  cytoplasmic membrane, 100
  electrostatic interactions, 104
  heavy-metal-resistant fungi, 100
  individual-spore bacteria, 103
  nanoparticles synthesis, 105
  NCIMB 8307, 102
  oxidoreductase, 104
  silver-resistant mechanism, 101
  zygomycete, 104
Transmission electron microscopy (TEM), 178, 268
Tribolium castaneum, 266, 301
Trichoderma, 185
Trichoderma harzianum, 310
Trichoderma reesei, 34, 325, 328
Trichoderma viride, 247, 248
Tricholoma crassum, 4
Trichophyton mentagrophytes, 275
Trichophyton rubrum, 54
Tuberculosis (TB), 226

U
Unconstrained free diffusion system, 204
US Environmental Protection Agency, 297

V
V. longisporum, 176
Vaccine development, 231
Verticillium sp, 51, 100, 176, 242–244
Volatile immigration, 202
Volatile organic compounds (VOCs), 295
Volatile organic solvents (VOCs), 292
Volvariella volvacea, 7

W
Wettable powders, 299
White rot fungi, 34–36
Wood pulp, 260
World Packaging Organization (WPO), 194

X
X-ray diffraction (XRD), 104, 178, 268

Y
Yarrowia lipolytica, 75

Z
Zinc, 162–163
Zinc oxide (ZnO), 146, 181
Zirconia ions, 34
Zinc oxide nanoparticles (ZnO NPs), 248, 269, 275, 279
  hydrothermal technique, 158–159
  luminescent materials, 158
  near-UV emission, 158
Zinc titanium oxide ($ZnTiO_3$), 181

MIX
Papier aus verantwortungsvollen Quellen
Paper from responsible sources
FSC® C105338

If you have any concerns about our products,
you can contact us on
ProductSafety@springernature.com

In case Publisher is established outside the EU,
the EU authorized representative is:
Springer Nature Customer Service Center GmbH
Europaplatz 3, 69115 Heidelberg, Germany

Printed by Libri Plureos GmbH
in Hamburg, Germany